D1534725

Fractal Image Compression

Springer

New York
Berlin
Heidelberg
Barcelona
Budapest
Hong Kong
London
Milan
Paris
Santa Clara
Singapore
Tokyo

Yuval Fisher
Institute for Nonlinear Science
University of California, San Diego
9500 Gilman Drive
La Jolla, CA 92093-0402
USA

Library of Congress Cataloging-in-Publication Data
Fractal image compression : theory and application /
    [edited by] Yuval Fisher.
        p.   cm.
    Includes bibliographical references and index.
    ISBN 0-387-94211-4 (New York). — ISBN 3-540-94211-4 (Berlin)
    1. Image processing — Digital techniques.   2. Image compression.
3. Fractals.   I. Fisher, Yuval.
TA1637.F73   1994
006.6 — dc20   94-11615

Printed on acid-free paper.

Production managed by Hal Henglein; manufacturing supervised by Jacqui Ashri.
Photocomposed copy prepared from the editor's LaTeX file.
Printed and bound by Braun-Brumfield, Ann Arbor, MI.
Printed in the United States of America.

9 8 7 6 5 4 3 2 (Corrected second printing, 1996)

ISBN 0-387-94211-4 Springer-Verlag New York Berlin Heidelberg
ISBN 3-540-94211-4 Springer-Verlag Berlin Heidelberg New York   SPIN 10544268

Yuval Fisher
Editor

Fractal Image Compression

Theory and Application

With 139 Illustrations

Springer

Preface

What is "Fractal Image Compression," anyway? You will have to read the book to find out everything about it, and if you read the book, you really will find out almost everything that is currently known about it. In a sentence or two: fractal image compression is a method, or class of methods, that allows images to be stored on computers in much less memory than standard ways of storing images. The "fractal" part means that the methods have something to do with fractals, complicated looking sets that arise out of simple algorithms.

This book contains a collection of articles on fractal image compression. Beginners will find simple explanations, working C code, and exercises to check their progress. Mathematicians will find a rigorous and detailed development of the subject. Non-mathematicians will find a parallel intuitive discussion that explains what is behind all the "theorem–proofs." Finally, researchers – even researchers in fractal image compression – will find new and exciting results, both theoretical and applied.

Here is a brief synopsis of each chapter:

Chapter 1 contains a simple introduction aimed at the lay reader. It uses almost no math but explains all the main concepts of a fractal encoding/decoding scheme, so that the interested reader can write his or her own code.

Chapter 2 has a rigorous mathematical description of iterated function systems and their generalizations for image encoding. An informal presentation of the material is made in parallel in the chapter using sans serif font.

Chapter 3 contains a detailed description of a quadtree-based method for fractal encoding. The chapter is readily accessible, containing no mathematics. It does contain almost everything anyone would care to know about the quadtree method.

The following chapters are contributed articles.

Chapter 4 details an important optimization which can reduce encoding times significantly. It naturally follows the previous chapter, but the methods can be applied in more general settings.

Chapter 5 contains a theoretical development of fractal data encoding using a pyramidal approach. The results include an ultra-fast decoding method and a description of the relationship between the finite- and infinite-dimensional representation of the compressed data.

Chapter 6 describes the details of a fractal encoding scheme that matches or exceeds results obtainable using JPEG and some wavelet methods.

Chapter 7 and the next three chapters form a subsection of the book dedicated to results obtainable through a linear algebraic approach. This chapter sets up the model and gives simple, but previously elusive, conditions for convergence of the decoding process in the commonly used rms metric.

Chapter 8 derives a different ultrafast decoding scheme with the advantage of requiring a fixed number of decoding steps. This chapter also describes ways of overcoming some of the difficulties associated with encoding images as fractals.

Chapter 9 contains a theoretical treatment of a method to significantly reduce encoding times. The theoretical framework relates to other image compression methods (most notably VQ).

Chapter 10 contains a new approach to encoding images using the concepts of Chapters 7 and 8. This method overcomes the difficulty that standard fractal methods have in achieving very high fidelity.

Chapter 11 contains a theoretical treatment of fractal encoding with an emphasis on convergence.

Chapter 12 gives both a new model and an implementation of a fast encoding/decoding fractal method. This method is a direct IFS based solution to the image coding problem.

Chapter 13 contains a formulation of an image encoding method based on finite automata. The method generates highly compressed, resolution-independent encodings

The following appendices contain supplementary material.

Appendix A contains a listing of the code used to generate the results in Chapter 3, as well as an explanation of the code and a manual on its use.

Appendix B contains exercises that complement the main text. For the most part, these exercises are of the useful "show that such-and-such is true" rather than the uninformative "find something-or-other."

Appendix C contains a list of projects including video, parallelization, and new encoding and decoding methods.

Appendix D contains a brief comparison of the results in the book with JPEG and other methods.

Appendix E consists of the original images used in the text.

If the list of contributors has any conspicuous omissions, they are Michael Barnsley and Arnaud Jacquin. Barnsley (and his group, including D. Hardin, J. Elton, and A. Sloan) and Jacquin have probably done more innovative research in fractal image compression than anyone else in the field. Dr. Barnsley has his own book on the topic, and Dr. Jacquin declined to contribute to this book. Too bad.

Here is a brief editorial about fractal compression: Does fractal image compression have a role to play in the current rush to standardize video and still image compression methods? The fractal scheme suffers from two serious drawbacks: encoding is computationally intensive, and there is no "representation" theorem. The first means that even near-real time applications will require specialized hardware (for the foreseeable future); this is not the end of the world. The second is more serious; it means that unlike Fourier or wavelet methods, for example, the size of fractally encoded data gets very large as we attempt to approach perfect reconstruction. For example, a checkerboard image consisting of alternating black and white pixels cannot be encoded by any of the fractal schemes discussed in this book, except by the trivial (in the mathematical sense) solution of defining a map into each pixel of the image, leading to fractal image *expansion*.

Does this mean that fractal image compression is doomed? Probably not. In spite of the problems above, empirical results show that the fractal scheme is at least as good as, and better at some compression ranges, than the current standard, JPEG. Also, the scheme does possess several intriguing features. It is resolution independent; images can be reconstructed at any resolution, with the decoding process creating artificial data, when necessary, that is commensurate with the local behavior of the image data. This is currently something of a solution in search of a problem, but it may be useful. More importantly, the fractal scheme is computationally simple to decode. Software decoding of video, as well as still images, may be its saving grace.

The aim of this book is to show that a rich and interesting theory exists with results that are applicable. Even in the short amount of time devoted to this field, results are comparable with compression methods that have received hundreds of thousands, if not millions, more man-hours of research effort.

Finally, this book wouldn't have come into being without the support of my wife, Melinda. She said "sounds good to me," when anyone else would have said "what's that rattling sound," or "I smell something funny." She often says "sounds good to me" (as well as the other two things, now that I think of it), and I appreciate it.

I would also like to express my gratitude to the following people: my co-authors, whose contributions made this book possible; Barbara Burke, for editing my portion of the manuscript; and Elizabeth Sheehan, my calm editor at Springer-Verlag. My thanks also go to Henry Abarbanel, Hassan Aref, Andrew Gross, Arnold Mandel, Pierre Moussa, Rama Ramachandran, Dan Rogovin, Dan Salzbach, and Janice Shen, who, in one way or another, helped me along the way.

This book was writen in LaTeX, a macro package written by Leslie Lamport for Donald Knuth's TeX typesetting package. The bibliography and index were compiled using BibTeX and makeindex, both also motivated by Leslie Lamport. In its final form, the book exists as a single 36 Megabyte postscript file.

Yuval Fisher, August 1994

The Authors

Izhak Baharav received a B.Sc. in electrical engineering from Tel-Aviv University, Israel, in 1986. From 1988 to 1991 he was a research engineer at Rafael, Israel. Since 1992 he has been a graduate student at the electrical engineering department in the Technion - Israel Institute of Technology, Haifa, Israel.
address:

        Department of Electrical Engineering
        Technion-Israel Institute of Technology
        Haifa 32000, Israel

Ben Bielefeld was born in Ohio. He received a B.S. in mathematics from Ohio State University and an M.A. and Ph.D. in mathematics from Cornell University. His dissertation was in complex analytic dynamical systems. He had a three-year research/teaching position at the Institute for Mathematical Sciences in Stony Brook where he continued to do research in dynamical systems. He then had a postdoc for 1 year in the applied math department at Stony Brook where he did research in electromagnetic scattering and groundwater modeling. Dr. Bielefeld currently works for the National Security Agency.

Roger D. Boss received his B.S. from Kent State University and his Ph.D. in Analytical Chemistry from Michigan State University in 1980 and 1985, respectively. He has worked in the Materials Research Branch of the NCCOSC RDT&E Division since 1985. His past research interests have included non-aqueous solution chemistry; spectroelectrochemistry of electron transfer; conducting polymers; high-temperature superconducting ceramics; chaotic and stochastic effects in neurons; and fractal-based image compression. His current research involves macromolecular solid-state chemistry.
address:

        NCCOSC RDT&E Division 573
        49590 Lassing Road
        San Diego, CA 92152-6171

Karel Culik II got his M.S. degree at the Charles University in Prague and his Ph.D. from the Czechoslovak Academy of Sciences in Prague. From 1969 to 1987 he was at the computer science department at the University of Waterloo; since 1987 he has been the Bankers' Trust Chair Professor of Computer Science at the University of South Carolina.
address:

> Department of Computer Science
> University of South Carolina
> Columbia, SC 29208

Frank Dudbridge gained the B.Sc. degree in mathematics and computing from Kings College, London, in 1988. He was awarded the Ph.D. degree in computing by Imperial College, London, in 1992, for research into image compression using fractals. He is currently a SERC/NATO research fellow at the University of California, San Diego, conducting further research into fractal image compression. His other research interests include the calculus of fractal functions, statistical iterated function systems, and global optimization problems.
address:

> Institute for Nonlinear Science
> University of California, San Diego
> La Jolla, CA 92093-0402

Yuval Fisher has B.S. degrees from the University of California, Irvine, in mathematics and physics. He has an M.S. in computer science from Cornell University, where he also completed his Ph.D. in Mathematics in 1989. Dr. Fisher is currently a research mathematician at the Institute for Nonlinear Science at the University of California, San Diego.
address:

> Institute for Nonlinear Science
> University of California, San Diego
> La Jolla, CA 92093-0402

Bill Jacobs received his B.S. degree in physics and M.S. degree in applied physics from the University of California, San Diego, in 1981 and 1986, respectively. He has worked in the Materials Research Branch of the NCCOSC RDT&E Division since 1981, and during that time he has studied a variety of research topics. Some of these included properties of piezoelectric polymers; properties of high-temperature superconducting ceramics; chaotic and stochastic effects in nonlinear dynamical systems; and fractal-based image compression.
address:

> NCCOSC RDT&E Division 573
> 49590 Lassing Road
> San Diego, CA 92152-6171

Jarkko Kari received his Ph.D. in mathematics from the University of Turku, Finland, in 1990. He is currently working as a researcher for the Academy of Finland.
address:

Mathematics Department
University of Turku
20500 Turku, Finland

Ehud D. Karnin received B.Sc. and M.S. degrees in electrical engineering from the Technion - Israel Institute of Technology, Haifa, Israel, in 1973 and 1976, respectively, and an M.S. degree in statistics and a Ph.D. degree in electrical engineering from Stanford University in 1983. From 1973 to 1979 he was a research engineer at Rafael, Israel. From 1980 to 1982 he was a research assistant at Stanford University. During 1983 he was a visiting scientist at the IBM Research Center, San Jose, CA. Since 1984 he has been a research staff member at the IBM Science and Technology Center, Haifa, Israel, and an adjunct faculty member of the electrical engineering department, Technion - Israel Institute of Technology. In 1988-1989 he was a visiting scientist at the IBM Watson Research Center, Yorktown Heights, NY. His past research interests included information theory, cryptography, and VLSI systems. His current activities are image processing, visualization, and data compression.
address:

IBM Science and Technology
MATAM-Advanced Technology Center
Haifa 31905, Israel

Skjalg Lepsøy received his Siv.Ing. degree in electrical engineering from the Norwegian Institute of Technology (NTH) in 1985, where he also received his Dr.Ing. in digital image processing in 1993. He has worked on source coding and pattern recognition at the research foundation at NTH (SINTEF) 1987-1992, and he is currently working on video compression at Consensus Analysis, an industrial mathematics R&D company.
address:

Consensus Analysis
Postboks 1391
1401 Ski, Norway

Lars M. Lundheim received M.S. and Ph.D. degrees from the Norwegian Institute of Technology, Trondheim, Norway, in 1985 and 1992, respectively. From February 1985 to May 1992 he was a research scientist at the Electronics Research Laboratory (SINTEF-DELAB), Norwegian Institute of Technology, where he worked with digital signal processing, communications, and data compression techniques for speech and images. Since May 1992 he has been with Trondheim College of Engineering.
address:

Trondheim College of Engineering
Department of Electrical Engineering
N-7005 Trondheim, Norway

David Malah received his B.S. and M.S. degrees in 1964 and 1967, respectively, from the Technion - Israel Institute of Technology, Haifa, Israel, and the Ph.D. degree in 1971 from the University of Minnesota, Minneapolis, MN, all in electrical engineering. During 1971-1972 he was an Assistant Professor at the Electrical Engineering Department of the University of New Brunswick, Fredericton, N.B., Canada. In 1972 he joined the Electrical Engineering Department of the Technion, where he is presently a Professor. During 1979-1981 and 1988-1989, as well as the summers of 1983, 1986, and 1991, he was on leave at AT&T Bell Laboratories, Murray Hill, NJ. Since 1975 (except during the leave periods) he has been in charge of the Signal and Image Processing Laboratory at the EE Department, which is active in image and speech communication research. His main research interests are in image, video, and speech coding; image and speech enhancement; and digital signal processing techniques. He has been a Fellow of the IEEE since 1987.
address:

Department of Electrical Engineering
Technion-Israel Institute of Technology
Haifa 32000, Israel

Spencer Menlove became interested in fractal image compression after receiving a B.S. in cognitive science from the University of California, San Diego. He researched fractal compression and other compression techniques under a Navy contract while working in San Diego. He is currently a graduate student in computer science at Stanford University doing work in image processing and artificial intelligence.
address:

Department of Computer Science
Stanford University
Palo Alto, CA 94305

Geir Egil Øien graduated with a Siv.Ing. degree from the Department of Telecommunications at the Norwegian Institute of Technology (NTH) in 1989. He was a research assistant with the Signal Processing Group at the same department in 1989-1990. In 1990 he received a 3-year scholarship from the Royal Norwegian Council of Scientific Research (NTNF) and started his Dr.Ing. studies. He received his Dr.Ing. degree from the Department of Telecommunications, NTH, in 1993. The subject of his thesis was L^2-optimal attractor image coding with fast decoder convergence. Beginning in 1994 he will be an associate professor at Rogaland University Centre, Stavanger, Norway. His research interests are within digital signal/image processing with an emphasis on source coding.
address:

The Norwegian Institute of Technology
Department of Telecommunications
O. S. Bragstads Plass 2
7034 Trondheim-NTH, Norway

Dietmar Saupe received the Dr. rer. nat. degree in mathematics from the University of Bremen, Germany, in 1982. He was Visiting Assistant Professor of Mathematics at the University of California at Santa Cruz, 1985–1987, and Assistant Professor at the University of Bremen, 1987–1993. Since 1993 he has been Professor of Computer Science at the University of Freiburg, Germany. His areas of interest include visualization, image processing, computer graphics, and dynamical systems. He is coauthor of the book *Chaos and Fractals* by H.-O. Peitgen, H. Jürgens, D. Saupe, Springer-Verlag, 1992, and coeditor of *The Science of Fractal Images,* H.-O. Peitgen, D. Saupe, (eds.), Springer-Verlag, 1988.

address:

Institut für Informatik
Rheinstrasse 10–12
79104 Freiburg
Germany

Greg Vines was born in Memphis, Tennessee, on June 13, 1960. He received his B.S. from the University of Virginia in 1982, and his M.S. and Ph.D. degrees in electrical engineering from the Georgia Institute of Technology in 1990 and 1993, respectively. While at the Georgia Institute of Technology, he was a graduate research assistant from 1988 until 1993. He is presently working at General Instrument's Vider Cipher Division. His research interests include signal modeling, image processing, and image/video coding.

address:

General Instrument Corporation
6262 Lusk Boulevard
San Diego, CA 92121

Contents

Chapter 1

Introduction

Y. Fisher

A picture may be worth a thousand words, but it requires far more computer memory to store. Images are stored on computers as collections of bits representing pixels, or points forming the picture elements. (A bit is a binary unit of information which can answer one "yes" or "no" question.) Since the human eye can process large amounts of information, many pixels – some 8 million bits' worth – are required to store even moderate-quality images. These bits provide the "yes" or "no" answers to 8 million questions that determine what the image looks like, though the questions are not the "is it bigger than a bread-box?" variety but a more mundane "what color is this or that pixel?"

Although the storage cost per bit is (in 1994 prices) about half a millionth of a dollar, a family album with several hundred photos can cost more than a thousand dollars to store! This is one area where image compression can play an important role. Storing images in less memory cuts cost. Another useful feature of image compression is the rapid transmission of data; fewer data requires less time to send.

So how can images be compressed? Most images contain some amount of redundancy that can sometimes be removed when the image is stored and replaced when it is reconstructed, but eliminating this redundancy does not lead to high compression. Fortunately, the human eye is insensitive to a wide variety of information loss. That is, an image can be changed in many ways that are either not detectable by the human eye or do not contribute to "degradation" of the image. If these changes lead to highly redundant data, then the data can be greatly compressed when the redundancy can be detected. For example, the sequence $2, 0, 0, 2, 0, 2, 2, 0, 0, 2, 0, 2, \ldots$, is (in some sense) similar to $1, 1, 1, 1, 1 \ldots$, with random fluctuations of ± 1. If the latter sequence can serve our purpose as well as the first, we would benefit from storing it in place of the first, since it can be specified very compactly.

Standard methods of image compression come in several varieties. The currently most popular method relies on eliminating high-frequency components of the signal by storing only

1

the low-frequency Fourier coefficients. This method uses a discrete cosine transform (DCT) [17], and is the basis of the so-called JPEG standard, which comes in many incompatible flavors. Another method, called vector quantization [55], uses a "building block" approach, breaking up images into a small number of canonical pieces and storing only a reference to which piece goes where. In this book, we will explore several distinct new schemes based on "fractals."

A fractal scheme has been developed by M. Barnsley, who founded a company based on fractal image compression technology but who has released only some details of his scheme. A. Jacquin, a former student of Barnsley's, was the first to publish a fractal image compression scheme in [45], and after this came a long list of variations, generalizations, and improvements. Early work on fractal image compression was also done by E.W. Jacobs and R.D. Boss of the Naval Ocean Systems Center in San Diego who used regular partitioning and classification of curve segments in order to compress measured fractal curves (such as map boundary data) in two dimensions [10], [43].

The goal of this introductory chapter is to explain an approach to fractal image compression in very simple terms, with as little mathematics as possible. The later chapters will review the same subjects in depth and with rigor, but for now we will concentrate on the general concepts. We will begin by describing a simple scheme that can generate complex-looking fractals from a small amount of information. We will then generalize this scheme to allow the encoding of images as "fractals," and finally we will discuss some ways this scheme can be implemented.

1.1   What Is Fractal Image Compression?

Imagine a special type of photocopying machine that reduces the image to be copied by a half and reproduces it three times on the copy, as in Figure 1.1. What happens when we feed the output of this machine back as input? Figure 1.2 shows several iterations of this process on several input images. What we observe, and what is in fact true, is that all the copies seem to be converging to the same final image, the one in 1.2c. We also see that this final image is not changed by the process, and since it is formed of three reduced copies of itself, it must have detail at every scale – it is a fractal. We call this image the *attractor* for this copying machine. Because the copying machine reduces the input image, the copies of any initial image will be reduced to a point as we repeatedly feed the output back as input; there will be more and more copies, but each copy gets smaller and smaller. So, the initial image doesn't affect the final attractor; in fact, it is only the position and the orientation of the copies that determines what the final image will look like.

Since the final result of running the copy machine in a feedback loop is determined by the way the input image is transformed, we only describe these transformations. Different transformations lead to different attractors, with the technical limitation that the transformations must be *contractive* – that is, a given transformation applied to any two points in the input image must bring them closer together in the copy. This technical condition is very natural, since if points in the copy were spread out, the attractor might have to be of infinite size. Except for this condition, the transformations can have any form. In practice, choosing transformations of the form

$$w_i \begin{bmatrix} x \\ y \end{bmatrix} = \begin{bmatrix} a_i & b_i \\ c_i & d_i \end{bmatrix} \begin{bmatrix} x \\ y \end{bmatrix} + \begin{bmatrix} e_i \\ f_i \end{bmatrix}$$

is sufficient to yield a rich and interesting set of attractors (see Exercise 8). Such transformations

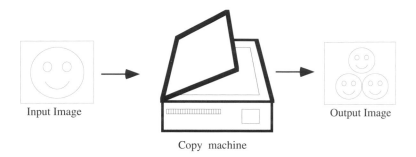

Copy machine

Figure 1.1: A copy machine that makes three reduced copies of the input image.

are called *affine* transformations of the plane, and each can skew, stretch, rotate, scale, and translate an input image.

Figure 1.3 shows some affine transformations, the resulting attractors, and a zoom on a region of the attractor. The transformations are displayed by showing an initial square marked with an "⊔" and its image by the transformations. The ⊔ shows how a particular transformation flips or rotates the square. The first example shows the transformations used in the copy machine of Figure 1.1. These transformations reduce the square to half its size and copy it at three different locations, each copy with the same orientation. The second example is very similar, except that one transformation flips the square, resulting in a different attractor (see Exercise 1). The last example is the Barnsley fern. It consists of four transformations, one of which is squashed flat to yield the stem of the fern (see Exercise 2).

A common feature of these and all attractors formed this way is that in the position of each of the images of the original square there is a transformed copy of the whole image. Thus, each image is formed from transformed (and reduced) copies of itself, and hence it must have detail at every scale. That is, the images are *fractals*. This method of generating fractals is due to John Hutchinson [36]. More information about many ways to generate such fractals can be found in books by Peitgen, Saupe, and Jürgens [67],[68], [69], and by Barnsley [4].

M. Barnsley suggested that perhaps storing images as collections of transformations could lead to image compression. His argument went as follows: the fern in Figure 1.3 looks complicated and intricate, yet it is generated from only four affine transformations. Each affine transformation w_i is defined by six numbers, a_i, b_i, c_i, d_i, e_i and f_i, which do not require much memory to store on a computer (they can be stored in 4 transformations × 6 numbers per transformation × 32 bits per number = 768 bits). Storing the image of the fern as a collection of pixels, however, requires much more memory (at least 65,536 bits for the resolution shown in Figure 1.3). So if we wish to store a picture of a fern, we can do it by storing the numbers that define the affine transformations and simply generating the fern whenever we want to see it. Now suppose that we were given any arbitrary image, say a face. If a small number of affine transformations could generate that face, then it too could be stored compactly. This is what this book is about.

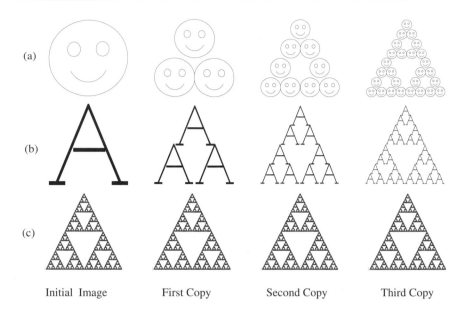

| | | | |
| Initial Image | First Copy | Second Copy | Third Copy |

Figure 1.2: The first three copies generated on the copying machine of Figure 1.1.

Why Is It *Fractal* Image Compression?

The schemes discussed in this book can be said to be fractal in several senses. Some of the schemes encode an image as a collection of transforms that are very similar to the copy machine metaphor. This has several implications. For example, just as the fern is a set which has detail at every scale, an image reconstructed from a collection of transforms also has detail created at every scale. Also, if one scales the transformations defining the fern (say by multiplying everything by 2), the resulting attractor will be scaled (also by a factor of 2). In the same way, the decoded image has no natural size; it can be decoded at any size. The extra detail needed for decoding at larger sizes is generated automatically by the encoding transforms. One may wonder (but hopefully not for long) if this detail is "real"; if we decode an image of a person at a larger and larger size, will we eventually see skin cells or perhaps atoms? The answer is, of course, no. The detail is not at all related to the actual detail present when the image was digitized; it is just the product of the encoding transforms, which only encode the large-scale features well. However, in some cases the detail is realistic at low magnification, and this can be a useful feature of the method. For example, Figure 1.4 shows a detail from a fractal encoding of an image, along with a magnification of the original. The whole original image can be seen in Figure 1.6; it is the now famous image of Lenna which is commonly used in image compression literature.

The magnification of the original shows pixelization; the dots that make up the image are clearly discernible. This is because it is magnified by a factor of 4 by local replication of the pixels. The decoded image does not show pixelization since detail is created at all scales.

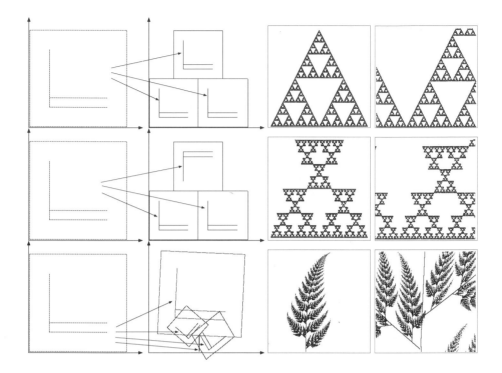

Figure 1.3: Transformations, their attractor, and a zoom on the attractors.

Why Is It Fractal Image *Compression*?

Standard image compression methods can be evaluated using their compression ratio: the ratio of the memory required to store an image as a collection of pixels and the memory required to store a representation of the image in compressed form. As we saw before, the fern could be generated from 768 bits of data but required 65,536 bits to store as a collection of pixels, giving a compression ratio of $65,536/768 = 85.3$ to 1.

The compression ratio for the fractal scheme is easy to misunderstand, since the image can be decoded at any scale. For example, the decoded image in Figure 1.4 is a portion of a 5.7 to 1 compression of the whole Lenna image. It is decoded at 4 times its original size, so the full decoded image contains 16 times as many pixels and hence its compression ratio can be considered to be 91.2 to 1. In practice, it is important to either give the initial and decompressed image sizes or use the same sizes (the case throughout this book) for a proper evaluation. The schemes we will discuss significantly reduce the memory needed to store an image that is similar (but not identical) to the original, and so they compress the data. Because the decoded image is not exactly the same as the original, such schemes are said to be *lossy*.

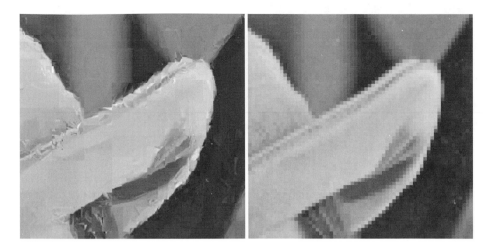

Figure 1.4: A portion of Lenna's hat decoded at 4 times its encoding size (left), and the original image enlarged to 4 times its size (right), showing pixelization.

Iterated Function Systems

Before we describe an image compression scheme, we will discuss the copy machine example with some notation. Later we will use the same notation in the image compression case, but for now it is easier to understand in the context of the copy machine example.

Running the special copy machine in a feedback loop is a metaphor for a mathematical model called an iterated function system (IFS). The formal and abstract mathematical description of IFS is given in Chapter 2, so for now we will remain informal. An iterated function system consists of a collection of contractive transformations $\{w_i : \mathbb{R}^2 \to \mathbb{R}^2 \mid i = 1, \ldots, n\}$ which map the plane R^2 to itself. This collection of transformations defines a map

$$W(\cdot) = \bigcup_{i=1}^{n} w_i(\cdot).$$

The map W is not applied to the plane, it is applied to sets – that is, collections of points in the plane. Given an input set S, we can compute $w_i(S)$ for each i (this corresponds to making a reduced copy of the input image S), take the union of these sets (this corresponds to assembling the reduced copies), and get a new set $W(S)$ (the output of the copier). So W is a map on the space of subsets of the plane. We will call a subset of the plane an image, because the set defines an image when the points in the set are drawn in black, and because later we will want to use the same notation for graphs of functions representinging actual images, or pictures.

We now list two important facts:

- When the w_i are contractive in the plane, then W is contractive in a space of (closed and bounded[1]) subsets of the plane. This was proved by Hutchinson. For now, it is not

---

[1]The "closed and bounded" part is one of several technicalities that arise at this point. What are these terms and what

necessary to worry about what it means for W to be contractive; it is sufficient to think of it as a label to help with the next step.

- If we are given a contractive map W on a space of images, then there is a special image, called the attractor and denoted x_W, with the following properties:

    1. If we apply the copy machine to the attractor, the output is equal to the input; the image is fixed, and the attractor x_W is called the *fixed point* of W. That is,

    $$W(x_W) = x_W = w_1(x_W) \cup w_2(x_W) \cup \cdots \cup w_n(x_W).$$

    2. Given an input image S_0, we can run the copying machine once to get $S_1 = W(S_0)$, twice to get $S_2 = W(S_1) = W(W(S_0)) \equiv W^{\circ 2}(S_0)$, and so on. The superscript "$\circ$" indicates that we are talking about iterations, not exponents: $W^{\circ 2}$ is the output of the second iteration. The attractor, which is the result of running the copying machine in a feedback loop, is the limit set

    $$x_W \equiv S_\infty = \lim_{n \to \infty} W^{\circ n}(S_0)$$

    which is not dependent on the choice of S_0.

    3. x_W is unique. If we find *any* set S and an image transformation W satisfying $W(S) = S$, then S is the attractor of W; that is, $S = x_W$. This means that only one set will satisfy the fixed-point equation in property 1 above.

    In their rigorous form, these three properties are known as the *Contractive Mapping Fixed-Point Theorem*.

Iterated function systems are interesting in their own right, but we are not concerned with them specifically. We will generalize the idea of the copy machine and use it to encode grey-scale images; that is, images that are not just black and white but contain shades of grey as well.

1.2 Self-Similarity in Images

In the remainder of this chapter, we will use the term *image* to mean a grey-scale image.

Images as Graphs of Functions

In order to discuss image compression, we need a mathematical model of an image. Figure 1.5 shows the graph of a function $z = f(x, y)$. This graph is generated by taking the image of Lenna (see Figure 1.6) and plotting the grey level of the pixel at position (x, y) as a height, with white being high and black low. This is our model for an image, except that while the graph

---

are they doing here? The terms make the statement precise and their function is to reduce complaints by mathematicians. Having W contractive is meaningless unless we give a way of determining distance between two sets. There is such a distance function (or metric), called the Hausdorff metric, which measures the difference between two closed and bounded subsets of the plane, and in this metric W is contractive on the space of closed and bounded subsets of the plane. This is as much as we will say about this now; Chapter 2 contains the details.

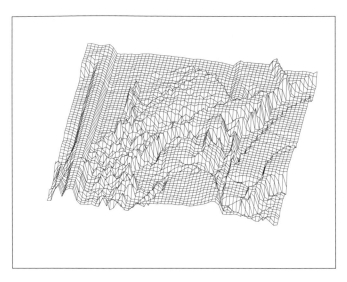

Figure 1.5: A graph generated from the Lenna image.

in Figure 1.5 is generated by connecting the heights on a 64×64 grid, we generalize this and assume that every position (x, y) can have an independent height. That is, our image model has infinite resolution.

Thus, when we wish to refer to an image, we refer to the function $f(x, y)$ that gives the grey level at each point (x, y). In practice, we will not distinguish between the *function f* and the *graph* of the function (which is a set in $\mathbb{R}^3$ consisting of the points in the surface defined by f). For simplicity, we assume we are dealing with square images of size 1; that is, $(x, y) \in \{(u, v) : 0 \le u, v \le 1\} \equiv I^2$, and $f(x, y) \in I \equiv [0, 1]$. We have introduced some convenient notation here: I means the interval $[0, 1]$ and I^2 is the unit square.

A Metric on Images

Imagine the collection of all possible images: clouds, trees, dogs, random junk, the surface of Jupiter, etc. We will now find a map W that takes an input image and yields an output image, just as we did before with subsets of the plane. If we want to know when W is contractive, we will have to define a distance between two images.

A metric is a function that measures the distance between two things. For example, the things can be two points on the real line, and the metric can then be the absolute value of their difference. The reason we use the word "metric" rather than "difference" or "distance" is because the concept is meant to be general. There are metrics that measure the distance between two images, the distance between two points, or the distance between two sets, etc.

There are many metrics to choose from, but the simplest to use are the supremum metric

$$d_{sup}(f, g) = \sup_{(x, y) \in I^2} |f(x, y) - g(x, y)|, \tag{1.1}$$

Figure 1.6: The original 256 × 256 pixel Lenna image.

and rms (root mean square) metric

$$d_{rms}(f, g) = \sqrt{\int_{I^2} (f(x, y) - g(x, y))^2 \, dx \, dy}.$$ (1.2)

The sup metric finds the position (x, y) where two images f and g differ the most and sets this value as the distance between f and g. The rms metric is more convenient in applications.[2]

Natural Images Are Not Exactly Self-Similar

A typical image of a face, for example Figure 1.6, does not contain the type of self-similarity found in the fractals of Figure 1.3. The image does not appear to contain affine transformations of itself. But, in fact, this image does contain a different sort of self-similarity. Figure 1.7 shows sample regions of Lenna that are similar at different scales: a portion of her shoulder overlaps a smaller region that is almost identical, and a portion of the reflection of the hat in the mirror is similar (after transformation) to a smaller part of her hat. The difference is that in Figure 1.3 the image was formed of copies of its *whole* self (under appropriate affine transformation), while here the image will be formed of properly transformed *parts* of itself. These transformed parts do not fit together, in general, to form an exact copy of the original image, and so we must allow some error in our representation of an image as a set of self-transformations. This means that an image that we encode as a set of transformations will not be an identical copy but an approximation.

---

[2]There are other possible choices for image models and other possible metrics. In fact, the choice of metric determines whether a transformation is contractive or not. These details appear in Chapters 2, 7, 8, and 11.

Figure 1.7: Self-similar portions of the Lenna image.

What kind of images exhibit this type of self-similarity? Experimental results suggest that most naturally occurring images can be compressed by taking advantage of this type of self-similarity; for example, images of trees, faces, houses, mountains, clouds, etc. This restricted self-similarity is the redundancy that fractal image compression schemes attempt to eliminate.

1.3   A Special Copying Machine

In this section we describe an extension of the copying machine metaphor that can be used to encode and decode grey-scale images.

Partitioned Copying Machines

The copy machine described in Section 1.1 has the following features:

- the number of copies of the original pasted together to form the output,

- a setting of position and scaling, stretching, skewing, and rotation factors for each copy.

We upgrade the machine with the following features:

- a contrast and brightness adjustment for each copy,

- a mask that selects, for each copy, a part of the original to be copied.

These extra features are sufficient to allow the encoding of grey-scale images. The last capability is the central improvement. It partitions an image into pieces which are each transformed separately. By partitioning the image into pieces, we allow the encoding of many shapes that are impossible to encode using an IFS.

Let us review what happens when we copy an original image using this machine. A portion of the original, which we denote by D_i, is copied (with a brightness and contrast transformation) to a part of the produced copy, denoted R_i. We call the D_i *domains* and the R_i *ranges*.[3] We denote this transformation by w_i. This notation does not make the partitioning explicit; each w_i comes with an implicit D_i. This way, we can use almost the same notation as with an IFS. Given an image f, a single copying step in a machine with N copies can be written as $W(f) = w_1(f) \cup w_2(f) \cup \cdots \cup w_N(f)$. As before, the machine runs in a feedback loop; its own output is fed back as its new input again and again.

Partitioned Copying Machines Are PIFS

The mathematical analogue of a partitioned copying machine is called a partitioned iterated function system (PIFS). As before, the definition of a PIFS is not dependent on the type of transformations, but in this discussion we will use affine transformations. There are two spatial dimensions and the grey level adds a third dimension, so the transformations w_i are of the form,

$$w_i \begin{bmatrix} x \\ y \\ z \end{bmatrix} = \begin{bmatrix} a_i & b_i & 0 \\ c_i & d_i & 0 \\ 0 & 0 & s_i \end{bmatrix} \begin{bmatrix} x \\ y \\ z \end{bmatrix} + \begin{bmatrix} e_i \\ f_i \\ o_i \end{bmatrix} \tag{1.3}$$

where s_i controls the contrast and o_i controls the brightness of the transformation. It is convenient to define the spatial part v_i of the transformation above by

$$v_i(x, y) = \begin{bmatrix} a_i & b_i \\ c_i & d_i \end{bmatrix} \begin{bmatrix} x \\ y \end{bmatrix} + \begin{bmatrix} e_i \\ f_i \end{bmatrix}.$$

Since an image is modeled as a function $f(x, y)$, we can apply w_i to an image f by $w_i(f) \equiv w_i(x, y, f(x, y))$. Then v_i determines how the partitioned domains of an original are mapped to the copy, while s_i and o_i determine the contrast and brightness of the transformation. We think of the pieces of the image D_i and R_i as lying in the plane, but it is implicit, and important to remember, that each w_i is restricted to $D_i \times I$, the vertical space above D_i. That is, w_i applies only to the part of the image that is above the domain D_i. This means that $v_i(D_i) = R_i$. See Figure 1.8.

Since we want $W(f)$ to be an image, we must insist that $\cup R_i = I^2$ and that $R_i \cap R_j = \emptyset$ when $i \neq j$. That is, when we apply W to an image, we get some (single-valued) function above each point of the square I^2. In the copy machine metaphor, this is equivalent to saying that the copies cover the whole square page, and that they are adjacent but not overlapping.

Running the copying machine in a loop means iterating the map W. We begin with an initial image f_0 and then iterate $f_1 = W(f_0)$, $f_2 = W(f_1) = W(W(f_0))$, and so on. We denote the n-th iterate by $f_n = W^{\circ n}(f_0)$.

Fixed Points for Partitioned Iterated Function Systems

In the PIFS case, a fixed point, or attractor, is an image f that satisfies $W(f) = f$; that is, when we apply the transformations to the image, we get back the original image. The Contractive Mapping Theorem says that the fixed point of W will be the image we get when we compute

---

[3]A **domain** is where a transformation maps *from*, and a **range** is where it maps *to*.

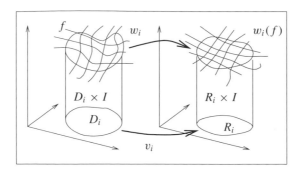

Figure 1.8: The maps w_i map the graph above D_i to a graph above R_i.

the sequence $W(f_0)$, $W(W(f_0))$, $W(W(W(f_0)))$, ..., where f_0 is *any* image. So if we can be assured that W is contractive in the space of all images, then it will have a unique fixed point that will then be some image.

Since the metric we chose in Equation (1.1) is only sensitive to what happens in the z direction, it is not necessary to impose contractivity conditions in the x or y directions. The transformation W will be contractive when each $s_i < 1$; that is, when z distances are scaled by a factor less than 1. In fact, the Contractive Mapping Theorem can be applied to $W^{\circ m}$ (for some m), so it is sufficient for $W^{\circ m}$ to be contractive. It is possible for $W^{\circ m}$ to be contractive when some $s_i > 1$, because $W^{\circ m}$ "mixes" the scalings (in this case W is called eventually contractive). This leads to the somewhat surprising result that there is no condition on any *specific* s_i either. In practice, it is safest to take $s_i < 1$ to ensure contractivity. But experiments show that taking $s_i < 1.2$ is safe and results in slightly better encodings.

Suppose that we take all the $s_i < 1$. This means that the copying machine always reduces the contrast in each copy. This seems to suggest that when the machine is run in a feedback loop, the resulting attractor will be an insipid, homogeneous grey. But this is wrong, since contrast is created between ranges that have different brightness levels o_i. Is the only contrast in the attractor between the R_i? No, if we take the v_i to be contractive, then the places where there is contrast between the R_i in the image will propagate to smaller and smaller scales; this is how detail is created in the attractor. This is one reason to require that the v_i be contractive.

We now know how to decode an image that is encoded as a PIFS. Start with any initial image and repeatedly run the copy machine, or repeatedly apply W until we get close to the fixed point x_W. The decoding is easy, but it is the encoding which is interesting. To encode an image we need to figure out R_i, D_i and w_i, as well as N, the number of maps w_i we wish to use.

1.4   Encoding Images

Suppose we are given an image f that we wish to encode. This means we want to find a collection of maps $w_1, w_2 \ldots, w_N$ with $W = \cup_{i=1}^{N} w_i$ and $f = x_W$. That is, we want f to be the

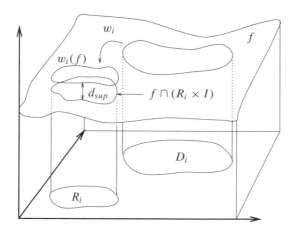

Figure 1.9: We seek to minimize the difference between the part of the graph $f \cap (R_i \times I)$ above R_i and the image $w_i(f)$ of the part of the graph above D_i.

fixed point of the map W. The fixed-point equation

$$f = W(f) = w_1(f) \cup w_2(f) \cup \cdots w_N(f)$$

suggests how this may be achieved. We seek a partition of f into pieces to which we apply the transforms w_i and get back f; this was the case with the copy machine examples in Figure 1.3c in which the images are made up of reduced copies of themselves. In general, this is too much to hope for, since images are not composed of pieces that can be transformed to fit exactly somewhere else in the image. What we can hope to find is another image $f' = x_W$ with $d_{rms}(f', f)$ small. That is, we seek a transformation W whose fixed point $f' = x_W$ is close to, and hopefully looks like, f. In that case,

$$f \approx f' = W(f') \approx W(f) = w_1(f) \cup w_2(f) \cup \cdots w_N(f).$$

Thus it is sufficient to approximate the parts of the image with transformed pieces. We do this by minimizing the following quantities

$$d_{rms}(f \cap (R_i \times I), w_i(f)) \quad i = 1, \ldots, N. \tag{1.4}$$

Figure 1.9 shows this process. That is, we find pieces D_i and maps w_i, so that when we apply a w_i to the part of the image over D_i, we get something that is very close to the part of the image over R_i. The heart of the problem is finding the pieces R_i (and corresponding D_i).

A Simple Illustrative Example

The following example suggests how this can be done. Suppose we are dealing with a 256×256 pixel image in which each pixel can be one of 256 levels of grey (ranging from black to white). Let $R_1, R_2, \ldots, R_{1024}$ be the 8×8 pixel nonoverlapping sub-squares of the image, and let **D** be the collection of all 16×16 pixel (overlapping) sub-squares of the image. The collection **D** contains $241 \cdot 241 = 58,081$ squares. For each R_i, search through all of **D** to find a $D_i \in$ **D** which minimizes Equation (1.4); that is, find the part of the image that most looks like the image above R_i. This domain is said to *cover* the range. There are 8 ways[4] to map one square onto another, so that this means comparing $8 \cdot 58,081 = 464,648$ squares with each of the 1024 range squares. Also, a square in **D** has 4 times as many pixels as an R_i, so we must either subsample (choose 1 from each 2×2 sub-square of D_i) or average the 2×2 sub-squares corresponding to each pixel of R_i when we minimize Equation (1.4).

Minimizing Equation (1.4) means two things. First, it means finding a good choice for D_i (that is the part of the image that most looks like the image above R_i). Second, it means finding good contrast and brightness settings s_i and o_i for w_i. For each $D \in$ **D** we can compute s_i and o_i using least squares regression (see Section 1.6), which also gives a resulting root mean square (rms) difference. We then pick as D_i the $D \in$ **D** with the least rms difference.

A choice of D_i, along with a corresponding s_i and o_i, determines a map w_i of the form of Equation (1.3). Once we have the collection $w_1, \ldots, w_{1024}$ we can decode the image by estimating x_W. Figure 1.10 shows four images: an initial image f_0 chosen to show texture; the first iteration $W(f_0)$, which shows some of the texture from f_0; $W^{\circ 2}(f_0)$; and $W^{\circ 10}(f_0)$.

The result is surprisingly good, given the naive nature of the encoding algorithm. The original image required 65,536 bytes of storage, whereas the transformations required only 3968 bytes,[5] giving a compression ratio of 16.5:1. With this encoding the rms error is 10.4 and each pixel is on average only 6.2 grey levels away from the correct value. Figure 1.10 shows how detail is added at each iteration. The first iteration contains detail at size 8×8, the next at size 4×4, and so on.

Jacquin [45] originally encoded images with fewer grey levels using a method similar to this example but with two sizes of ranges. In order to reduce the number of domains searched, he also classified the ranges and domains by their edge (or lack of edge) properties. This is very similar to the scheme used by Boss et al. [43] to encode contours.

A Note About Metrics

We have done something sneaky with the metrics. For a simple theoretical motivation, we use the supremum metric, which is very convenient for this. But in practice we are happier using the rms metric, which allows us to make least square computations. (We could have developed a theory with the rms metric, of course, but checking contractivity in this metric is much harder. See Chapter 7.)

1.5   Ways to Partition Images

The example in Section 1.4 is naive and simple, but it contains most of the ideas of a practical fractal image encoding scheme: first partition the image by some collection of ranges R_i; then for each R_i, seek from some collection of image pieces a D_i that has a low rms error when mapped

---

[4]The square can be rotated to 4 orientations or flipped and rotated into 4 other orientations.

[5]A byte consists of 8 bits. Each transformation requires 8 bits in each of the x and y directions to determine the position of D_i, 7 bits for o_i, 5 bits for s_i and 3 bits to determine a rotation and flip operation for mapping D_i to R_i. The position of R_i is implicit in the ordering of the transformations.

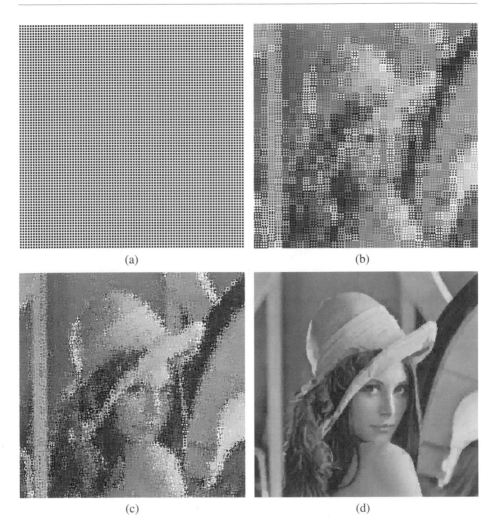

Figure 1.10: The initial image (a), and the first (b), second (c), and tenth (d) iterates of the encoding transformations.

to R_i. If we know R_i and D_i, then we can determine s_i and o_i as well as a_i, b_i, c_i, d_i, e_i and f_i in Equation (1.3). We then get a transformation $W = \cup w_i$ that encodes an approximation of the original image. There are many possible partitions that can be used to select the R_i; examples are shown in Figure 1.11. Some of these are discussed in greater detail later in this book.

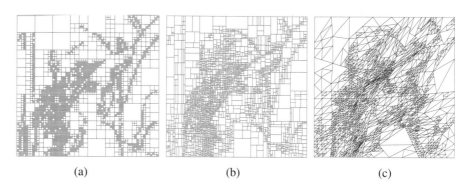

(a)   (b)   (c)

Figure 1.11: (a) A quadtree partition (5008 squares), (b) an HV partition (2910 rectangles), and (c) a triangular partition (2954 triangles).

Quadtree Partitioning

A weakness of the example of Section 1.4 is the use of fixed-size R_i, since there are regions of the image that are difficult to cover well this way (for example, Lenna's eyes). Similarly, there are regions that could be covered well with larger R_i, thus reducing the total number of w_i maps needed (and increasing the compression of the image). A generalization of the fixed-size R_i is the use of a quadtree partition of the image. In a quadtree partition (Figure 1.11a), a square in the image is broken up into four equal-sized sub-squares when it is not covered well enough by some domain. This process repeats recursively starting from the whole image and continuing until the squares are small enough to be covered within some specified rms tolerance. Small squares can be covered better than large ones because contiguous pixels in an image tend to be highly correlated.

Here is an algorithm based on these ideas which works well. Lets assume the image size is 256 × 256 pixels. Choose for the collection **D** of permissible domains all the sub-squares in the image of size 8, 12, 16, 24, 32, 48 and 64. Partition the image recursively by a quadtree method until the squares are of size 32. Attempt to cover each square in the quadtree partition by a domain that is larger. If a predetermined tolerance rms value e_c is met, then call the square R_i and the covering domain D_i. If not, subdivide the square and repeat. This algorithm works well. It works even better if the domain pool **D** also includes diagonally oriented squares, as in [27]. Figure 1.12 shows an image of a collie compressed using this scheme, which is discussed in detail in Chapter 3.

Figure 1.12: A collie image (256×256) compressed with the quadtree scheme at a compression of 28.95:1 with an rms error of 8.5.

HV-Partitioning

A weakness of quadtree-based partitioning is that it makes no attempt to select the domain pool **D** in a content-dependent way. The collection must be chosen to be very large so that a good fit to a given range can be found. A way to remedy this, while increasing the flexibility of the range partition, is to use an HV-partition. In an HV-partition (Figure 1.11b) a rectangular image is recursively partitioned either horizontally or vertically to form two new rectangles. The partitioning repeats recursively until a covering tolerance is satisfied, as in the quadtree scheme.

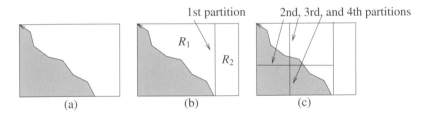

Figure 1.13: The HV scheme attempts to create self-similar rectangles at different scales.

This scheme is more flexible, since the position of the partition is variable. We can then try to make the partitions in such a way that they share some self-similar structure. For example, we can try to arrange the partitions so that edges in the image will tend to run diagonally through them. It is then possible to use the larger partitions to cover the smaller ones with a reasonable expectation of a good cover. Figure 1.13 demonstrates this idea. The figure shows a part of an image (a); in (b) the first partition generates two rectangles, R_1 with the edge running diagonally through it, and R_2 with no edge; and in (c) the next three partitions of R_1 partition it into four rectangles – two rectangles that can be well covered by R_1 (since they have an edge running diagonally) and two that can be covered by R_2 (since they contain no edge). Figure 1.14 shows an image of San Francisco encoded using this scheme.

Figure 1.14: San Francisco (256×256) compressed with the diagonal-matching HV scheme at 7.6:1 with an rms error of 7.1.

Other Partitioning

Partitioning schemes come in as many varieties as ice cream. Chapter 6 discusses a variation of the HV scheme, and in Appendix C we discuss, among other things, other partitioning methods which may yield better results. Figure 1.11c shows a triangular partitioning scheme. In this scheme, a rectangular image is divided diagonally into two triangles. Each of these is recursively subdivided into four triangles by segmenting the triangle along lines that join three partitioning points along the three sides of the triangle. This scheme has several potential advantages over the HV-partitioning scheme. It is flexible, so that triangles in the scheme can be chosen to share self-similar properties, as before. The artifacts arising from the covering do not run horizontally and vertically, which is less distracting. Also, the triangles can have any orientation, so we break away from the rigid 90 degree rotations of the quadtree- and HV-partitioning schemes. This scheme, however, remains to be fully developed and explored.

1.6   Implementation

In this section we discuss the basic concepts behind the implementation of a fractal image compression scheme. The reader should note that there are many variations on these methods, as well as schemes that are completely different. Our goal is to get the interested reader programming as soon as possible (see also the code in Appendix A).

To encode an image, we need to select an image-partitioning scheme to generate the range blocks $R_i \subset I^2$. For the purpose of this discussion, we will assume that the R_i are generated by a quadtree or HV partition, though they may also be thought of as fixed-size subsquares. We must also select a domain pool **D**. This can be chosen to be all subsquares in the image, or some subset of this rather large collection. Jacquin selected squares centered on a lattice with a spacing of one-half of the domain size, and this choice is common in the other chapters. It is convenient to select domains with twice the range size and then to subsample or average groups of 2×2 pixels to get a reduced domain with the same number of pixels as the range.

In the example of Section 1.4, the number of transformations is fixed. In contrast, the quadtree- and HV- partitioning algorithms are adaptive, in the sense that they use a range size that varies depending on the local image complexity. For a fixed image, more transformations lead to better fidelity but worse compression. This trade-off between compression and fidelity leads to two different approaches to encoding an image f – one targeting fidelity and one targeting compression. These approaches are outlined in the pseudo-code in Tables 1.1 and 1.2. In the tables, size(R_i) refers to the size of the range; in the case of rectangles, size(R_i) is the length of the longest side. The value r_{min} is a parameter that determines the smallest size range that will be allowed in the encoding.

Table 1.1: Pseudo-code targeting a fidelity e_c.

- Choose a tolerance level e_c.

- Set $R_1 = I^2$ and mark it uncovered.

- While there are uncovered ranges R_i do {

  - Out of the possible domains **D**, find the domain D_i and the corresponding w_i that best cover R_i (i.e., that minimize expression 1.4).

  - If $d_{rms}(f \cap (R_i \times I), w_i(f)) < e_c$ or size(R_i) $\leq r_{min}$ then

    - Mark R_i as covered, and write out the transformation w_i;

  - else

    - Partition R_i into smaller ranges that are marked as uncovered, and remove R_i from the list of uncovered ranges.

  }

The code in Table 1.1 attempts to target an error by finding a covering such that Equation (1.4) is below some criterion e_c. This does not mean that the resulting encoding will have this fidelity. However, encodings made with lower e_c will have better fidelity, and those with higher e_c will have worse fidelity. The method in Table 1.2 attempts to target a compression ratio by limiting the number of transforms used in the encoding. Again, the resulting encoding will not have the exact targeted compression ratio, since the number of bits per transform (typically) varies from one transformation to the next and since the number of transformations used cannot

Table 1.2: Pseudo-code targeting an encoding with N_r transformations. Since the average number of bits per transformation is roughly constant for different encodings, this code can target a compression ratio.

- Choose a target number of ranges N_r.

- Set a list to contain $R_1 = I^2$, and mark it as uncovered.

- While there are uncovered ranges in the list do {

    - For each uncovered range in the list, find and store the domain $D_i \in \mathbf{D}$ and the map w_i that covers it best, and mark the range as covered.

    - Out of the list of ranges, find the range R_j with size(R_j) > r_{min} with the largest

        $$d_{rms}(f \cap (R_j \times I), w_j(f))$$

        (i.e. which is covered worst).

    - If the number of ranges in the list is less than N_r then {

        - Partition R_j into smaller ranges which are added to the list and marked as uncovered.

        - Remove R_j, w_j and D_j from the list.

        }

    }

- Write out all the w_i in the list.

be exactly specified. However, since the number of transformations is high (ranging from several hundred to several thousand), the variation in memory required to store a transform tends to cancel, and so it is possible to target a compression ratio with relative accuracy.

Finally, decoding an image is simple. Starting from any initial image, we repeatedly apply the w_i until we approximate the fixed point. This means that for each w_i, we find the domain D_i, shrink it to the size of its range R_i, multiply the pixel values by s_i and add o_i, and put the resulting pixel values in the position of R_i. Typically, 10 iterations are sufficient. In the later chapters, other decoding methods are discussed. In particular, in some cases it is possible to decode exactly using a fixed number of iterations (see Chapter 8) or a completely different method (see Chapter 11 and Section C.13).

The RMS Metric

In practice, we compare a domain and range using the rms metric. Using this metric also allows easy computation of optimal values for s_i and o_i in Equation (1.3). Given two squares containing n pixel intensities, $a_1, \ldots, a_n$ (from D_i) and $b_1, \ldots, b_n$ (from R_i), we can seek s and o to minimize the quantity

$$R = \sum_{i=1}^{n} (s \cdot a_i + o - b_i)^2.$$

This will give us contrast and brightness settings that make the affinely transformed a_i values have the least squared distance from the b_i values. The minimum of R occurs when the partial derivatives with respect to s and o are zero, which occurs when

$$s = \frac{\left[n \sum_{i=1}^{n} a_i b_i - \sum_{i=1}^{n} a_i \sum_{i=1}^{n} b_i \right]}{\left[n \sum_{i=1}^{n} a_i^2 - \left(\sum_{i=1}^{n} a_i \right)^2 \right]}$$

and

$$o = \frac{1}{n} \left[\sum_{i=1}^{n} b_i - s \sum_{i=1}^{n} a_i \right]$$

In that case,

$$R = \frac{1}{n} \left[\sum_{i=1}^{n} b_i^2 + s \left(s \sum_{i=1}^{n} a_i^2 - 2 \sum_{i=1}^{n} a_i b_i + 2o \sum_{i=1}^{n} a_i \right) + o \left(no - 2 \sum_{i=1}^{n} b_i \right) \right] \qquad (1.5)$$

If $n \sum_{i=1}^{n} a_i^2 - (\sum_{i=1}^{n} a_i)^2 = 0$, then $s = 0$ and $o = \frac{1}{n} \sum_{i=1}^{n} b_i$. There is a simpler formula for R but it is best to use this one as we'll see later. The rms error is equal to $\sqrt{R}$.

The step "compute $d_{rms}(f \cap (R_i \times I), w_i(f))$" is central to the algorithm, and so it is discussed in detail for the rms metric in Table 1.3.

Storing the Encoding Compactly

To store the encoding compactly, we do not store all the coefficients in Equation (1.3). The contrast and brightness settings have a non-uniform distribution, which means that some form of entropy coding is beneficial. If these values are to be quantized and stored in a fixed number of bits, then using 5 bits to store s_i and 7 bits to store o_i is roughly optimal in general (see Chapter 3). One could compute the optimal s_i and o_i and then quantize them for storage. However, a *significant* improvement in fidelity can be obtained if only quantized s_i and o_i values are used when computing the error during encoding (Equation (1.5) facilitates this).

The remaining coefficients are computed when the image is decoded. Instead of storing them directly, we store the positions of R_i and D_i. In the case of a quadtree partition, R_i can be encoded by the storage order of the transformations, if we know the size of R_i. The domains D_i must be referenced by position and size. This is not sufficient, though, since there are eight ways to map the four corners of D_i to the corners of R_i. So we also must use three bits to determine this rotation and flip information.

Table 1.3: Details of the computation of $d_{rms}(f \cap (R_i \times I), w_i(f))$ in the case where D_i is twice the size of R_i.

- Let D_i be the domain of w_i.

- Take the pixels of D_i and average nonoverlapping 2×2 blocks to form a new collection of pixels F_i that has the same size as R_i.

- If w_i involves a rotation or flip, permute the pixels of F_i to the new orientation.

- Compute $\sum_{a \in F_i} a$ and $\sum_{a \in F_i} a^2$.

- Compute $\sum_{b \in R_i} b$ and $\sum_{b \in R_i} b^2$.

- Compute $\sum_{a \in F_i, b \in R_i} ab$. In this sum, only the elements a and b in the same pixel position are summed.

- These sums can be used to compute s_i, o_i and R. Note that all but the last sum can be done ahead of time. That is, it is not necessary to repeat the domain sums for different ranges.

- $d_{rms}(f \cap (R_i \times I), w_i(f)) = \sqrt{R}$.

In the case of the HV partitioning and triangular partitioning, the partition is stored as a collection of offset values. As the rectangles (or triangles) become smaller in the partition, fewer bits are required to store the offset value. The partition can be completely reconstructed by the decoding routine. One bit must be used to determine if a partition is further subdivided or will be used as an R_i, and a variable number of bits must be used to specify the index of each D_i in a list of all the partitions. For all three methods, and without too much effort, it is possible to achieve a compression of roughly 31–34 bits per w_i.

Optimizing Encoding Time

Another concern is encoding time, which can be significantly reduced by classifying the ranges and domains. Both ranges and domains are classified using some criteria such as their edgelike nature, or the orientation of bright spots, etc. Considerable time savings result from only using domains in the same class as a given range when seeking a cover, the rationale being that domains in the same class as a range should cover it best. Chapters 4 and 9 are devoted to this topic, and other methods can be found in Chapters 3 and 6.

1.7 Conclusion

The power of fractal encoding is shown by its ability to outperform[6] (or at least match) the DCT, a method which has had the benefit of hundreds of thousands, if not millions, of man-hours of research, optimization, and general tweaking. While the fractal scheme currently has more of a cult following than the respectful attention of the average engineer, today's standards and fashions become tomorrow's detritus, and at least some of today's new ideas flower into popular use. For the gentle reader interested in nurturing this scheme through implementation or theory, the remainder of this book presents other theoretical and experimental results, as well as refinements of the ideas in this chapter and a list of unsolved problems and further research.

Acknowledgments

This work was partially supported by DOE contract DE-FG03-90ER418. Other support was provided by the San Diego Super Computer Center; the Institute for Nonlinear Science at the University of California, San Diego; and the Technion Israel Institute of Technology. This chapter is based on Appendix A of [69] and [24].

---

[6]See Appendix D for a comparison of results.

Chapter 2

Mathematical Background

Y. Fisher

Hutchinson [36] introduced the theory of iterated function systems (a term coined by M. Barnsley) to model collections of contractive transformations in a metric space as dynamical systems. His idea was to use the Contractive Mapping Fixed-Point Theorem to show the existence and uniqueness of fractal sets that arise as fixed points of such systems. It was Barnsley's observation, however, that led to the idea of using iterated function systems (IFS's) to encode images. He noted that many fractals that can be very compactly specified by iterated function systems have a "natural" appearance. Given an IFS, it is easy to generate the fractal that it defines, but Barnsley posed the opposite question: given an image, is it possible to find an IFS that defines it?

After the appearance of the acronym "IFS," a slew of others appeared on the scene. These include (but are probably not limited to) RIFS, RFIF, PIFS, WFA, HIFS, and MRCM. The details of the evolution of the topic are interesting (and possibly sordid); in this chapter we will simply present a synopsis. This chapter has the misfortune of being aimed at readers with widely varying background; trivialities and technicalities mingle. Every attempt at rigor is made, but to help separate the text into thick and thin, most topics are also presented informally in sans serif font. Finally, this chapter is not completely general nor generally complete; for an undergraduate-level presentation of the IFS material, the interested reader should refer to [4] or [69].

2.1 Fractals

Unfortunately, a good definition of the term fractal is elusive. Any particular definition seems to either exclude sets that are thought of as fractals or to include sets that are not thought of as fractals. In [23], Kenneth Falconer writes:

My personal feeling is that the definition of a "fractal" should be regarded in the same way as the biologist regards the definition of "life." There is no hard and fast definition, but just a list of properties characteristic of a living thing.... In the same way, it seems best to regard a fractal as a set that has properties such as those listed below, rather than to look for a precise definition which will almost certainly exclude some interesting cases.

If we consider a set F to be a fractal, we think of it as having (some) of the following properties:

1. F has detail at every scale.

2. F is (exactly, approximately, or statistically) self-similar.

3. The "fractal dimension" of F is greater than its topological dimension. Definitions for these dimensions are given below.

4. There is a simple algorithmic description of F.

Of these properties, the third is the most rigorous, and so we define it here. Our interest in these definitions is lukewarm, however, because there are few results on the fractal dimension of fractally encoded images.

Definition 2.1 *The topological dimension of a totally disconnected set is always zero. The topological dimension of a set F is n if arbitrarily small neighborhoods of every point of F have boundary with topological dimension $n - 1$.*

The topological dimension is always an integer. An interval, for example, has topological dimension 1 because at each point we can find a neighborhood, which is also an interval, whose boundary is a disconnected set and hence has topological dimension 0.

There are many definitions for non-integral dimensions. The most commonly used fractal dimension is the **box dimension**, which is defined as follows.

Definition 2.2 *For $F \subset \mathbb{R}^n$, let $N_\epsilon(F)$ denote the smallest number of sets with diameter no larger than ϵ that can cover F. The box dimension of F is*

$$\lim_{\epsilon \to 0} \frac{\log N_\epsilon(F)}{-\log \epsilon}.$$

when this limit exists.

The fractal dimension can be thought of as a scaling relationship. Figure 2.1 shows four examples of sets and the scaling relationship for each (i.e., the way the number of boxes it takes to cover the set scales with the size of the box). For each example, we describe the scaling relationship below:

a. A curve of length ℓ can be covered by ℓ/ϵ boxes of size ϵ and $2\ell/\epsilon$ boxes of size $\epsilon/2$.

b. A region of area A can be covered by A/ϵ^2 boxes of size ϵ and $2^2 A/\epsilon^2$ boxes of size $\epsilon/2$.

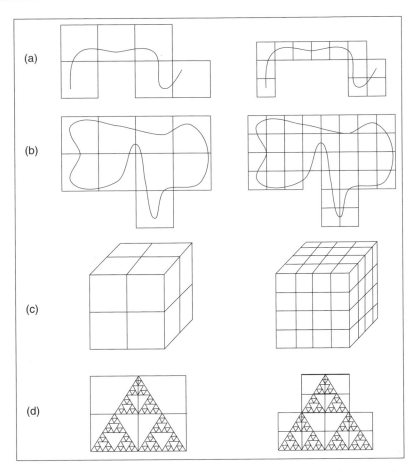

Figure 2.1: Four examples of sets and the number of ϵ boxes required to cover them.

c. A set with volume V can be covered by V/ϵ^3 boxes of size ϵ and $2^3 V/\epsilon^3$ boxes of size $\epsilon/2$.

d. If the Sierpinski triangle is covered with N boxes of size ϵ, then it takes $3N$ boxes of size $\epsilon/2$ to cover it. This is shown for ϵ equal to half the width of the set in the figure.

We can see that for sets whose dimension is 1, 2, and 3, halving box size corresponds to increasing the number of boxes required to cover the set by a factor of 2^1, 2^2, and 2^3, at least in the limit when ϵ is very small. For the Sierpinski triangle, however, the number of boxes increases by 2^d, where $d = \log 3/\log 2$, and so we call this the (fractal) dimension of the set. It is a number between 1 and 2, which is reasonable since the set has no area[1] but is more 'meaty' than a curve. See also Exercise 3.

2.2   Iterated Function Systems

In this section we will formally define iterated function systems. IFS's served as the motivating concept behind the development of fractal image compression, and most of the work on fractal image compression is based on IFS's or their generalization.

An Example

Let's begin with an example similar to the copy machine of Chapter 1. Figure 2.2 shows three transformations in the plane:

$$w_1 \begin{bmatrix} x \\ y \end{bmatrix} = \begin{bmatrix} \frac{1}{2} & 0 \\ 0 & \frac{1}{2} \end{bmatrix} \begin{bmatrix} x \\ y \end{bmatrix}$$

$$w_2 \begin{bmatrix} x \\ y \end{bmatrix} = \begin{bmatrix} \frac{1}{2} & 0 \\ 0 & \frac{1}{2} \end{bmatrix} \begin{bmatrix} x \\ y \end{bmatrix} + \begin{bmatrix} 0 \\ \frac{1}{2} \end{bmatrix}$$

$$w_3 \begin{bmatrix} x \\ y \end{bmatrix} = \begin{bmatrix} \frac{1}{2} & 0 \\ 0 & \frac{1}{2} \end{bmatrix} \begin{bmatrix} x \\ y \end{bmatrix} + \begin{bmatrix} \frac{1}{2} \\ 0 \end{bmatrix}.$$

The copying process consists of taking the input (which is assumed to lie in the square region between the points (0,0) and (1,1) in the plane), applying the three transformations to it, and pasting the result together. This output is then used again as the input. Figure 2.3 shows this process beginning with both a square and a circle. The process appears to converge to the image in Figure 2.4.

Note that the maps w_i bring points closer together. That is, the distance between $P = (x, y)$ and $Q = (u, v)$ is always more than the distance between $w_i(P)$ and $w_i(Q)$ (by a factor greater than one). This is the same as saying that the copy machine makes reductions before pasting the copies together. What would happen if one (or more) of the copies was an expansion? In

---

[1]The Sierpinski triangle can be generated by recursively removing the middle inverted triangle from a partition of an equilateral triangle into 4 equilateral triangles of half the size. Since at each removal 3/4 of the area is left, the limit leaves $\lim_{n\to\infty}(3/4)^n = 0$ area.

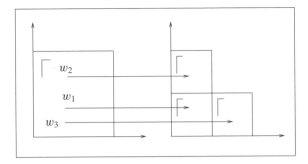

Figure 2.2: Three transformations in the plane.

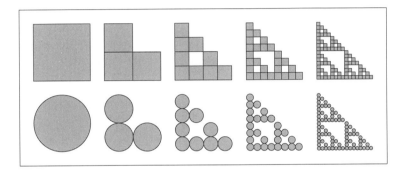

Figure 2.3: Applying the transformations of Figure 2.2 to a square and a circle.

that case, each successive copy would contain a bigger copy of the whole image, and the limit copy would have to be infinitely big.[2] So we arrive at the following rule:

Rule: The copies produced in the copy machine should be reductions.

Now let's examine what happens when we start the copying machine with different initial images. The copies of the input image are reduced at each step, so in the limit copy the initial image will be reduced to a point. Thus, it doesn't matter what image we start with; we will approach the same limit image. Therefore, we arrive at the following observation:

Observation: The limit image obtained by running the copy machine in a feedback loop is independent of the initial image.

An analogy to keep in mind is the following: The iterates of the map $x \mapsto x/2$ will converge to 0 from any initial point on the real line. In the same way, the copying process can be considered as a map in a space of images. This map converges to a point (which is an image) in this space,

---

[2]This reasoning is not quite right. It is possible to have non-contractive copies that nevertheless have limit copies of finite extent. However, if we want to be certain that the limit copy is finite, then having reductions is sufficient. See, for example, Section C.18.

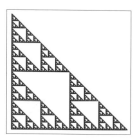

Figure 2.4: The limit image resulting from running the copy machine infinitely many times.

independent of the starting image.

The following sections are devoted to making these notions precise. We first discuss the space of images (and associated technical conditions). Next, we define "reduction" and restate the above rule. We then state the Contractive Mapping Fixed-Point Theorem, which corresponds to the above observation. The remaining sections generalize this theory and apply it to the image compression problem.

Complete Metric Spaces

Our goal is to define a subset of the plane (which will often be a fractal) as the fixed point of a map from a space of subsets of the plane to itself. Following Hutchinson [36], we will first define a metric on such a space, and then a contractive map that will determine a unique fixed point. The definition of an IFS makes use of the notions of a complete metric space, and contractive mappings.

Definition 2.3 *A* metric space *is a set X on which a real-valued distance function $d : X \times X \to \mathbb{R}$ is defined, with the following properties:*

1. $d(a, b) \geq 0$ *for all $a, b \in X$.*
2. $d(a, b) = 0$ *if and only if $a = b$, for all $a, b \in X$.*
3. $d(a, b) = d(b, a)$ *for all $a, b \in X$.*
4. $d(a, c) \leq d(a, b) + d(b, c)$ *for all $a, b, c \in C$ (triangle inequality).*

Such a function d is called a metric.

A metric space is a set of points along with a function that takes two elements of the set and gives the distance between them. The set can be a set of points, or a set of images. One of our metric spaces will be the collection $\mathcal{H}$ of all (closed and bounded) subsets of the plane. This set can be imagined as the collection of all black-and-white pictures, in which a subset of the plane is represented by a picture that is black at the points of the subset and white elsewhere. The space $\mathcal{H}$ is very large, containing, for example, a line drawing of you, gentle reader, addressing the United Nations (as well as many other pictures).

The Hausdorff Metric

Our first goal is to define the Hausdorff metric on this space. This metric will tell us the "distance" between any two pictures. The Hausdorff metric is defined formally below. To find the Hausdorff distance $h(A, B)$ between two subsets of the plane, A and B, we carry out the following procedure. For each point x of A, find the closest[3] point y in B. Measure these minimal distances (both starting from A and staring from B) and choose the largest one. This is the Hausdorff distance. Figure 2.5 shows three examples of sets A and B and the Hausdorff distance (based on Euclidean distance) between them, marked with a line. The Hausdorff metric is not sensitive to content; it can't decide that the distance between two pictures of people is less than the distance between a picture of a person and a picture of a fern.

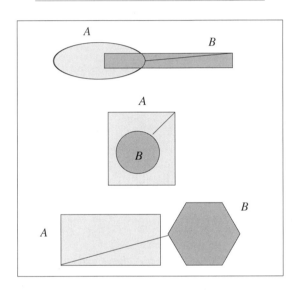

Figure 2.5: Sets A and B and the Hausdorff distance between them, indicated by a line. In the top example, the longest minimal distance is measured from B to A; in the other two, it is measured from A to B.

Here are some examples of metrics (see Exercise 4):

1. In the plane, the function

$$d\left((x_1, y_1), (x_2, y_2)\right) = \sqrt{(x_1 - x_2)^2 + (y_1 - y_2)^2}$$

   is the standard Euclidean metric which gives the distance between the points (x_1, y_1) and (x_2, y_2).

---

[3] The astute reader will notice that "close" already makes use of the notion of distance. Here we mean the normal Euclidean distance.

2. Another metric between these points is $\max\{|x_1 - x_2|, |y_1 - y_2|\}$.

3. For the (measurable) functions $f, g : \mathbb{R}^2 \to \mathbb{R}$, the supremum metric is

$$d_{sup}(f, g) = \sup_{x \in \mathbb{R}^2} |f(x) - g(x)|,$$

and the L^p metrics are

$$\|f - g\|_p = \left(\int_{\mathbb{R}^2} |f - g|^p dx dy \right)^{1/p}.$$

4. If x is a point in the plane and A is a subset of the plane, we can define $\delta_d(x, A) = \inf_{y \in A} d(x, y)$, where $d(\cdot, \cdot)$ is some metric in the plane. If B is also a subset of the plane, then we can define the metric

$$h_d(A, B) = \max \left\{ \sup_{x \in A} \delta_d(x, B), \ \sup_{y \in B} \delta_d(y, A) \right\}.$$

This is the Hausdorff metric. We will give a different (possibly simpler) definition of this metric later.

Definition 2.4 *A sequence of points $\{x_n\}$ in a metric space is called a* Cauchy *sequence if for any $\epsilon > 0$, there exists an integer N such that*

$$d(x_m, x_n) < \epsilon \quad \text{for all } n, m > N.$$

Definition 2.5 *A metric space X is* complete *if every Cauchy sequence in X converges to a limit point in X.*

Definition 2.6 *A metric space X is* compact *if it is closed and bounded.*

We now define the space in which we will work. Let $\mathcal{H}(X) = \{S \subset X \mid S \text{ is compact}\}$. If $A \in \mathcal{H}(X)$, then we write $A_d(\epsilon) = \{x \mid d(x, y) \le \epsilon \text{ for some } y \in A\}$; $A_d(\epsilon)$ is the set of points that are of maximal d distance ϵ from A (See Figure 2.6.) We can then define the *Hausdorff distance* between two elements $A, B \in \mathcal{H}(X)$ to be

$$h_d(A, B) = \max \{\inf\{\epsilon \mid B \subset A_d(\epsilon)\}, \inf\{\epsilon \mid A \subset B_d(\epsilon)\}\}.$$

Notice that h_d depends on the metric d. In practice, we drop the subscript d when the context is clear. The compactness condition in the definition of $\mathcal{H}(X)$ is necessary for the following theorem to hold.

Theorem 2.1 *Let X be a complete metric space with metric d, then $\mathcal{H}(X)$ with the Hausdorff metric h_d is a complete metric space.*

Proof: See [4] or [36].   ∎

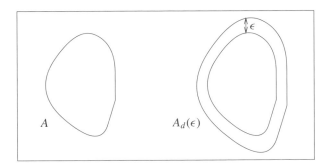

Figure 2.6: A set A and its thickened version $A_d(\epsilon)$.

A Cauchy sequence is a sequence of points that converges to a limit point. Our definition of a Cauchy sequence is dependent on measuring distances, which is why we must have a metric before we decide if the space is complete. For example, the sequence $1, \frac{3}{2}, \frac{7}{5}, \frac{17}{12}, \ldots, \frac{p}{q}, \frac{p+2q}{p+q}, \ldots$ is convergent (see Exercise 5), while $1, 2, 3, 4, \ldots$ is not. It is sufficient to think of complete metric spaces as spaces that don't have "holes" or values missing. For example, the set of rational numbers can be metrized using $d(x, y) = |x - y|$. But it is not a complete metric space, because the first sequence above converges to $\sqrt{2}$, which is not rational.

The "compact" part of the definition of $\mathcal{H}(X)$ is required in order to make $\mathcal{H}(X)$ be a complete space. We are concerned with completeness because we will want to know that a convergent sequence of images will converge to an image. That is, we want our space of images to be complete.

Contractive Maps and IFS's

Definition 2.7 *Let X be a metric space with metric d. A map $w : X \to X$ is* Lipschitz *with Lipschitz factor s if there exists a positive real value s such that*

$$d(w(x), w(y)) \le sd(x, y)$$

for every $x, y \in X$. If the Lipschitz constant satisfies $s < 1$, then w is said to be contractive *with contractivity s.*

Examples:

1. The function $f(x) = 1/x$ is Lipschitz on the closed interval $[1, 2]$ but not on the open interval $(0, 1)$. On $[1, 2]$, $|f(x) - f(y)| = |1/x - 1/y| = |(x - y)/(xy)| \le |x - y|$, but on $(0, 1)$ the slope of f becomes unbounded and so even when $|x - y|$ is small, $|1/x - 1/y|$ can be very large. This example suggests that the technicalities, such as closure, are at least relevant.

2. If f is discontinuous, then f is not Lipschitz because if there is a discontinuity between x and y, then when x and y get arbitrarily close, $f(x)$ and $f(y)$ can remain separated.

3. If $f : \mathbb{R}^n \to \mathbb{R}^n$ is given by $f(\vec{x}) = A\vec{x} + \vec{b}$, then f is Lipschitz. In fact, any linear operator on a finite-dimensional vector space is Lipschitz.

Lemma 2.1 *If $f : X \to X$ is Lipschitz, then f is continuous.*

Proof: It is sufficient to think of continuous functions as maps that have $f(x)$ and $f(y)$ arbitrarily close as x gets close to y. But as x and y get close, $d(x, y)$ gets arbitrarily small so that $d(f(x), f(y)) \leq s d(x, y)$ must also. ∎

A map is contractive if it brings points closer together. The contractivity s measures how much closer two points are brought together. As an example, consider the map $f(x) = x/2$ on the real line (with the usual metric $d(x, y) = |x - y|$). This map is contractive, because the distance between $f(x)$ and $f(y)$ is never more than half the distance between x and y. If we pick any initial x and iterate f, computing $f(x)$, $f(f(x))$, $f^{\circ 3}(x), \ldots$, then we will converge to a fixed point $0 = f(0)$. This is a general property of contractive maps: if we iterate them from *any* initial point, we converge to a unique fixed point.

While all contractive maps have unique fixed points, not all maps with unique fixed points are contractive. For example, $x \mapsto 2x$ only fixes 0, but it is an expansive map (see also Exercise 6).

Here are some examples of contractive maps:

1. If $f : \mathbb{R}^n \to \mathbb{R}^n$ is defined by $f(\vec{x}) = A\vec{x} + \vec{B}$, where A is a $n \times n$ matrix and $\vec{x}$, $\vec{B}$ are vectors, then f is contractive if $\vec{x} \mapsto A\vec{x}$ is contractive. This map will be contractive if $\sup_{\vec{x} \in \mathbb{R}^n} \|A\vec{x}\| / \|\vec{x}\| < 1$. This value is called the *norm* of A.

2. Let $w_i : \mathbb{R}^2 \to \mathbb{R}^2$, $i = 1, \ldots, n$ be a collection of contractive transformations. Then we can define a map $W : \mathcal{H}(\mathbb{R}^2) \to \mathcal{H}(\mathbb{R}^2)$ by

$$W(S) = \bigcup_{i=1}^{n} w_i(S). \tag{2.1}$$

The next theorem tells us that this map is contractive also.

Theorem 2.2 *If $w_i : \mathbb{R}^2 \to \mathbb{R}^2$ is contractive with contractivity s_i for $i = 1, \ldots, n$, then $W = \cup_{i=1}^{n} w_i : \mathcal{H}(\mathbb{R}^2) \to \mathcal{H}(\mathbb{R}^2)$ is contractive in the Hausdorff metric with contractivity $s = \max_{i=1,\ldots,n}\{s_i\}$.*

Proof: We will prove this for the case $n = 2$. The general case can be proved by induction. Select $A, B \in \mathcal{H}(\mathbb{R}^2)$, and let $\epsilon = h(A, B)$. Then $A \subset B(\epsilon)$ and $B \subset A(\epsilon)$. Let

$s = \max\{s_1, s_2\}$,

$A_i = w_i(A)$,

$B_i = w_i(B)$,

$A' = W(A) = A_1 \cup A_2$, and

$B' = W(B) = B_1 \cup B_2$.

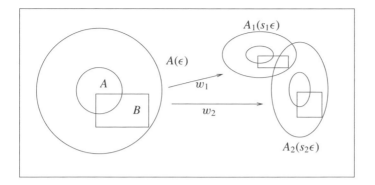

Figure 2.7: $w_i(A(\epsilon)) \subset A_i(s_i\epsilon)$, for $i = 1, 2$

Then we want to show that $h(W(A), W(B)) < s\, h(A, B)$, or $B' \subset A'(s\epsilon)$ and $A' \subset B'(s\epsilon)$. But this follows from $B_i \subset w_i(A(\epsilon)) \subset A_i(s_i\epsilon) \subset A_i(s\epsilon)$ and $A_i \subset w_i(B(\epsilon)) \subset B_i(s_i\epsilon) \subset B_i(s\epsilon)$, for $i = 1, 2$. See Figure 2.7. ∎

Recall the copy machine example of Chapter 1. Theorem 2.2 says that if the individual copies are reductions (that is, if the transformations composing the copies are contractive) then the resulting copying process W is contractive in the space of all possible images. Why do we care about this? Because just as the contractive map $x \mapsto x/2$ has a unique fixed point 0 which is the limit of iterating starting from any initial point, so will the contractive map W have a unique fixed picture, the limit of iterating W from any initial picture.

The Contractive Mapping Fixed-Point Theorem

We are now ready to state one of the main tools of this subject:

Theorem 2.3 (The Contractive Mapping Fixed-Point Theorem) *Let X be a complete metric space and $f : X \to X$ be a contractive mapping. Then there exists a unique point $x_f \in X$ such that for any point $x \in X$*

$$x_f = f(x_f) = \lim_{n \to \infty} f^{\circ n}(x).$$

Such a point is called a fixed point *or the* attractor *of the mapping f.*

Proof: Select $x \in X$. Then for $n > m$,

$$d(f^{\circ m}(x), f^{\circ n}(x)) < s\, d(f^{\circ m-1}(x), f^{\circ n-1}(x)) < s^m\, d(x, f^{\circ n-m}(x)). \tag{2.2}$$

Now, using this and the triangle inequality repeatedly,

$$d(x, f^{\circ k}(x)) \leq d(x, f^{\circ k-1}(x)) + d(f^{\circ k-1}(x), f^{\circ k}(x))$$

$$\leq \ d(x, f(x)) + d(f(x), f(f(x))) + \cdots +$$
$$d(f^{\circ k-1}(x), f^{\circ k}(x))$$
$$\leq \ (1 + s + \cdots + s^{k-2} + s^{k-1}) d(x, f(x))$$
$$\leq \ \frac{1}{1-s} d(x, f(x)). \tag{2.3}$$

We can then write Equation (2.2) as

$$d(f^{\circ m}(x), f^{\circ n}(x)) < \frac{s^m}{1-s} d(x, f(x)),$$

and since $s < 1$, the left side can be made as small as we like if n and m are sufficiently large. This means that the sequence $x, f(x), f(f(x)), \ldots$ is a Cauchy sequence, and since X is complete, the limit point $x_f = \lim_{n \to \infty} f^{\circ n}(x)$ is in X. By Lemma 2.1, contractivity of f implies that f is continuous, and so $f(x_f) = f(\lim f^{\circ n}(x)) = \lim f^{\circ n+1}(x) = x_f$.

To prove uniqueness: suppose x_1 and x_2 are both fixed points. Then $d(f(x_1), f(x_2)) = d(x_1, x_2)$, but we also have $d(f(x_1), f(x_2)) \leq s \, d(x_1, x_2)$, a contradiction. ∎

Corollary 2.1 (Collage Theorem) *With the hypothesis of the Contractive Mapping Fixed-Point Theorem,*

$$d(x, x_f) \leq \frac{1}{1-s} d(x, f(x)).$$

Proof: This is a consequence of taking the limit as k goes to infinity in Equation 2.3. ∎

Note that it is not necessary for f to be contractive for it to have a fixed point which attracts all of X. For example, it is sufficient for some iterate of f to be contractive. This leads to the following generalization of the previous theorem.

Definition 2.8 *Let f be a Lipschitz function. If there is a number n such that $f^{\circ n}$ is contractive, then we call f eventually contractive. We call n the* exponent of eventual contractivity.

Corollary 2.2 (Generalized Collage Theorem) *Let f be eventually contractive with exponent n; then there exists a unique fixed point $x_f \in X$ such that for any $x \in X$*

$$x_f = f(x_f) = \lim_{k \to \infty} f^{\circ k}(x).$$

In this case,

$$d(x, x_f) \leq \frac{1}{1-s} \frac{1-\sigma^n}{1-\sigma} d(x, f(x)),$$

where s is the contractivity of $f^{\circ n}$ and σ is the Lipschitz factor of f.

Proof: Let $g = f^{\circ n}$. We want to show that $f^{\circ k}$ converges to x_g; that is, $f^{\circ k}(x)$ is arbitrarily close to x_g for all sufficiently large k. For any k, we can write $k = q \, n + r$, with $0 \leq r < n$. So,

$$
\begin{aligned}
d(f^{\circ k}(x), x_g) &= d(f^{\circ qn+r}, x_g) \\
&\leq d(f^{\circ qn+r}(x), f^{\circ qn}(x)) + d(f^{\circ qn}(x), x_g) \\
&= d(g^{\circ q}(f^{\circ r}(x)), g^{\circ q}(x)) + d(g^{\circ q}(x), x_g) \\
&\leq s^q d(f^{\circ r}(x), x) + d(g^{\circ q}(x), x_g).
\end{aligned}
$$

But both of these terms can be made arbitrarily small for $0 \leq r < n$ and q sufficiently large. The fixed-point condition follows from the continuity of f, and uniqueness follows from the uniqueness of x_g.

For the inequality, we know from the Collage Theorem that:

$$d(x, x_g) \leq \frac{1}{1-s} d(x, g(x)) \tag{2.4}$$

and

$$
\begin{aligned}
d(x, g(x)) \;=\; & d(x, f^{\circ n}(x)) \leq \sum_{i=1}^{n} d(f^{\circ i}(x), f^{\circ i-1}(x)) \\
\leq \; & d(x, f(x)) \sum_{i=1}^{n} \sigma^{i-1} \\
\leq \; & \frac{1 - \sigma^n}{1 - \sigma} d(x, f(x)).
\end{aligned}
\tag{2.5}
$$

The result follows from Equations (2.4) and (2.5). ∎

Remarks: Note that it is not necessary for f to be contractive for all n that are sufficiently large. It is sufficient that f be contractive for *some* n. Also, it is possible for $W = \cup w_i$ to be eventually contractive if some of the w_i are not contractive; see Section C.18.

Fractals As Fixed Points

Now we are ready to apply the Contractive Mapping Fixed-Point Theorem. Let $w_1, \ldots, w_n$ be contractive transformations. Then the map $W = \cup w_i$ is contractive and defines a unique fixed point in $\mathcal{H}$. For example, we may choose w_i to be affine transformations of the plane, in which case the fixed point x_W is a subset of the plane which is often a fractal.

Definition 2.9 *Let X be a complete metric space. An* iterated function system *is a collection of contractive maps $w_i : X \to X$, for $i = 1, \ldots, n$.*

So an IFS is just a collection of maps that, by the Contractive Mapping Fixed-Point Theorem, defines a unique attractor. The point is that since the attractor is unique, it is completely specified by the map W. The image encoding problem, or the *inverse problem*, as it has come to be called, is: if we are given some set S, can we find an IFS whose attractor is S? No completely general, or even mildly satisfying, solution to this problem exists to date. However, some insight can be found in the fixed-point equation

$$x_W = W(x_W) = w_1(x_W) \cup \cdots \cup w_n(x_W)$$

and the Collage Theorem

$$d(S, x_W) \leq \frac{1}{1-s} d(S, W(S)).$$

The first says that the fixed point is constructed out of transformed copies of itself, so we should take the set S, transform it by contractive transformations and attempt to paste these together to reconstruct S. The uniqueness of the fixed point is important, because if, given S, we can find a W (or, if given W, we can find a set S) satisfying $S = W(S)$, then we will know that $S = x_W$; that is, S is the attractor of W. The second equation says that even if we can't paste the pieces to fit exactly, the better the fit between the original set S and the pasted "collage" $W(S)$, the closer the attractor x_W will be to S (barring the use of some transformations that are barely contractive, that is, with s close to 1).

An obvious question is: why not take W to slightly contract S, in which case $d(S, W(S))$ will be small, and hopefully $d(S, x_W)$ will be small too? This will fail because for such a W, the $\frac{1}{1-s}$ term is very large, so we don't know that $d(S, x_W)$ must be small. (In fact, it isn't small because if it were, there wouldn't be books on the subject.)

Figure 2.8 shows a simple example of using the fixed-point equation. Here the set S is a square, which can be exactly covered by four copies of itself reduced by a half. The transformations w_1, w_2, w_3, and w_4 are shown in Figure 2.8a, and it is clear that $S = w_1(S) \cup w_2(S) \cup w_3(S) \cup w_4(S) = W(S)$, which means that the fixed point x_W for W is S. The first three iterates of W starting from a circle are shown in Figure 2.8b. We can see that the iterates converge to S, as the Contractive Mapping Fixed-Point Theorem states.

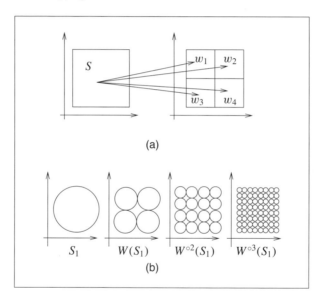

(a)

(b)

Figure 2.8: (a) The transformations w_1, w_2, w_3, w_4 that cover a square, and (b) the first three iterates of $W = \cup_{i=1}^4 w_i$.

Figure 2.9 shows two examples of IFS's, showing both the fixed point and the transformations. The transformations are shown as the parallelograms which are the images of the large square. The orientation of the images is shown by the ⊢. Note that each parallelogram contains a transformed (reduced) copy of the whole image, and that the union of the copies in each parallelogram forms the whole attractor. In the spiral, for example, the small parallelogram contains a copy of the whole image; the orientation is reversed, however. The transformation corresponding to the large parallelogram maps the whole attractor into itself, shifted by one sub-spiral. The union of this spiral and the small spiral forms the whole attractor. This is exactly what the fixed-point equation says. See Exercise 7 for other examples.

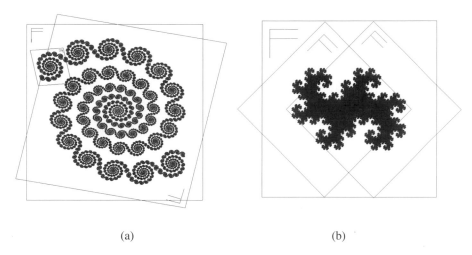

(a)   (b)

Figure 2.9: Two examples of IFS's with their attractors. The transformations of the IFS's are shown as the images of the square around the attractor.

2.3   Recurrent Iterated Function Systems

In this section, we review some of the work of Barnsley, Elton, and Hardin [5] on recurrent iterated function systems (RIFS). A RIFS is a collection $w_1, \ldots, w_n$ of Lipschitz maps in a complete metric space X, and an $n \times n$ matrix (p_{ij}) satisfying $\sum_j p_{ij} = 1$ for all i. An attractor for a RIFS is computed in the following way. Select an initial x_0 and an initial $i_0 \in \{1, \ldots, n\}$. Choose a number $i_{n+1} \in \{1, \ldots, n\}$ with the probability of $i_{n+1} = j$ equal to $p_{i_n j}$. Set $x_{n+1} = w_{i_{n+1}}(x_n)$, and in this way define a sequence $\{x_n\}$. The attractor A is defined as the set of points x such that every neighborhood of x contains infinitely many x_n for almost all x_0. Reference [5] is concerned with the existence, uniqueness and characterization of the limit set A. This is the most general definition of a RIFS, and we will not pursue it further. We will restrict ourselves to an alternative development, given below.

Suppose that for some reason one wished to generate a fern in which the leaves were constructed out of Sierpinski triangles. Is there a simple generalization to an IFS that would allow this? Yes. To do this, we could think of two IFS's working in parallel. One would construct an IFS for a Sierpinski triangle as before, and this IFS would feed its output into another IFS which would copy the triangle where the leaves would belong. The second IFS would also be responsible for making copies of the leaves to fill out the fern. This is shown in Figure 2.10, with the resulting attractor shown in Figure 2.11. RIFS are a general framework for specifying such processes.

In a RIFS, we imagine some number of disjoint planes, each containing an IFS. We allow maps between the planes, and these allow us to mix and match the attractors. See Figure 2.12.

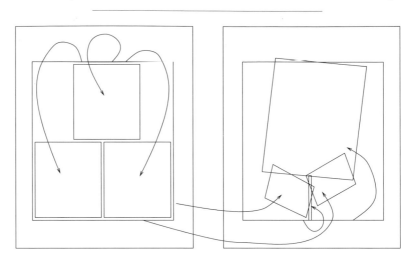

Figure 2.10: Two IFS's which are mixed to form a fern with Sierpinski triangle leaves.

Figure 2.11: The attractor resulting on the right of Figure 2.10.

The theory of recurrent IFS, as far as we are concerned, can fit into IFS theory. Let $X_1, \ldots, X_n$ be complete metric spaces, and let H_i be the set of non-empty compact subsets of X_i, with its associated Hausdorff metric h_i. Define

$$H = H_1 \times H_2 \cdots \times H_n.$$

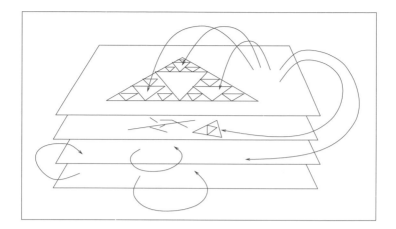

Figure 2.12: A RIFS consists of IFS's on disjoint planes with maps between the planes.

Then an element of H is an n-tuple of non-empty subsets $(A_1, \ldots, A_n)$. We metrize H with

$$h((A_1, \ldots, A_n), (B_1, \ldots, B_n)) = \max_i \{h_i(A_i, B_i)\}.$$

With this metric, H is a complete metric space.[4]

We now define a set of transformations $W_{ij} : H_j \to H_i$, some of which may map to the empty set.[5] The W_{ij} are constructed as a union of transformations $W_{ij} = \cup_k w_{ijk}$ where $w_{ijk} : X_j \to X_i$ is contractive. Because H has non-empty components, we require that for each i there is some W_{ij} with non-empty image.

We can now define $W : H \to H$ by

$$W(A_1, \ldots, A_n) = \left(\cup_j W_{1j}(A_j), \ldots, \cup_j W_{nj}(A_j) \right).$$

The reader is encouraged to show that the contractivity of W is the maximum of the contractivities of the w_{ijk}. By the Contractive Mapping Fixed-Point Theorem, W has a unique fixed point $x_W = (A_1, \ldots, A_n) \in H$.

The generality of the theory allows us to apply it to various types of data. For example, Barnsley and Jacquin discussed the compression of contour data in [8]. In this scheme, contour data was partitioned (in some unspecified way) into domain D_i and range R_i contour pieces, as in Figure 2.13. For each domain piece, a map w_i was chosen that mapped the endpoints of D_i onto the endpoints of R_i and that minimized the distance between $w_i(D_i)$ and R_i. The resulting map $W = \cup w_i$ has a fixed point (in a space of curves) that approximates the original contour.

---

[4]This is an immediate consequence of the completeness of the X_i.

[5]We could specify an index set I of pairs, such that $(i, j) \in I$ implies that we have some map W_{ij}. Instead, we let all pairs have some map, and we allow the case $W_{ij}(\cdot) = \emptyset$.

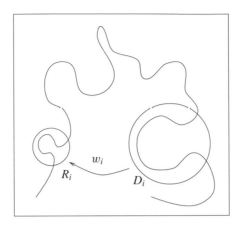

Figure 2.13: A map w_i mapping a contour domain D_i onto a contour range R_i.

The details of the partitioning or selection of domain-range pairs were not included in [8], but they did appear in the work of Boss and Jacobs [43]. In this scheme, the original contour was segmented and each segment was classified by its curvature at different scales. The segmentation was regular and classification was used to speed the massive domain-range comparison. Although the scheme actually resulted in image expansion – that is, more data was required to store the RIFS than to cleverly store the original contour – the scheme was one of the first publically available automated schemes to encode measured data using fractals. We will apply these ideas to encoding images, after we have settled on an appropriate model for an image.

2.4   Image Models

We are interested in encoding monochrome images, not black-and-white images that are representations of subsets of the plane. Color images are typically extensions of the grey-scale representation of an image, so we will focus only on grey-scale models. Three image models are generally used: measure spaces, pixelized data, and functions. We discuss these below.

Measures As Image Models

An image can be represented as a measure μ in the plane, in which case the intensity can be measured on subsets A of the plane by

$$\mu(A) = \int_A d\mu.$$

In this model, it makes no sense to speak of the intensity at a point, which is reasonable. Our eye, for example, measures the light flux through an area, not at a point. Barnsley and Hurd's book [7] stresses this model, but it is not used in this book.

Pixelized Data

In this model, an image is represented as a collection of discrete picture elements or pixels. Each pixel value takes on discrete values (typically ranging from 0 to 255) representing a grey level.[6] The number of bits per pixel used to store the pixel values determines the grey-scale resolution, while the total number of pixels determines the image resolution. Computer implementations can only work on this type of representation. Chapter 7 contains an analysis and results based on this representation, as do Chapters 8, 9, and 10.

An image containing $M \times N$ pixels can be thought of as a vector in an $n = MN$-dimensional space. Typically, the discrete nature of the pixel values is ignored, so the space is $\mathbb{R}^n$. Common norms on $\mathbb{R}^n$ are the p-norms, defined by

$$||x||_p = \left(|x_1|^p + |x_2|^p + \cdots + |x_n|^p\right)^{\frac{1}{p}},$$

which define metrics by

$$d_p(x, y) = ||x - y||_p.$$

The ∞-norm is given by

$$||x||_\infty = \max_{i=1,\ldots,n} |x_i|.$$

The 2-norm is also called the ℓ^2 norm, and its induced metric is often called the rms metric. Thus, if $\mathbf{x} = (x_1, \ldots, x_n)$ and $\mathbf{y} = (y_1, \ldots, y_n)$ are images, the ℓ^2, or rms, difference between them is given by

$$d_{rms}(\mathbf{x}, \mathbf{y}) = ||\mathbf{x} - \mathbf{y}||_2 = \sqrt{\frac{1}{n} \sum_{i=1}^{n}(x_i - y_i)^2}.$$

The ℓ^2 metric is far more convenient to use than other metrics because if we denote the standard inner product by $\langle \cdot, \cdot \rangle$, then $d_2(\mathbf{x}, \mathbf{y}) = \sqrt{\langle \mathbf{x} - \mathbf{y}, \mathbf{x} - \mathbf{y} \rangle}$. This means that certain minimization problems become easy to solve. For example, we can find a, b which minimize $d_2(a\mathbf{x} + b\mathbf{y}, \mathbf{z})$ by minimizing

$$
\begin{aligned}
d_2^2(a\mathbf{x} + b\mathbf{y}, \mathbf{z}) &= \langle a\mathbf{x} + b\mathbf{y} - \mathbf{z}, a\mathbf{x} + b\mathbf{y} - \mathbf{z} \rangle \\
&= a^2 \langle \mathbf{x}, \mathbf{x} \rangle + 2ab\langle \mathbf{x}, \mathbf{y} \rangle + b^2 \langle \mathbf{y}, \mathbf{y} \rangle - \\
&\quad 2a \langle \mathbf{x}, \mathbf{z} \rangle - 2b\langle \mathbf{y}, \mathbf{z} \rangle + \langle \mathbf{z}, \mathbf{z} \rangle.
\end{aligned}
$$

Differentiating with respect to a and b to find the maximum and solving the resulting linear equations gives

$$a = \frac{\langle \mathbf{y}, \mathbf{z} \rangle \langle \mathbf{x}, \mathbf{y} \rangle - \langle \mathbf{y}, \mathbf{y} \rangle \langle \mathbf{x}, \mathbf{z} \rangle}{\langle \mathbf{x}, \mathbf{y} \rangle^2 - \langle \mathbf{x}, \mathbf{x} \rangle \langle \mathbf{y}, \mathbf{y} \rangle} \tag{2.6}$$

and

$$b = \frac{\langle \mathbf{x}, \mathbf{y} \rangle \langle \mathbf{x}, \mathbf{z} \rangle - \langle \mathbf{x}, \mathbf{x} \rangle \langle \mathbf{y}, \mathbf{z} \rangle}{\langle \mathbf{x}, \mathbf{y} \rangle^2 - \langle \mathbf{x}, \mathbf{x} \rangle \langle \mathbf{y}, \mathbf{y} \rangle}. \tag{2.7}$$

This is the least squares solution.

---

[6]This value can also be an index into a table of colors. Alternatively, pixels may contain three values representing red, green, and blue levels.

In practice, the Peak signal-to-noise ratio (PSNR) is used to measure the difference between two images. It is defined as

$$PSNR = 20 \log_{10} \left(\frac{b}{rms} \right),$$

where b is the largest possible value of the signal (typically 255), and rms is the rms difference between two images. The PSNR is given in decibel units (dB) which measure the ratio of the peak signal and the difference between two images. An increase of 20 dB corresponds to a ten-fold decrease in the rms difference between two images. There are many versions of signal-to-noise ratios, but the PSNR is very common in image processing, probably because it gives better-sounding numbers than other measures.

Functions on $\mathbb{R}^2$

A function $f : \mathbb{R}^2 \to \mathbb{R}$ can be thought of as an image of infinite resolution. Because real images are finite in extent, we take the domain of f to be the unit square $I^2 = \{(x, y) \mid 0 \le x, y \le 1\}$ and the range to be $I = \{x \mid 0 \le x \le 1\}$, with the value of f representing the grey level. This is the model we adopt in this chapter.

For functions, the metric is inherited from the $L^p(I^2)$ norms. The L^p metrics are

$$d_p(f, g) = \|f - g\|_p = \left(\int_{I^2} |f - g|^p \right)^{\frac{1}{p}},$$

so that the rms metric corresponds to L^2. Equations (2.6) and (2.7) hold in this case too, with $\langle f, g \rangle = \int_{I^2} fg$.

There are two other metrics of interest: the L^∞ metric and the supremum metric. The supremum metric of two functions f and g is

$$d_{sup}(f, g) = \sup_{(x,y) \in I^2} |f(x, y) - g(x, y)|. \qquad (2.8)$$

It finds the point $(x, y) \in I^2$ where f and g differ the most and sets this difference as the distance between f and g. This metric is very easy to work with, but it is impractical. It is not reasonable to have the distance between two functions depend on just one point (for example, if the functions are identical except at one point).

The L^∞ metric is less sensitive than the supremum metric. To define it, we define the essential supremum of f to be $\inf\{\beta \mid \mu(f^{-1}((\beta, \infty))) = 0\}$, where μ is the Lebesgue measure. The number β is the smallest value such that $f(x) > \beta$ on a set of measure 0. We define the L^∞ distance between f and g to be the essential supremum of $|f - g|$. This metric avoids the problems of the supremum metric, but is harder to work with.

Note that we are actually concerned with the convergence of an iterative process and not so much with the distance. For functions, the notions of *pointwise convergence* and *almost everywhere (a.e.) convergence* are important. These notions do not depend on a metric in the space of functions, just a metric on the range of the functions. We say that f_n converges to f pointwise if for each x we have $f_n(x) \to f(x)$. Almost everywhere convergence means pointwise convergence except on a set of measure zero. It is possible, thus, to find a sequence of convergent images f_n without discussing a particular metric at all. In fact, it is possible to find examples for which $f_n \to f$ pointwise but $f_n \not\to f$ for any L^p norm (see Chapter 11).

Finally, given a function $f : \mathbb{R}^2 \to \mathbb{R}$, the *graph* of f is a subset of $\mathbb{R}^3$ consisting of all the triplets $(x, y, f(x, y))$. Formally, the graph of f is different from f. However, the two contain the same information, and so we sometimes use them interchangeably.

Color

It is fortunate that the human visual system uses only three color channels to encode color information that is sent to the brain. This means that colors, which correspond to different wavelengths of light, can be simulated by a superposition of three primary colors, typically red, green, and blue. When a color image is digitized, three filters are used to extract the red, green, and blue (RGB) intensities in the image, and when these are recombined we perceive them as some color. This simplifies important things, such as color TV, and less important things, such as this section. See [12] for more information on the human visual system.

A method that can encode a monochrome image can be used to encode color images. We describe one scheme here, but first, here is what *not* to do: encoding the RGB components separately is not reasonable. This is because the human visual system is not particularly sensitive to color information, and there are ways of compressing color information to take advantage of this insensitivity.

Given a RGB triplet (R_i, G_i, B_i) for pixel i, we compute the YIQ values[7]

$$\begin{bmatrix} Y_i \\ I_i \\ Q_i \end{bmatrix} = \begin{bmatrix} 0.299 & 0.587 & 0.114 \\ 0.596 & -0.274 & -0.322 \\ 0.211 & -0.523 & 0.312 \end{bmatrix} \begin{bmatrix} R_i \\ G_i \\ B_i \end{bmatrix}.$$

This yields the three YIQ channels: the Y, or luminance, signal measures the perceived brightness of the color; the the I, or hue, signal is the actual color; and the Q, or saturation, signal measures the vividness or depth of the color. The I and Q channels are also called the chrominance signals. The original RGB signals can be computed using the inverse transformation

$$\begin{bmatrix} R_i \\ G_i \\ B_i \end{bmatrix} = \begin{bmatrix} 1.000 & 0.956 & 0.621 \\ 1.000 & -0.273 & -0.647 \\ 1.000 & -1.104 & 1.701 \end{bmatrix} \begin{bmatrix} Y_i \\ I_i \\ Q_i \end{bmatrix}.$$

It is possible to significantly compress the I and Q signals with little to no apparent degradation.[8] The I and Q channels are typically decimated to one-half or one-quarter of their original size and compressed (usually at a lower bit rate than the Y channel). This means that the overall compression ratio for color images is significantly greater. When we compress

---

[7]These matrices have the curious property of never being identical in any two references. The differences are inconsequential, being in the third digit; they are due to roundoff errors and the selection of one matrix as definitive and the other as its inverse.

[8]This is, roughly, why monochrome and color television sets can operate using the same broadcast signals. In North America, which uses the NTSC television broadcasting standard, color and monochrome television sets receive YIQ signals. Only the Y signal is displayed on a black and white set, while the I and Q color information is added to the Y signal after being amplitude modulated at around 3.5 Mhz in two phases. (A brief pulse used to tell the television to begin scanning the next line contains a synchronization "color burst" signal that was needed before stable oscillators could be made.) Color television sets combine the I and Q color signal with the Y signal to form a color signal. Because the I and Q signals can be highly compressed, there is enough time to send them during this rapid horizontal sync pulse. This allowed the introduction of color televisions without the need for a new broadcasting method. The two color-control knobs on color television sets essentially control the I and Q channels.

three channels, there is three times as much data to compress, and this is reduced by a factor of about 1.1 to 1.5 for compression of the YIQ signals. So we can expect a color image to have a compression ratio that is larger by a factor of 2 to 2.7 than the compression ratio of the same monochrome image at about the same perceived quality. See, for example, [16] for more information.

The decoding process consists of computing the Y and the reduced I and Q signals; expanding the I and Q signals to the original image size (the resolution independence of the fractal schemes is convenient here, although not necessary); and computing the RGB components from the Y and expanded IQ components. This book contains no results on color, but the scheme has been successfully implemented in conjunction with a fractal scheme. with the expected results: it works.

2.5   Affine Transformations

An affine transformation $w : \mathbb{R}^n \to \mathbb{R}^n$ can always be written as $w = Ax + b$, where $A \in \mathbb{R}^{n \times n}$ is an $n \times n$ matrix and $b \in \mathbb{R}^n$ is an offset vector. Such a transformation will be contractive exactly when its linear part is contractive, and this depends on the metric used to measure distances. If we select a norm $|| \cdot ||$ in $\mathbb{R}^n$ then $x \mapsto Ax$ is contractive when

$$||A|| = \sup_{\vec{x} \in \mathbb{R}^n} ||A\vec{x}|| / ||\vec{x}|| < 1.$$

We list without proof the norm of A for $p = 1, 2, \infty$.

$$||A||_1 \quad = \max_{1 \le j \le n} \sum_{i=1}^{n} |a_{ij}|$$

$$||A||_2 \quad = \text{square root of the largest eigenvalue of } A^T A$$

$$||A||_\infty \quad = \max_{1 \le i \le n} \sum_{j=1}^{n} |a_{ij}|$$

See Exercise 8 for an expression for the norm of A in the 2×2 case.

When, as is most common, we have an IFS in $\mathbb{R}^2$, we can understand the action of A geometrically, as is shown in Figure 2.14. A linear transformation can scale with A_s, stretch with A_t, skew with A_u, and rotate with A_θ. Combinations of these maps can map in a variety of ways, though we can see that linear transformations always map parallelograms to parallelograms. See also Exercise 9.

2.6   Partitioned Iterated Function Systems

Our goal now is to repeat the "IFS recipe" for generating fractals but in a more general setting. We will find maps $w_1, \ldots, w_n$, collect them into a transformation $W = \cup w_i$, show that under certain conditions W is contractive, deduce that it has a fixed point, and attempt to select W so that its attractor is close to some given image.

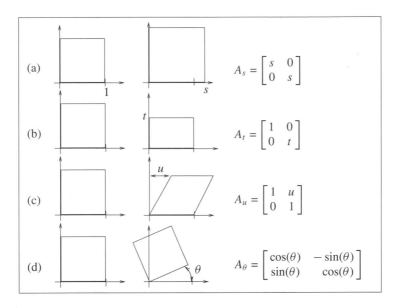

Figure 2.14: A linear transformation in $\mathbb{R}^2$ can (a) scale, (b) stretch, (c) skew, and (d) rotate.

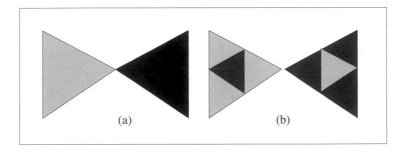

Figure 2.15: (a) A "bow tie" with the domains of a PIFS indicated by shading, and (b) the ranges of the transformations indicated by shading.

The partitioned iterated function system (PIFS) is a generalization of an IFS. In a PIFS, the domains of the w_i are restricted. This simplifies the encoding of more general, non-self-similar sets. For example, it is not easy to see how the "bow tie" in Figure 2.15 can be encoded by an IFS (see Exercise 10), but it is easy to encode it using a PIFS: with eight transformations, four restricted to the right half and four restricted to the left half, we can generate the two triangles of the bow tie separately. The formal definition is:

Definition 2.10 *Let X be a complete metric space, and let $D_i \subset X$ for $i = 1, \ldots, n$. A partitioned iterated function system is a collection of contractive maps $w_i : D_i \to X$, for $i = 1, \ldots, n$.*

We wish to use the Contractive Mapping Fixed-Point Theorem, as before, to define a unique fixed point for a PIFS. But this is not possible to do in complete generality. One problem is that since the domains are restricted, the initial point is very important; if we are not careful, we may end up with an empty set after one iteration. In the specific case of image representation, however, this is not a problem. A complete theory of PIFS remains to be developed, but many special cases are understood; they are discussed in later chapters.

2.7   Encoding Images

In this section we will use the same notation as in the IFS case. This is an advantage for the novice who understands IFS theory but a disadvantage for a more sophisticated approach. The notation is cumbersome and hides what is really happening. We discuss better notation in Section 2.8.

We will work with the space $F = \{f : I^2 \to \mathbb{R}\}$ of images defined as the graphs of (measurable) functions over the unit square I^2 with values in $I = [0, 1]$. We allow the range of f to be in $\mathbb{R}$, however, so that sums and differences of images are still well defined. We will use the supremum metric, given in Equation (2.8).

Claim 2.1 *The space F with the metric d_{sup} is complete.*

Proof: See [75].   ∎

Let $D_1, \ldots, D_n$ and $R_1, \ldots, R_n$ be subsets of I^2 which we will call *domains* and *ranges*, even though these are not exactly domains and ranges. It is natural (or at least it is common usage) to confuse a subset of I^2 and the function that is defined over that subset. Thus, although we call D_i and R_i the domain and range of w_i, the actual domain and range are the Cartesian products $D_i \times I$ and $R_i \times I$.

Let $v_1, \ldots, v_n : I^3 \to I^3$ be some collection of maps. Define w_i as the restriction

$$w_i = v_i|_{D_i \times I}.$$

The maps $w_1, \ldots, w_n$ will form the PIFS. The idea is that w_i have restricted domains, but can otherwise be very general. The map w_i operates on f by $w_i(f) = w_i(x, y, f(x, y))$, and we think of the result as the graph of a function on I^2. Not every w_i can do this, which is what we discuss next.

Definition 2.11 *The maps $w_1, \ldots, w_n$ are said to* tile *I^2 if for all $f \in F$, $\bigcup_{i=1}^{n} w_i(f) \in F$.*

This means the following: when w_i is applied to the part of the image $f \cap (D_i \times I)$ above D_i, the result must be a graph of a function over R_i, and $I^2 = \cup_{i=1}^n R_i$. See Figure 1.8, page 12. Tiling implies that the union $\cup_{i=1}^n w_i(f)$ yields a graph of a function over I^2, and that the R_i's are disjoint.

Encoding Images with Recurrent Iterated Function Systems

The RIFS model can be used for image encoding. Since a RIFS is just a glorified IFS on a large space, it is possible to claim (for those who are concerned with such details) that this represents an IFS solution to the inverse problem. However, we must make the simplifying assumption, that domains are composed of unions of ranges.

If we are given $w_1, \ldots, w_n : I^3 \to I^3$, then we can let $D_i \subset I^2$ denote the domain of w_i and $R_i \subset I^2$ its range.[9] We also assume that w_i tile I^2, so in particular, $\cup_i R_i = I^2$. Our simplifying assumption is that each D_i is exactly equal to a union of some ranges.

If we consider each range R_i to be a space disjoint from the other ranges, we can compose a RIFS on $H = (R_1 \times I, \ldots, R_n \times I)$ by defining

$$W_{ij} = \begin{cases} w_i & \text{if } R_j \subset D_i \\ \emptyset & \text{otherwise.} \end{cases}$$

In an abuse of notation, we think of w_i both as a mapping $w_i : D_i \times I \to R_i \times I$ and as a mapping on a component of H. We think of the components of H as being spaces of graphs of functions over each R_i, and we use the supremum metric on these spaces. Thus, the metric on H is

$$d_1((A_1, \ldots, A_n), (B_1, \ldots, B_n)) = \sup_i \{d_{sup}(A_i, B_i)\},$$

and we claim that H with this metric is a complete metric space. We now define W as in Section 2.3

$$W(A_1, \ldots, A_n) = \left(\bigcup_j W_{1j}(A_j), \ldots, \bigcup_j W_{nj}(A_j) \right).$$

This can also be written

$$W(A_1, \ldots, A_n) = \left(\bigcup_{\{j \mid R_j \subset D_1\}} w_1(A_j), \ldots, \bigcup_{\{j \mid R_j \subset D_n\}} w_n(A_j) \right).$$

The Contractive Mapping Fixed-Point Theorem now assures us that we have a fixed point in H, and this fixed point can be embedded in $\mathbb{R}^3$ by the embeddings of the components of H into $\mathbb{R}^3$.

Since the w_i tile I^2, the fixed point can be thought of as the graph of a function $f : I^2 \to \mathbb{R}$. This follows because we know that if our initial point is a function, then each application of W is a function, and so is the limit. However, we must choose a range to dominate in the case when the attractor differs on the boundary of adjacent R_i. See Exercise 11 for an alternative method of representing an image in a RIFS model.

---

[9] As before, we will refer to D_i as the domain and R_i as the range, even though the actual domains and ranges are $D_i \times I$ and $R_i \times I$.

Remark 1: Note that the Hausdorff metric we use is h_{d_1}; it is the metric induced by the supremum metric on each component of H. In this metric, a convergent sequence of graphs of functions converges to a graph of a function. If we take $H = (H_1, H_2, \ldots, H_n)$, where H_i consists of the compact subsets of $R_i \times I$ with the Hausdorff metric h_d (d being the Euclidean metric), then the limit of graphs of functions may not be a graph of a function (although it is otherwise a perfectly healthy attractor). For example, $\lim_{n \to \infty} \arctan(nx)$ is not convergent using h_{d_1}, but if we use h_d, the limit converges to $\{(0, y) \mid -\pi/2 \le y \le \pi/2\} \cup \{(x, y) \mid x \in \mathbb{R}, \ y = \pi/2 \text{ when } x \ge 0, \ y = -\pi/2 \text{ when } x \le 0\} \subset \mathbb{R}^2$, which is not the graph of a function.

If we use h_d, then it is not necessary to start iterating with the graph of a function. We can begin iterating with any element of H and reach a unique fixed point. The condition that the w_i be contractive changes, in this case, since the contractivity is measured in a different metric. We don't have to insist that the w_i tile I^2, either. At this point, however, there is little relation between images and the fixed points of such systems.

Remark 2: If we eliminate the simplifying assumption, the definition of W_{ij} no longer makes sense; in that case, R_j may be mapped into two different components of H.

Remark 3: Since the supremum metric is only sensitive to what happens in the z direction, there is no condition on the spatial part – that is, the xy component – of each w_i. This part may be expansive. In fact, we will reach the fixed point by iteration if W is eventually contractive. Thus, it is possible to choose some w_i to be non-contractive without sacrificing the model.

Remark 4: RIFS theory can be used to gain some knowledge about the attractor and the encoding. The main theoretical results in this book use this model because it is far more tractable than the more general PIFS model. For example, if all the w_i are similitudes with contractivity s_i,[10] then the fractal dimension of the attractor can be computed in the following way (see [5]). Let $S(t)$ be the $n \times n$ diagonal matrix composed of the s_i^t,

$$S(t) = \mathrm{diag}(s_1^t, \ldots, s_n^t).$$

Then the fractal dimension of the attractor is the unique positive number d such that $S(d)C$ has a maximal eigenvalue of modulus 1, where C is the connectivity matrix

$$C_{ij} = \begin{cases} 1 & \text{if } R_i \subset D_j \\ 0 & \text{otherwise.} \end{cases}$$

Encoding Images with PIFS

In this section, we eliminate the simplifying assumption of the previous section while maintaining the IFS notation.

Definition 2.12 *If $w : \mathbb{R}^3 \to \mathbb{R}^3$ is a map with $(x', y', z_1') = w(x, y, z_1)$ and $(x', y', z_2') = w(x, y, z_2)$, then w is called z contractive if there exists a positive real number $s < 1$ such that*

$$|z_1' - z_2'| \le s|z_1 - z_2|$$

and if x' and y' are independent of z_1 or z_2, for all x, y, z_1, z_2.

---

[10] A similitude w is a map for which equality always holds in the Lipschitz inequality.

Lemma 2.2 *If $w_1, \ldots, w_n$ are z-contractive, then*

$$W = \bigcup_{i=1}^{n} w_i$$

is contractive in F with the supremum metric.

Proof: Let $s = \max_i s_i$, where s_i are the z contractivities of the w_i.

$$
\begin{aligned}
d_{sup}(W(f), W(g)) &= \sup\{|W(f)(x, y) - W(g)(x, y)| \; : \; (x, y) \in I^2\} \\
&= \sup\{z\text{-component of } |w_i(x, y, f(x, y)) - w_i(x, y, g(x, y))| \; : \\
&\qquad (x, y) \in D_i, i = 1, \ldots, n\} \\
&\leq \sup\{s_i|f(x, y) - g(x, y)| \; : \; i = 1, \ldots, n\} \\
&\leq \sup\{s|f(x, y) - g(x, y)|\} \\
&\leq s \sup\{|f(x, y) - g(x, y)|\} \\
&\leq s d_{sup}(f, g)
\end{aligned}
$$

∎

Notice that the z independence of the x and y coordinates of w_i is used in going from the equality to the first inequality above. We have shown that if we select the w_i to be contractive in the z direction, for example if w_i is of the form

$$
w_i \begin{bmatrix} x \\ y \\ z \end{bmatrix} = \begin{bmatrix} a_i & b_i & 0 \\ c_i & d_i & 0 \\ 0 & 0 & s_i \end{bmatrix} \begin{bmatrix} x \\ y \\ z \end{bmatrix} + \begin{bmatrix} e_i \\ f_i \\ o_i \end{bmatrix},
\tag{2.9}
$$

with $s_i < 1$, then $W = \cup_i w_i$ is contractive in d_{sup}. The tiling condition assures us that when we evaluate $W = \cup w_i$ we get a function again. This is necessary so that we can apply W the next time (since it takes a function as input). Contractivity of W in F then determines a unique fixed point in F by the Contractive Mapping Fixed-PointTheorem.

Note that z contractive maps may be expanding in the x and y directions. In fact, as with the RIFS model, we are happy with an eventually contractive W, so that in practice there are no a priori conditions on any particular w_i.

We are now ready to encode a given image.

Encoding a Given Image

Given a collection of z contractive mappings $w_1, \ldots, w_n$ that tile I^2, we know that $W = \cup_i w_i$ defines a unique attractor x_W in the space of images. Given W, it is easy to find the image that it encodes: begin with any image f_0 and successively compute $W(f_0), W(W(f_0)), \ldots$ until the images converge to x_W. The converse is considerably more difficult: given an image f, how can a mapping W be found such that $x_W = f$? A general non-trivial solution to this problem remains elusive. Instead, an image $f' \in F$ is sought such that $d(f, f')$ is minimal with $f' = x_W$. Since $x_W = W(x_W) = \bigcup_{i=1}^{n} w_i(x_W)$, it is reasonable to seek domains $D_1, \ldots, D_n$ and corresponding transformations $w_1, \ldots, w_n$ such that

$$f \approx W(f) = \bigcup_{i=1}^{n} w_i(f). \tag{2.10}$$

This equation says: cover f with parts of itself; the parts are defined by the D_i and the way those parts cover f is determined by the w_i. Equality in Equation (2.10) would imply that $f = x_W$. Being able to exactly cover f with parts of itself is not likely, so the best possible cover is sought with the hope that x_W and f will not look too different, i.e. that $d(f, x_W)$ is small. The hope for this comes from the Collage Theorem. The goal is to find a W such that $d(f, W(f))$ is minimized and such that s is small. In that case, x_W will be close (in d) to f. However, as was shown in [27], the bound in the corollary is not very good; it provides motivation only, and it is not a useful bound in practice. In fact, it is possible to generate examples in which the bound in the corollary is arbitrarily large while $d(f, x_W)$ is bounded. Also, empirical results show that restricting s to be small results in a *larger* error $d(f, x_W)$. So we lack a good motivation, but nevertheless, the procedure works. A more motivating Collage Theorem can be found in [65].

Thus, the encoding process is: partition I^2 by a set of ranges R_i. For each R_i, a $D_i \subset I^2$ and $w_i : D_i \times I \to I^3$ are sought such that $w_i(f)$ is as close to $f \cap (R_i \times I)$ as possible; that is,

$$d(f \cap (R_i \times I), w_i(f)) \tag{2.11}$$

is minimized. The map W is specified by the maps w_i. The w_i's must be chosen such that W or $W^{\circ m}$ is contractive. Specific partitioning methods and ways of storing W compactly are presented in later chapters, including the next one.

Eventually Contractive Maps

A brief explanation of how a transformation W can be eventually contractive but not contractive is in order. The map W is composed of a union of maps w_i operating on disjoint parts of an image. The iterated transform $W^{\circ m}$ is composed of a union of compositions of the form

$$w_{i_1} \circ w_{i_2} \circ \cdots w_{i_m}.$$

The product of the contractivities bounds the contractivity of the compositions, so the compositions will be contractive if each contains sufficiently contractive w_{i_j}. Thus, W will be eventually contractive (in the supremum metric) if it contains sufficient "mixing" so that the contractive w_i eventually dominate the expansive ones. In practice, given a PIFS, this condition is relatively simple to check in the supremum metric.

To check whether W is eventually contractive, we define new maps w_i', which have the same spatial (xy) component as w_i, but whose z component is $s_i z$, where s_i is the contractivity of w_i. Then we define $W' = \cup_i w_i'$ and iterate W' starting with $f(x, y) = 1$. If $W'^{\circ m}(f) < 1$, then W is eventually contractive with exponent m.

2.8   Other Models

The IFS notation for PIFS is ugly. Better notation and theory are discussed later in the book; thus, we only briefly list some of the other approaches.

Chapter 11 discusses the following approach. Consider an image as an element of $f \in L^p(I^2)$. The goal is to find functions $m : I^2 \to I^2$, $s : I^2 \to \mathbb{R}$ and $o : I^2 \to \mathbb{R}$ so that a given function f will be a fixed point for the operator $\tau_{m,s,o} : L^p(I^2) \to L^p(I^2)$ given by

$$\tau_{m,s,o}(f) = s(x)f(m(x)) + o(x).$$

In that case, m, s and o compose an encoding of f. Here $s(x)$ is a generalized version of the s_i in w_i. The function $m(x)$ represents the spatial part of the w_i transformation. Given $w_1, \ldots, w_n$, we can select $m(x)$ and $s(x)$ to be piecewise constant (on the ranges of the w_i) with $m(x)$ mapping the range of each w_i to its domain. (This has the unfortunate disadvantage of reversing the meaning of range and domain. Oh well!)

If we consider an image as a vector $f \in \mathbb{R}^n$ then we can write τ as an affine matrix operator,

$$\tau(f) = Sf + O,$$

where $O \in \mathbb{R}^n$ is a vector that plays the role of $o(x)$ and S does several things. First, S permutes the values in f according to the action of $m(x)$ (or the spatial part of each w_i); second, S scales the values of f according to the action of $s(x)$ (or the s_i in each w_i). This model is discussed in Chapters 7 – 10, as well as briefly in Chapter 11.

Weighted Finite Automata

The weighted finite automata (WFA) appears to be, at first glance, a completely different method of generating fractals and encoding images. Chapter 13 contains a description of the method, so we restrict ourselves here to a brief intuitive discussion. In some loose sense, a WFA restricts the type of transformations used in an IFS and allows more IFS's to "mix."

Roughly, a WFA consists of four $n \times n$ matrices W_0, W_1, W_2, W_3, and two n vectors $F = (F_1, \ldots, F_n)$ and $I = (I_1, \ldots, I_n)$. To this we add a collection of square images $\psi_1, \ldots, \psi_n$. These are not part of the definition, but for now it is instructive to think of them as given.

An image f encoded by the WFA is a linear combination of the (grey levels of the) images ψ_i,

$$f = \sum_{i=1}^{n} I_i \psi_i.$$

The components F_i of F consist of the average grey values for the image ψ_i. Finally, the images ψ_i are defined on each of their quadrants by linear combinations of each other. If we divide each image into four equal squares and label these quadrants by 0, 1, 2, 3, then the a-th quadrant of ψ_i is given by $\sum_j (W_a)_{ij} \psi_j$. The ψ_i can be generated from F and these relations, which is why they are not part of the definition.

Because the ψ_i are defined in terms of themselves, the WFA can easily generate fractals. The transformations in the WFA can map each image only into a quadrant, but the linear combination of images compensates for this restriction. The success of WFA encoding of images suggests that some hybrid consisting of WFA with more generalized transformations may yield even better results.

Chapter 3

Fractal Image Compression with Quadtrees

Y. Fisher

This chapter describes a quadtree-based fractal encoding scheme. The scheme is an extension of the one presented in [45] and is similar to that discussed in [26], [27], and in Chapters 7, 8, and 9. While the results are not optimal, this scheme serves as a reference point for other fractal schemes. It is also a "typical" fractal scheme and a good first step for those wishing to implement their own. The code used to generate the results presented in this chapter is given in Appendix A, along with a detailed explanation of its inner workings.

3.1 Encoding

The encoding step follows the pseudo-code in Table 1.1, page 19. In this case, the collection of ranges comes from a quadtree partition of the image, and the collection of domains consists of subsquares of the image that are twice the range size.

The Ranges

A quadtree partition is a representation of an image as a tree in which each node, corresponding to a square portion of the image, contains four subnodes, corresponding to the four quadrants of the square. The root of the tree is the initial image. See Figure 3.1.

The ranges are selected as follows: after some initial number of quadtree partitions are made (corresponding to a minimum tree depth), the squares at the nodes are compared with domains

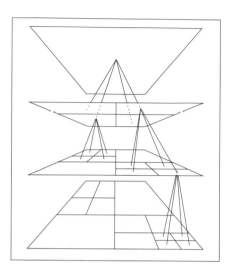

Figure 3.1: A quadtree partition is a representation of an image as a tree in which each node potentially has four subnodes. Each node of the tree corresponds to a square which is a quadrant of its parent's square.

from the domain library (or domain pool) **D**, which are twice the range size. The pixels in the domain are averaged in groups of four so that the domain is reduced to the size of the range, and the affine transformation of the pixel values (that is, a scaling and offset) is found that minimizes the rms difference between the transformed domain pixel values and the range pixel values (see Section 1.6). All the potential domains are compared with a range. If the resulting optimal rms value is above a preselected threshold and if the depth of the quadtree is less than a preselected maximum depth, then the range square is subdivided into four quadrants (which means adding four subnodes to the node corresponding to the range), and the process is repeated. If the rms value is below the threshold, the optimal domain and the affine transformation on the pixel values are stored – this constitutes one map w_i. The collection of all such maps $W = \cup w_i$ constitutes the encoding.

The above description must be revised a bit in order to fit the actual algorithm. First, the affine transformation on the pixel values consists of a multiplication by a scaling factor s_i and an addition of an offset value o_i (these can be thought of as a contrast and brightness adjustment). These values must be quantized for efficient storage, and it is these quantized values that are used when computing the rms difference. Second, if the magnitude of the optimal scaling factor $|s_i|$ is greater than 1 for any of the w_i, there is a chance that the resulting map W will not be eventually contractive. Thus, $|s_i|$ larger than some maximal value s_{max} are truncated.

Other ways of selecting the ranges are discussed in Chapter 6. Conditions on s_i for contractivity of W are discussed in Chapters 7 and 11. See also Sections C.12 and C.16.

The Domains

There are three major types of domain libraries – D_1, D_2, and D_3. The first has a roughly equal number of domains of each size. The second domain library has more larg domains and fewer small domains, the idea being that it is more important to find a good domain-range fit for larger ranges, since then the encoding will require fewer transformations. The third domain library has more small domains than large domains. The domains are selected as subsquares of the image whose upper-left corners are positioned on a lattice determined by a parameter l. The lattice spacing is selected as follows:

D_1 has a lattice with a fixed spacing equal to l.

D_2 has a lattice whose spacing is the domain size divided by l. Thus, there will be more small domains than large. This is the domain pool used by Jacquin in [45].

D_3 has a lattice as above but with the opposite spacing–size relationship. The largest domains have a lattice corresponding to the smallest domain size divided by l, and vice versa.

It is important to note that each domain can be mapped onto a range in eight different ways – four rotations and a flip with four more rotations. Thus, the domain pool can be thought of as containing the domains in eight possible orientations. In practice, the classification scheme discussed below defines a fixed rotation, so that the domains are only considered in one or two orientations.

An alternative method of selecting the domains is discussed in Section 6.6. In Chapter 12, the domain pool contains just one domain, leading to fast encoding times and a different contractivity condition.

The Classification

The domain-range comparison step of the encoding is very computationally intensive. We use a classification scheme in order to minimize the number of domains compared with a range. Before the encoding, all the domains in the domain library are classified; this avoids reclassification of domains. During the encoding, a potential range is classified, and only domains with the same (or near) classification are compared with the range. This significantly reduces the number of domain-range comparisons. By "near" classifications we mean squares that would have been classified differently if their pixel values were slightly different. The idea of using a classification scheme was developed independently in [45] and [43].

Many classification schemes are possible (see Section C.10): for example, Jacquin used a scheme that classified a sub-image into flat, edge, and texture regions. Here we discuss a scheme similar to the one presented in [26]. A square sub-image is divided into upper left, upper right, lower left, and lower right quadrants, numbered sequentially. On each quadrant we compute values proportional to the average and the variance: if the pixel values in quadrant i are $r_1^i, \ldots, r_n^i$ for $i = 1, 2, 3, 4$, we compute

$$A_i = \sum_{j=1}^{n} r_j^i$$

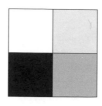

Figure 3.2: A square image can always be oriented so that the brightness, or average values, of its quadrants fall into one of these three canonical positions.

and

$$V_i = \sum_{j=1}^{n} \left(r_j^i\right)^2 - A_i^2.$$

It is always possible to orient the sub-image so that the A_i are ordered in one of the following three ways:

Major Class 1: $A_1 \geq A_2 \geq A_3 \geq A_4$.

Major Class 2: $A_1 \geq A_2 \geq A_4 \geq A_3$.

Major Class 3: $A_1 \geq A_4 \geq A_2 \geq A_3$.

These correspond to the brightness levels shown in Figure 3.2. Once the rotation of the square has been fixed, each of the 3 major classes has 24 subclasses consisting of the 24 orderings of the V_i. Thus, there are 72 classes in all. If the scaling value s_i is negative, the orderings in the classes above are rearranged. Therefore, each domain is classified in two orientations, one orientation for positive s_i and the other for negative s_i.

Chapters 4 and 9 give both an implementation and a theoretical motivation for such methods of classification.

The Encoding Parameters

Thus, the parameters for this algorithm are:

- The rms tolerance threshold e_c.
- The maximum depth of the quadtree partition.
- The minimum depth of the quadtree partition.
- The type of domain pool and the lattice spacing l.
- The maximum allowable scaling factor s_{max}.
- The number of classes compared with a range.

An encoding of an image consists of the following data:

- The final quadtree partition of the image.
- The scaling and offset values s_i and o_i for each range.

- For each range, a domain that is mapped to it.

- The symmetry operation (orientation) used to map the domain pixels onto the range pixels.

3.2 Decoding

Decoding an image consists of iterating W from any initial image. The quadtree partition is used to determine all the ranges in the image. For each range R_i, the domain D_i that maps to it is shrunk by two in each dimension by averaging nonoverlapping groups of 2×2 pixels. The shrunken domain pixel values are then multiplied by s_i, added to o_i, and placed in the location in the range determined by the orientation information. This constitutes one decoding iteration. The decoding step is iterated until the fixed point is approximated; that is, until further iteration doesn't change the image or until the change is below some small threshold value.

Fast Decoding

Somewhat surprisingly, there are many other methods for decoding an image. It is possible to find the exact fixed point directly by inverting a large but sparse matrix; see Section 11.4. Sections 5.4 and 6.3 discuss how the fixed point can be approximated in a lower-dimensional space, significantly reducing the number of operations required to approximate it. The hierarchical model described in Chapter 5 makes this point very nicely. Chapter 8 discusses a modification of the above scheme that allows decoding in a finite number of steps, sometimes just one! See also Section C.13.

Decoding at a Larger Size

The transformations that encode an image are resolution independent. The underlying model uses functions of infinite resolution, and the fixed point we reach is such a function. Therefore, when we decode, we may pick any resolution we wish. Figure 3.3 shows portions of both the original Lenna image and a version generated by decoding an encoding of Lenna at 8 times the original size. For comparison, the original version is shown, magnified by a factor of 8 also (by reproducing each pixel 64 times, and not, for example, by using bilinear interpolation, which would lead to less blocky artifacts). The original encoding results were: 512×512 Lenna, compressed by a factor of 10.96 at 35.31 dB; the image was encoded with a 24 class search in 794 CPU seconds. Since the large decoded version requires 64 times the memory of the original, in some (very loose) sense the compression ratio of the decoded image can be said to be $64 \times 10.96 = 701.4$. Continuing in this (very loose) vein, it is possible to suggest that 63/64th of the pixels are artificial data created by the transformations. Section 5.5 discusses the theory of superresolution.

Postprocessing

Since the ranges are encoded independently, there is no guarantee that the pixel values will be smooth at the block boundaries. The eye is sensitive to such discontinuities, even when they are small. It is possible to minimize these artifacts by postprocessing the image (see also Section

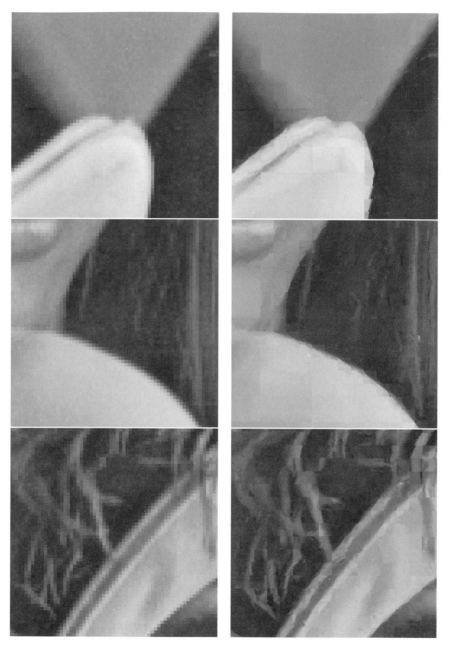

Figure 3.3: Several portions of Lenna (original, left; decoded, right) showing the artificially created detail.

C.16). In this implementation, we modify only the pixels at the boundary of the range blocks, using a weighted average of their values. If the pixel values on either size of a boundary are a and b, then these are replaced by $w_1 a + w_2 b$ and $w_2 a + w_1 b$, with $w_1 + w_2 = 1$. Ranges that occur at the maximum depth of the quadtree are averaged with weights of $w_1 = 5/6$ and $w_2 = 1/6$, and those above this depth are averaged with weights of $w_1 = 2/3$ and $w_2 = 1/3$. These values are largely heuristic (as is the method itself), but the results seem satisfactory. Section 6.3 discusses an extension of this method in more detail. Also see Exercise 12.

Efficient Storage

To increase the compression, the following bit allocation schemes are used.

Quadtree. One bit is used at each quadtree level to denote a further recursion or ensuing transformation information. At the maximum depth, however, no such bit is used, since the decoder knows that no further division is possible.

Scaling and Offset. Five bits are used to store the scaling and seven for the offset. No form of adaptive coding is done on these coefficients, even though their distributions show considerable structure.

Domains. The domains are indexed and referenced by this index. However, when the scaling value is zero, the domain is irrelevant, and so no domain or orientation information is stored in this case.[1]

Orientation. Three bits are used to store the orientation of the domain-range mapping.

3.3   Sample Results

This section consists of various data characterizing the quadtree scheme. Many similar results were reported first in [44] and [27]. (Appendix D contains a comparison of compression results for the quadtree method and for other image compression schemes, some of which are described later in this book.) Unless otherwise stated, the following can be assumed:

- Five bits were used to quantize the scaling coefficient, and seven were used for the offset.

- Domains from three classes were compared with each range; that is, three for positive scale factors and three for negative scale factors.

- For 512 × 512 Lenna, the maximum range size was 32 × 32 (minimum quadtree depth 4), and the minimum was 4 × 4 (maximum quadtree depth 7).

- The domain pool used was $\mathbf{D}_2$ with a lattice parameter $l = 1$. That is, the domains consisted of nonoverlapping subsquares of the image that were twice the range size.

- $s_{max} = 1$.

---

[1]On average, more than half the bits in the transformation data are used to store the domain position. This means that a "smarter" selection of domains could lead to a considerable saving in memory.

- Results in which the compression is varied are achieved by letting the rms tolerance criterion e_c take on (at least some of) the values 0.5, 1, 2, 3, 4, 5, 6, 8, 10, 12, 14, 16, 18, 20, leading to results ranging from low to high compression.

- PSNR was computed after postprocessing.

- Time results are given in CPU seconds on a personal IRIS 4D/35 workstation.

Varying s_{max}

If we restrict the absolute value of the largest scaling value s_{max} to be less than 1, then we may ensure contractivity of the encoding. Several questions arise, however: Does a smaller s_{max} lead to faster convergence? Does $s_{max} > 1$ imply divergence? How does changing s_{max} change the fidelity of the encoding?

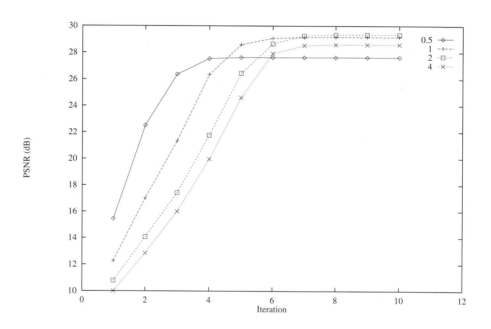

Figure 3.4: The PSNR versus the number of decoding steps for 256×256 Lenna encoded with $s_{max} = 0.5, 1.0, 2.0,$ and 4.0. The initial image was black.

Figure 3.4 shows the PSNR as a function of the number of decoding steps, starting from an initial black image. For these data, the image of Lenna was encoded with $s_{max} = 0.5, 1, 2,$ and 4. We see that smaller s_{max} does converge in fewer iterations, but to a fixed point which has a worse signal-to-noise ratio (even though the quantization remained constant, allowing for a finer resolution of the scaling parameter). The encodings with $s_{max} > 1$ still converged, because while they are not contractive, they are eventually contractive. In fact, $s_{max} = 2$ encodings reached a higher signal-to-noise ratio than encodings with $s_{max} \leq 1$ or $s_{max} = 4$. Some images, however, will converge very slowly when $s_{max} > 1$ so that the gain in fidelity may not be worth the uncertainty in decoding. This also depends on the decoding methods. Methods such as those in Sections 8.3 and 11.4 are not sensitive to the contractivity of the final encoding, so that $s_{max} > 1$ may not present a problem.

Scaling and Offset Bit Allocation

Figures 3.5 and 3.6 show plots of PSNR versus compression for various bit allocations used to store the offset and scaling parameters. The curves show results of using different numbers of bits (s_{bits} and o_{bits}) to uniformly quantize the scaling and offset parameters. These encodings were made with one search class and a minimal range size of 8×8. These curves show that $s_{bits} = 5$ and $o_{bits} = 7$ represents a reasonable compromise, although $s_{bits} = 4$ and $o_{bits} = 6$ (which is not shown) may also be a good value. What is clear is that a sophisticated scheme would use some form of adaptive bit allocation. Experiments with logarithmic quantization, which has a non-uniform resolution for the values of s_i, show no improvement.

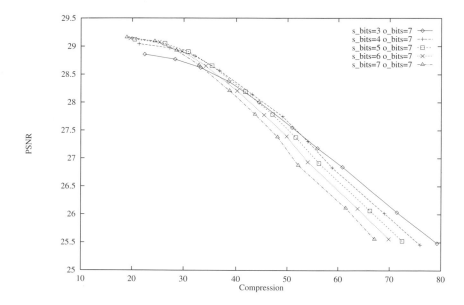

Figure 3.5: PSNR versus compression for 512×512 Lenna with $o_{bits} = 7$ and $s_{bits} = 3, 4, 5, 6, 7$.

Figures 3.7 and 3.8 show distributions of the scaling and offset coefficients from a typical encoding (compare with Figure 8.3). They show that both parameters would benefit from some adaptive coding, since their distributions show considerable structure. The large number of $|s_i| = 1$ transformations are due to clipping.[2] It is interesting to note that aside from $|s_i| = 1$,

---

[2]It is important to note, however, that the domain-range rms value is computed after the clipping. This means

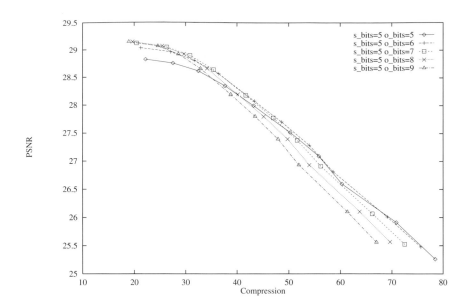

Figure 3.6: PSNR versus compression for 512×512 Lenna with $s_{bits} = 5$ and $o_{bits} = 5, 6, 7, 8, 9$.

the distribution of s_i values is rather uniform. This is not the case when $s_{max} > 1$, as is shown in Figure 3.9, in which $s_{max} = 2$. This means that another advantage to using $s_{max} > 1$ is that the scaling coefficients can potentially be better encoded.

Classification Matching

Figure 3.10 shows a typical histogram of the number of domains in each of the 72 classes (summed across all the domain sizes). This distribution leaves much to be desired, since it is far from uniform. A somewhat better distribution appears in Figure 3.11 in which the three major first classes have been combined. Somewhat surprisingly, this distribution is roughly comparable to the one in Figure 9.2 (p. 193).

We now attempt to answer the following questions: How fast can encodings be made? What is the gain in time and cost in fidelity associated with selecting a small proportion of the possible domains? PSNR and time versus compression ratio results for the 512×512 Lenna image are given in Figures 3.12 and 3.13. The data in both figures come from the same encodings, so that the points may be compared for fidelity and time results. The number of classes from which potential domains were chosen for comparison with a range was varied. The curve marked "Pos" consists of just one comparison class, using only positive scaling values. Since negative scaling of the domain leads to a different classification, the other curves represent a comparison

that transformations with $|s_i| = 1$ are optimal and do not, in general, consist of all the transformations which found a potential $|s_i| > 1$ and were clipped.

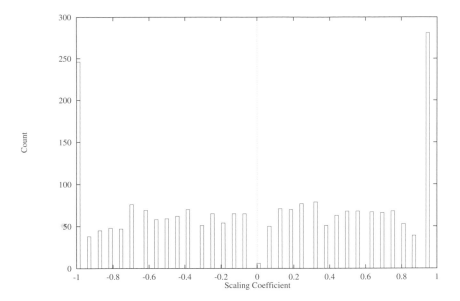

Figure 3.7: A distribution of the scaling coefficients for 256×256 Lenna with $s_{max} = 1$.

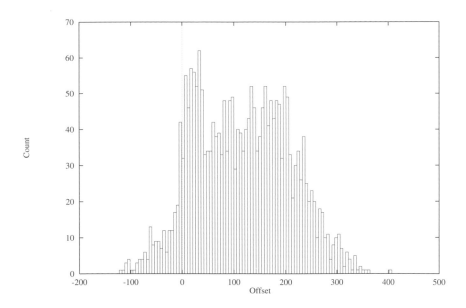

Figure 3.8: A distribution of the offset values for 256×256 Lenna with $s_{max} = 1$.

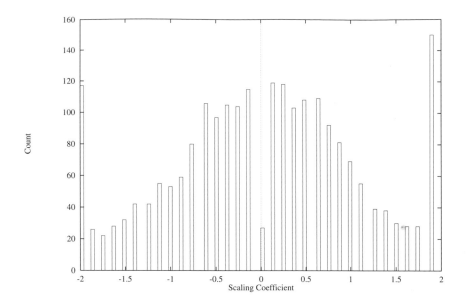

Figure 3.9: A distribution of the scaling coefficients for 256×256 Lenna with $s_{max} = 2$.

with twice the number of labeled classes – one for positive scaling and one for negative scaling. Finally, the curve marked "F 8×8" consists of using ranges of minimal size 8×8 (corresponding to a quadtree depth of 6) with a three class comparison.

As is to be expected, comparing a larger number of classes requires more time and leads to better fidelity. Referring to Figures 3.12 and 3.13, we see that a moderate compression of 20 at about 30 dB can be done in about 15 seconds – fairly fast for a fundamentally search-intensive scheme. The search times are very linear in the compression and progress linearly with the number of classes searched. A one-class search (meaning one for positive scaling and one for negative scaling) requires twice the search time; a three class search requires slightly less than six times the time; and so on. Given the non-uniformity of the class distribution in Figures 3.10 and 3.11, this is somewhat remarkable. It shows that when the whole image is encoded, some averaging takes place, eliminating this unevenness. The full 72-class search requires somewhat less time, proportionally, probably because the overhead involved with classifying the range and preclassifying all the domains doesn't scale with the number of classes.

Figure 3.12 also shows that the quadtree scheme encodes fine detail poorly. It is not able to achieve a fidelity better than about 37 dB (in this case, and about 38 dB in the next section). The encoding made with 8×8 minimum range size reaches a lower maximum of about 30 dB. However, this encoding requires considerably less time. In fact, most of the time is spent in attempting to cover the large blocks (which were not well matched), so that the method requires roughly the same time to encode images at all compression ratios. This means that for high compression ratios (when the fidelity can be compromised), the 8×8 minimum range size is

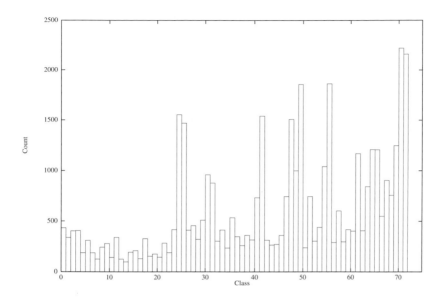

Figure 3.10: A typical histogram showing the number of domains in each of the 72 classes .

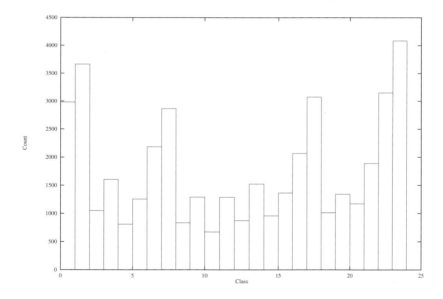

Figure 3.11: A typical histogram showing the number of domains in each of the 24 classes derived by combining the three major first classes in the 72-class classification.

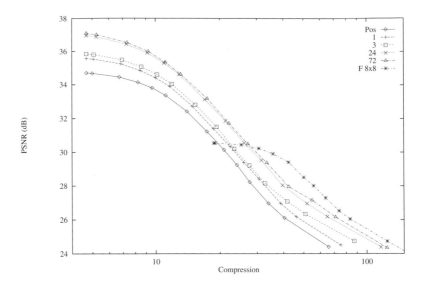

Figure 3.12: Lenna 512×512 PSNR versus compression results for various numbers of domain classes compared with each range.

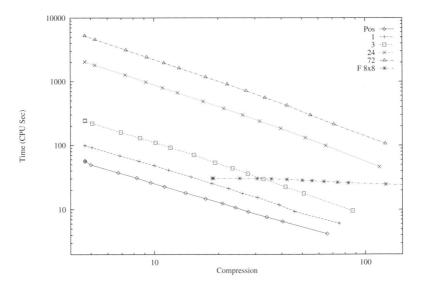

Figure 3.13: Lenna 512×512 time (CPU seconds) versus compression results for various numbers of domain classes compared with each range.

the method of choice. We can extrapolate from these data that using it with just one search class would require only about 10 seconds per encoding over the range of compression ratios with a loss of 0.25 to 0.5 dB in fidelity.

We can draw the following conclusions from these data:

- The improvement attained by using 72 rather than 24 classes is minimal and comes at great expense in time.

- The improvement of using both positive and negative scaling values over just positive scaling is significant, about 1 dB, but doubles the encoding time.

- Using three search classes in place of one gives little gain and triples the encoding time.

- Using a large minimum range size is beneficial when high fidelity is not important.

The Domain Pool

In this section we present results derived by varying the domain pool. We answer the following questions: Is there some form of local self-similarity in a "typical" image which we can exploit on to restrict the domain pool and hence increase the compression with minimal loss of fidelity? Is it a good idea to work hard to encode large ranges so that fewer ranges will be required overall, leading to high compression? Will increasing the total number of domains compared to a range, improve or worsen the results? One might argue that it will improve the results, since better domain-range matches may be found, and so either the fidelity will increase or larger ranges may be used resulting in higher compression. But the extra domains require more bits to store, and it may happen that ranges that were previously partitioned and fit well will now fit poorly at a larger size. We answer these questions and, as a consequence, we also find out what happens when the domain pool is selected in various ways.

Figure 3.14 shows a distribution of the domain-range distance derived from an encoding of Lenna. The figure also shows the theoretical probability distribution for the distance between two randomly chosen points (see Exercise 13). The distributions are almost identical, with a slight shift on the domain-range distance due (at least in part) to the fact that the smallest ranges are not just one point. This means that the domains are essentially random and that there is no preference for local domains.

We can examine this same question by plotting the distributions of the difference in the x and y positions of the domains and ranges. This is done in Figures 3.15 and 3.16. If (x_r, y_r) and (x_d, y_d) denote the range and domain positions, then the normalized distribution of $d_x = x_r - x_d$ looks very much like $\rho(d_x) = 1 - |d_x|$ (and similarly for d_y). The function $\rho(d_x)$ is the probability distribution for the difference in the x (similarly for y) coordinates of two points randomly chosen in the unit square with uniform probability. So even when the points are chosen randomly, it *appears* that there is a preference for local domains. However, this is an artifact. In practice, we can subtract this expected peak from the actual data to see if there is any effect. Figures 3.15 and 3.16 (in which the subtraction is left to the reader) show that there is a slight preference for local domains, but the effect is small. This may also partially account for the left shift of the distribution in Figure 3.14. This means that while there might be a very slight a benefit to adaptively coding (d_x, d_y), the domain for a fixed range is not more likely

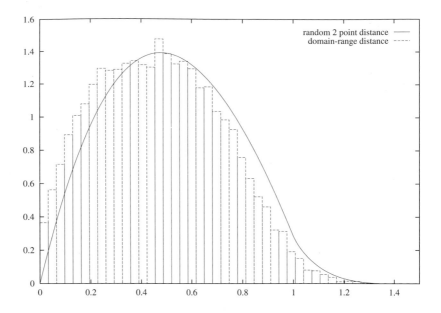

Figure 3.14: A distribution of the domain-range distance (for 512×512 Lenna) and a theoretical distribution of distance for two randomly chosen points in the unit square.

to be close to the range. Close domain positions appear more often because there are more of them overall, not because of local self-similarity.[3]

PSNR and time versus compression ratio results for the 512×512 Lenna image are given in Figures 3.17 and 3.18. Again, the data in both figures come from the same encodings, so that the points may be compared for fidelity and time results. The domain pool type is indicated in the plot, as well as the lattice spacing variable l. Table 3.1 shows the number of domains of each size for each of the domain pools and lattice parameters used.

These data show that increasing the size of the domain pool results in better compression and fidelity results. The time cost of having a large number of domains is very high, however. Also, the encodings done with $\mathbf{D}_3$ are comparable in PSNR to those done with $\mathbf{D}_1$ for high compression ratios, but the fidelity at that point is too low to be of use. At low compression ratios, the scheme performs worse than the other schemes in either encoding time or PSNR; this means that it is not reasonable to select this domain pool scheme. Note that the encoding time is roughly constant (and large) for this scheme, because the number of operations for a domain-range comparison is large when the range (and hence the domain) is large. The smaller ranges have a relatively small number of domains which require fewer operations to match, and

---

[3]To further elaborate this point, the question is: given a fixed range, is the domain likely to be close? The answer is no. The domain is equally likely to be anywhere. Averaged over all the ranges, however, there are more close domain positions and hence more close domains, and so the probability distribution of the x and y difference shows a peak.

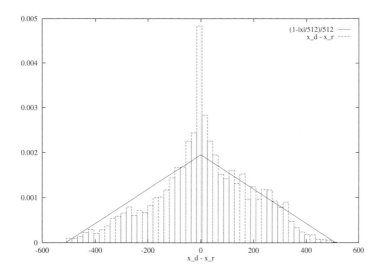

Figure 3.15: A distribution of the difference of the x positions of the domains and ranges (for an encoding of 512×512 Lenna). Also shown is the theoretical distribution of the difference of two randomly selected points.

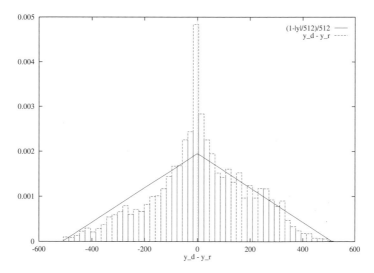

Figure 3.16: A distribution of the difference of the y positions of the domains and ranges (for an encoding of 512×512 Lenna). Also shown is the theoretical distribution of the difference of two randomly selected points.

Table 3.1: The number of domains in each of the domain pools for various values of the lattice parameter l.

Domain Pool	Domain Size	l	No. Domains
$\mathbf{D}_1$	32	2	50625
$\mathbf{D}_1$	16	2	58081
$\mathbf{D}_1$	8	2	62001
$\mathbf{D}_1$	4	2	64009
$\mathbf{D}_1$	32	4	12769
$\mathbf{D}_1$	16	4	14641
$\mathbf{D}_1$	8	4	15625
$\mathbf{D}_1$	4	4	16129
$\mathbf{D}_2$	32	2	225
$\mathbf{D}_2$	16	2	961
$\mathbf{D}_2$	8	2	3969
$\mathbf{D}_2$	4	2	16129
$\mathbf{D}_2$	32	1	64
$\mathbf{D}_2$	16	1	256
$\mathbf{D}_2$	8	1	1024
$\mathbf{D}_2$	4	1	4096
$\mathbf{D}_3$	32	2	12769
$\mathbf{D}_3$	16	2	3721
$\mathbf{D}_3$	8	2	1024
$\mathbf{D}_3$	4	2	256

hence they do not contribute significantly to the encoding time.

Postprocessing

The effect of postprocessing is shown in Figure 3.19 and in Table 3.2; the first two entries in the table refer to the data shown in the figure. In general, high-fidelity images have a minor reduction in PSNR (on the order of 0.1 dB), and all other encodings have an increase in PSNR. This is not the case when the weighting parameters at the bottom of the quadtree are selected to be equal to those at the inner nodes. A thorough study of postprocessing weights and methods, in general, remains to be done.

3.4   Remarks

It is possible to compare Figures 3.12 and 3.17 and Figures 3.13 and 3.18. The "3 class" curves in Figures 3.12 and 3.13 are identical with the "$\mathbf{D}_2$, $l = 1$" curve in Figures 3.17 and 3.18. We can see, for example, that searching all the domains of the smaller domain pool ($\mathbf{D}_2$ used in Figures 3.12 and 3.13) is roughly equivalent to searching some small subset (consisting of three classes searched in Figures 3.17 and 3.18) of the large domain pool $\mathbf{D}_1$, $l = 4$. This is

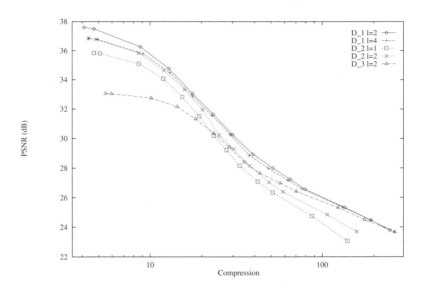

Figure 3.17: PSNR versus compression results for 512×512 Lenna for various domain pool selection schemes.

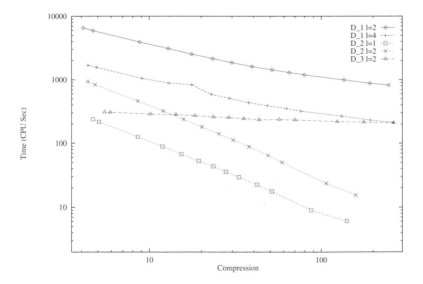

Figure 3.18: time (CPU seconds) versus compression results for 512×512 Lenna for various domain pool selection schemes.

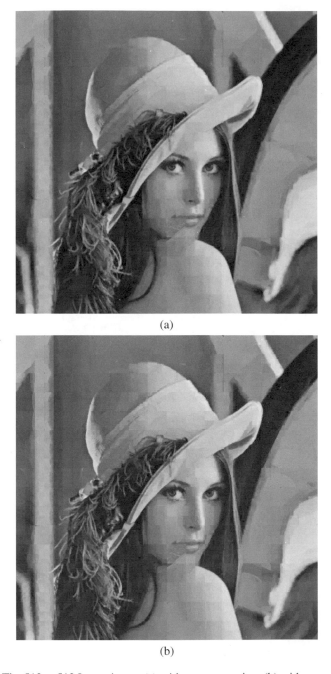

(a)

(b)

Figure 3.19: The 512×512 Lenna image (a) with postprocessing, (b) without postprocessing.

Table 3.2: Postprocessing results for 512×512 Lenna.

Figure	Postprocessed	dB PSNR	Comp. Ratio	Enc. Time
3.19a	Yes	31.72	22.09	903 sec
3.19b	No	31.29	22.09	903 sec
-	Yes	29.37	33.36	556 sec
-	No	28.85	33.36	556 sec
-	Yes	34.87	8.48	56.5 sec
-	No	34.56	8.48	56.6 sec
-	Yes	37.07	4.68	5244 sec
-	No	37.22	4.68	5244 sec

Table 3.3: Sample results for 512×512 and 256×256 Lenna.

Image Size	Comp. Ratio	PSNR	Enc. Time
256	13.56	29.20	38.9 sec
512	31.48	29.52	237.7 sec
256	7.74	31.19	188.9 sec
512	22.08	31.72	903.5 sec
256	6.59	29.65	5.1 sec
512	25.99	29.38	17.8 sec

also true locally, around the point of intersection of the parameters of these graphs: the gain in PSNR per change in ratio of computing time is roughly the same as we vary the number of classes (from three to one) and the density of the domains ($\mathbf{D}_2$, with $l = 1$ and $l = 2$).

Another issue is how the results scale with image size. Some sample comparative results are given in Table 3.3 and Figure 3.20. These encodings were made with one comparison class. The encoding time for the 512×512 Lenna ranges from three to six times the encoding time for the 256×256 version. The compression ratio for the 512×512 image, at a fixed fidelity, ranges from roughly two to four times the compression ratio of the 256×256 image.

3.5 Conclusion

We conclude with a list of "Do's," "Don'ts," and "Don't knows." Results here (or elsewhere) show that it is reasonable to:

- Select s_{max} large if the decoding method allows, if it is possible to check for eventual contractivity after the encoding is made, or if the risk (which seems minimal, empirically) is not overwhelming.

- Have a large domain pool. The larger the domain pool, the better the results.

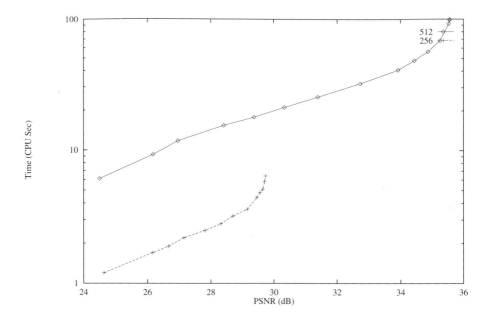

Figure 3.20: The compression time versus PSNR for both the 512×512 and 256×256 Lenna images.

- Encode the scaling coefficients using some form of adaptive bit encoding, Huffman coding or arithmetic coding. Quantize these using 5 and 7 bits, respectively, increasing both s_{bits} and o_{bits} for high-fidelity results.

- Allow small ranges when fine detail is important and use larger ranges when it is not.

- Use one search class for a good compromise between encoding time and fidelity.

- Select a large and uniform domain pool when encoding time is not a factor. Select a domain pool with more domains for the smaller ranges when time is a factor.

- Postprocess the image, except for high fidelity encodings.

The following are things that are *not* reasonable to do:

- Do not restrict the domains to be localized in the image. There is no local self-similarity that can be exploited to increase the chance of a good domain-range match.

- Do not use more domains for the larger ranges in the hope that larger ranges will have a sufficiently good domain-range rms difference.

- Do not increase the domain pool and search fewer classes. It is better to search all the domains in a smaller pool.

- Do not compute the optimal s_i and o_i and then quantize them for storage. First quantize them, and then compute the rms difference of the domain-range map.

The following issues remain to be sorted out:

- It is possible to stop searching as soon as a sufficiently good domain-range match is found. This will increase the encoding speed, but decrease the fidelity. It will also mean that the potential domains that occur early in the listing of the domain pool will occur more frequently as actual domains, and hence there may be a benefit in adaptively coding the domain used.

- What is the optimal number of classes? (See Chapter 9.)

- Can some simple model of image complexity suggest optimal parameter values over a range of compression ratios?

- Is there an automatic way to determine whether postprocessing is beneficial?

- At what point does adding domains worsen the results?

- When doing a 24 class search, there is an incremental improvement in quality derived by searching classes for both positive and negative scaling, but searching for both types of scaling takes twice the encoding time. Would it be better to search only one class ?

Chapter 4

Archetype Classification in an Iterated Transformation Image Compression Algorithm

R.D. Boss and E.W. Jacobs

Determining a good set of transformations that encode an image well is time consuming because for each range an extensive search through the candidate domains is required ([47]). The purpose of classification is to reduce the number of domains that have to be checked in order to find an acceptable covering. References [11] and [44] use a classification scheme based on the idea that by orienting blocks in a canonical form (based on brightness), and then subdividing these primary classes further by the location of strong edges, it should be possible to find good coverings with minimal computation. In fact, this type of classification method performs quite well. In this chapter, a different classification method using *archetypes* is presented (see [11]). The method for generating a set of archetypes is described, and the archetypes are then used to classify ranges and domains in an iterated transformation image compression encoder. Fidelity versus encoding time data are presented, and compared with a more conventional classification scheme.

4.1 Archetype Classification

Since the goal of the encoding scheme is to find good *coverings* (i.e., find a domain and corresponding transformation w which results in an accurate mapping to each range), it is reasonable to try to base the classification scheme on finding good coverings. This is what the archetype classification scheme is designed to do. An archetype is that member of a class that is best able to cover all the other members of the class. The determination of archetypes is similar

79

to the determination of a vector quantization (VQ) codebook. However, there are important differences. Vector quantization is a well-studied technique commonly used to compress digital images. Because it introduces some of basic ideas that will be used to construct archetypes, the following section describes, in some detail, a method used to construct a vector codebook. The specific method discussed is based upon the methodology described by Linde, Buzo, and Gray [55]. The following discussion will serve to better expose the similarities and differences between archetypes and VQ codebooks.

Vector Quantization Codebooks

In VQ, an image is subdivided into many smaller sections, called the *target vectors*. Each target vector is encoded as a *code vector*, which is drawn from a list of code vectors, called the *codebook*. The storage of the image is then simply a list of the codes that indicates which code vector in the codebook best matches the sequence of target vectors that make up the image. The subsequent reconstruction of the image is then simply a pasting together of the indicated code vectors.

Encoding an image is a simple procedure. For each target vector, a search through the code vectors is performed to find the one that (measured in a given metric) is closest. Therefore, the most interesting problem in such an encoding scheme, and the problem that has been the focus of much of the research in the VQ field, is the construction of the codebook.

A specific method to generate a codebook based on the general method of Linde, Buzo, and Gray is shown in Figure 4.1. The initial state is assumed to be a single code vector (of all zeros) and a set of *teaching* vectors (all of mean zero). A search is then performed to determine those two vectors that are members of the teaching set which, when used to encode the entire set, give a minimum square error. The search is *not* performed for all possible sets of any two vectors because such a search is too time consuming. For example, a full search for the first step, for a teaching set of 8×8 vectors drawn from five 512×512 images requires the computation of of 20,478 square errors for each of 209,704,960 different combinations. Consequently, the two vectors are determined from a search of a randomly chosen subset of 200 elements. Each of the 19,900 possible combinations is checked, with the minimum error pair being used as the initial code vectors.

The entire set of teaching vectors is then sorted into subsets by determining which of the new code vectors best encodes them. Each subset is then used to compute a new best code vector. In this case, the best code vector is simply determined by computing the average of all the vectors in the subset. This process is then repeated until a self-consistent set of code vectors is obtained.

If the number of self-consistent code vectors is less than the desired number of code vectors, then each code vector is bifurcated into two new code vectors by dividing the subset of teaching vectorsprecisely as described previously. This process of iterating to self-consistency then bifurcating is continued until the number of desired code vectors is obtained. (Note, the only problem encountered with this process occurs when a subset of the teaching vectors consists of exactly one vector, in which case the subset cannot be divided. To continue the bifurcation process in this case, the subset that has the most members is divided once more, after all other subsets have been divided. This extra division of the largest subset is performed for all subsets of size one; then, and only then, is the self-consistency iteration performed.)

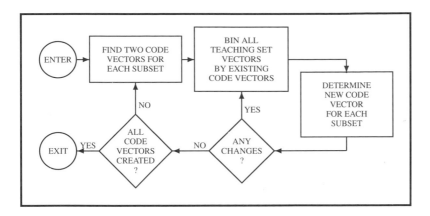

Figure 4.1: A flow chart showing how to make a codebook.

Archetypes

The archetype $\mathcal{A}$ for a set of vectors is determined by searching through the entire set to find that member of the set which can *cover* the other members best. That is, for each vector $\mathcal{E}_l$, $l \in \{1, 2, \ldots, n\}$, of length n_{pixels} the best transform w is found for every other member of the set $\mathcal{E}_k$, $k \in \{1, 2, \ldots, n\}$ and $k \neq l$, such that the rms distance $\delta_{rms_{lk}}$ given by

$$\delta_{rms_{lk}} = \left\{ \frac{\sum_{pixels} [w(\mathcal{E}_l) - \mathcal{E}_k]^2}{n_{pixels}} \right\}^{\frac{1}{2}}$$

is minimized. The archetype is then chosen as that vector for which the total error Δ_{rms_l}, given by

$$\Delta_{rms_l} = \sum_{k=1}^{n} \delta_{rms_{lk}},$$

is minimized. That is, $\mathcal{A} = \mathcal{E}_l$ if and only if $\Delta_{rms_l} \leq \Delta_{rms_k}$ for all k.

If some method exists to separate the entire set of vectors into classes (or bins), then the set of vectors in each class has a definable archetype. The determination of the archetypes for some set of classes is given by the flow chart in Figure 4.2a.

A more useful set of archetypes can be generated if, rather than stopping once the archetypes are determined, the archetypes are then used to perform the sorting of the vectors into classes (i.e., the archetype that best covers a given vector defines the class into which the vector is classified), with the reclassified vectors then being used to determine new archetypes. This process can then be iterated to self-consistency. A flow chart of this process is shown in Figure 4.2b.

The determination of archetypes is obviously similar to the determination of a vector quantization codebook; however, there are major differences. The most fundamental difference is that the transformation w is included in the process of determining archetypes. In addition, an archetype is a member of the teaching set. This is different from VQ algorithms, in which the

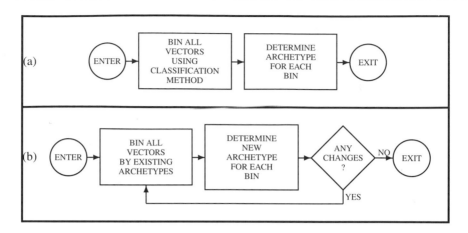

Figure 4.2: A flow chart showing how to make self-consistent archetypes.

teaching vectors are used as a starting point, but the resulting code vectors are not members of the set of teaching vectors.

To use the archetypes for classification in an iterated transformation encoder, a set of archetypes is found from an image or set of images by following the procedure outlined above. This set of archetypes is then used to define the classes into which ranges and domains are binned at the start of the iterated transform encoding procedure; i.e., each range and domain is assigned to the class of the archetype that covers it best. Once the classification has been done, the encoding of the image proceeds in the normal fashion.

As in VQ, in which the codebook is generated from a set of images which does not typically include the images that the algorithm will ultimately be used to encode, in the archetype classification method the classes are defined using images other than the ones that will ultimately be encoded. The idea of using predetermined archetypes (from images other than the one being encoded) is somewhat contrary to the basic concept of the iterated transform method. Therefore, it is necessary to emphasize that the archetypes are used only to classify domains and ranges in an efficient way so as to reduce encoding time, and they have no other role in the encoding procedure. If the images used to find the archetypes are sufficiently general, then (as will be shown in the next section), a set of archetypes which will be useful for encoding a broad variety of images will result. If, for a specific application, the set of images to be encoded have similar characteristics, archetypes could be found from a set of images with these same characteristics, and the classification method might work particularly well.

4.2   Results

The encoder used in this paper used 8×8 ranges. The set **D** was chosen to be 16×16 squares with corners on a lattice with vertical and horizontal spacing of 8. Two similar encoders, one using a conventional classification scheme based on the brightness and edge like character of each quadrant of the domain and range squares, and one using an archetype classification

Figure 4.3: Images from which archetype sets were generated, and the test images.

scheme, were used to encode five different 256×256, 8 bpp test images. These five test images are shown in row four of Figure 4.3. The compression ratio ($\approx 16{:}1$) of all the encodings was identical. In the archetype encoder, 72 archetypes were used to define the classes (see Exercise 14). To determine how results vary depending on the archetype sets, encodings were done using five different sets of archetypes. Each set of archetypes was generated from five images, using the self-consistent technique described in Section 4.1. Three of the archetype sets were generated from sets of five qualitatively dissimilar images. These sets are referred to as grand canonical ensemble (GCE) 1, 2, and 3, and are shown in Figure 4.3, rows one, two, and three, respectively. The remaining two archetype sets were generated from sets of five qualitatively similar images. One of these archetype sets (SIM 1) was generated from five different images of buildings and mechanical equipment, and the other one (SIM 2) was generated from five different images of faces. The SIM 1 set was made up of the five images, GCE 1 columns one and two, GCE 2 columns one and two, and GCE 3 column one, as shown in Figure 4.3. The SIM 2 set was made up of the images GCE 1 column three, GCE 2 columns three and four, and GCE 3 columns two and three. None of the test images were included in the sets of images used to generate the archetype sets.

In Figure 4.4 the archetypes for the sets GCE 1, GCE 2, and GCE 3 are shown. These archetypes show some strong similarities. The similarities are due to the underlying classi-

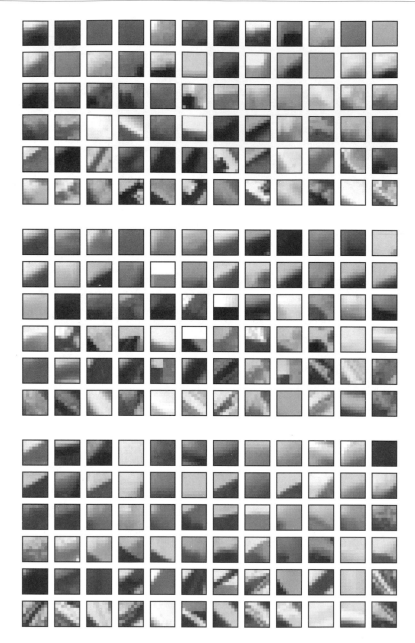

Figure 4.4: The archetypes generated from the sets GCE 1 (top), GCE 2 (center), and GCE 3 (bottom).

fication which was used to start the process. This can be easily seen by looking at the first member of each archetype set. The initial classification method begins by dividing all test vectors into one of three major classes based upon the orientation of the relative brightness of the quadrants. The initial classification further divided each of the major classes into 24 subsets based upon the amount of brightness variation (not variance, as in Chapter 3) in the sub-quadrants. To be specific the first class would have the brightest quadrant in the upper left, the next brightest in the upper right, the third brightest in the lower left, and the darkest in the lower right. It would also have the most variation of brightness in the upper left, next most variation in the upper right, third most variation in the lower right, and the least variation in the lower left. It is seen in Figure 4.4 that the first archetype in each archetype set has the same overall orientation of brightness as the original classification, and the variation of the brightness also is maintained. That these properties are maintained results in the strong similarity between corresponding archetypes. Even in those instances in which the corresponding archetypes do not appear similar, they still share the same basic underlying properties which were passed on from the initial classification process. In these cases the classes have relatively few members, thus resulting in a relatively poorly defined archetype, thereby allowing *apparently* dissimilar archetypes.

Figures 4.5 through 4.8 show results for some of the encodings. In Figure 4.5, PSNR versus the number of classes searched (n_c) is shown for encodings of the standard image Lenna (row 4, column 1 of Figure 4.3). Data are shown for the conventional classification scheme for $n_c = 1, 4, 6, 12, 24,$ and 72. Data are shown for the archetype classification scheme using GCE 1, 2, and 3 archetypes, with $n_c = 1, 2, 3, 4, 6, 12, 24,$ and 72. For both conventional and archetype classification schemes, the accuracy of the encodings increased as n_c increased. For a given $n_c < 72$, the archetype classification scheme results in more accurate encodings than the conventional classification scheme. In the case of the complete 72-class search, all the domains are checked; therefore encodings done with the two different classification schemes yield the same encoding, and thus the same PSNR.

Of more interest are the data of Figure 4.6 where PSNR is plotted as a function of encoding time for the Lenna and Tank Farm (row 4, column 2 of Figure 4.3) test images. The encoding times are given in seconds, as performed on an Apollo 4500 work station. (The time difference between encodings made with the conventional and archetype classifier depends on both the computer and the compiler being used. Tests were done using several different computers, with several different compiler optimization levels. The results presented here were typical, and correlate well with what one might predict based on the actual differences in the codes.)

For a given n_c, the total encoding time for the archetype classification encodings is longer than the conventional classification encodings. This is because sorting the domains and ranges in the archetype classification scheme requires more time than it does for the standard classification method. Therefore, to outperform the conventional classification method, the archetype method must yield more accurate encodings while searching through fewer classes. Indeed, Figure 4.6 shows that the curves for the archetype classification and standard classification crossover, thus indicating that the archetype classification can (by searching fewer classes) result in improvements in both encoding accuracy and encoding speed. The decoded Tank Farm image from the encoding done using GCE 1 archetypes with a 6-bin (or class) search is shown in Figure 4.7. The original Tank Farm image appears in Appendix E.

In Figures 4.5 and 4.6, it is seen that the archetype encodings done with GCE 1, 2, and 3 archetypes yielded almost identical results. This indicates that if a relatively diverse set of

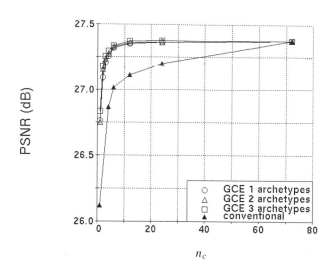

Figure 4.5: PSNR vs. number of bins searched for Lenna image.

images is used to generate the archetype sets, then a useful set of archetypes will be obtained. It is also of interest that all three sets of GCE archetypes yielded good results for Lenna and Tank Farm, two rather disparate images. In fact, results indicating that archetype classification works well as compared with conventional classification were obtained from all three of the other test images that were encoded. To further investigate the importance of the images used to make the archetype sets, the test images were encoded using the SIM 1 and SIM 2 archetypes classification sets. In Figure 4.8, PSNR versus encoding time for encodings made with SIM 1, SIM 2, and GCE 1 archetypes are shown for Lenna and Tank Farm. Figure 4.8a indicates that encodings of Lenna made using the archetype set made from the images of faces (SIM 2) resulted in slightly more accurate encodings than encodings made with the GCE 1 classification set. Encodings made with the archetype set made from images of buildings and machinery (SIM 1) resulted in slightly poorer encodings than GCE 1 archetype encodings. In Figure 4.8b, it is evident that the differences between encodings obtained using the different archetype sets on the Tank Farm image were less distinguishable than for the Lenna image. Fidelity results for the other three test images also indicated little dependence on the archetype set.

4.3   Discussion

Results have been presented that compare the fidelity and encoding times of two similar iterated transformation image compression encoders, one utilizing a conventional classification method and the other an archetype classification method. Because classifying a given range or domain

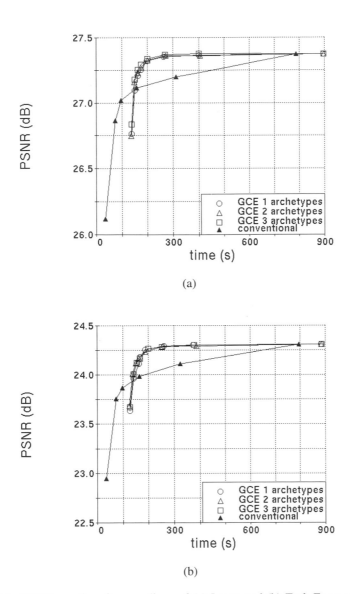

Figure 4.6: PSNR vs. time for encodings of (a) Lenna and (b) Tank Farm, using different number of bins searched.

Figure 4.7: Decoded Tank Farm with PSNR = 24.26 dB.

by archetypes requires more computing time than a conventional classifier, the archetype scheme requires that the classification be more accurate to ultimately yield improved fidelity and faster encoding times. The results indicate that to approach the maximum fidelity (attained with the complete 72-bin search), the archetype classifier required a search through fewer bins than did the conventional classifier. That this improved classifying accuracy resulted in an improved encoding performance is clearly indicated for the Lenna and Tank Farm images in Figure 4.6, where it is seen that the fidelity versus time curve for the archetype encodings cross the curve for the conventional encodings. The archetype data indicate that a 6-bin search resulted in close to the same fidelity as the complete search in roughly one-third of the encoding time. The fastest encodings are still attained with a conventional classifier, but to attain fidelity approaching that of a complete search, significant time can be saved by using an archetype classifier.

The iterated transformation method is based on the ability to encode an image using a self-referential mapping. The idea of using a classifier relying on archetypes generated from a set of independent images seems contrary to the basis of the method. This might lead one to expect that such a classification method may not be robust. The data presented here dispel this concern. The three GCE archetype sets yielded almost identical results. The data of Figure 4.8 indicate that using an archetype set generated from a set of images disparate from the target image results in only a slight decrease in performance. These same figures also indicate that the GCE archetypes performed essentially as well as the archetypes generated from images similar to the target image. The data for the three other test images yielded similar results. These results suggest that the method described in Section 4.1 requires only a reasonable choice for the set of teaching images to result in a robust set of archetypes.

Unlike the encoders described in [44] and [45], the encoder described in this chapter utilizes ranges of only one size. This was done because the results from such an encoder are more easily analyzed than results from encoders using variable-sized ranges. Applying archetype classification to an encoder that utilizes variable sized range squares would be direct, although it is not

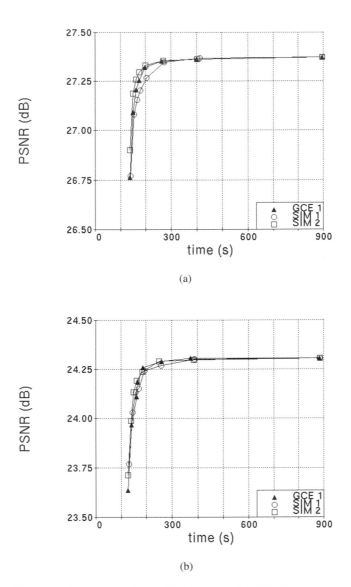

Figure 4.8: PSNR vs. time for encodings of (a) Lenna and (b) Tank Farm using GCE and SIM archetype sets.

clear that the improvement in encoding time performance would be greater than, or less than, the improvement demonstrated for the constant-sized range square encoder. In a variable-sized range square encoder, the number of possible domains in **D** is increased relative to a constant-sized range square encoder. Because an archetype encoder takes longer to classify each domain, increasing the number of domains makes it even more important for the archetype classifier to classify accurately in order to maintain its advantage over a conventional classification method. On the other hand, accurate classification achieved using an archetype classifier may result in good coverings being found for large range squares with a relatively short search. An encoder using a less accurate classifier would have to partition to small ranges more often, resulting in increased encoding time.

The method described above results in archetypes that are members of the teaching set. Because of this, it is possible to apply this archetype concept to the iterated transformation encoding method in another manner. Given an image to be encoded, a set of archetypes could be determined from the image using the process of Figure 4.2a. The image could then be encoded in the normal fashion using the set of archetypes as the set of possible domains (**D**). The location of the archetypes would therefore have to be stored as part the encoded image. The set **D** would have few, but carefully chosen members. As a result, a large number of relatively compactly specified w_i's could be used to encode an image.

This method is the opposite extreme of an algorithm in which the set **D** is large, the w_i's require more bits for storage, and fewer w_i's are used to encode the image. The problem with this method is that determining the archetypes entails performing almost the same calculations necessary in encoding the image with a more general set **D**. Therefore, even though **D** is small in number, the encoding time is not reduced. For reference, encodings using this method yielded fidelities approximately equivalent to that obtained using the conventional classifier with a one bin search. The necessary computation is approximately equivalent to that required for a 4-bin search, while the compression is improved by approximately a factor of 1.2.

In Section 4.1, it was pointed out that one of the difference between archetypes and VQ code-books is that archetypes are actually members of the set of teaching vectors. If the archetypes are to be used solely for classification, as is done in this chapter, this need not be the case. One could conceive of a similar scheme in which the generated archetypes were not members of the teaching set, but were derived in a more general way, resulting in other pseudo-archetype sets. Such pseudo-archetype sets may result in improved results. In light of the vast amount of work done to improve VQ codebooks, there are, no doubt, improvements that can be made to the archetype selection process presented here.

Chapter 5

Hierarchical Interpretation of Fractal Image Coding and Its Applications

Z. Baharav, D. Malah, and E. Karnin

Many interesting features associated with iterated function systems and standard fractals seem to fade away as the fractal coding of images evolves. For example, the property of self-similarity at different resolutions, inherent to fractals, does not show up in a simple form in the image coding problem. In this chapter we will demonstrate a hierarchical model for the fractal encoding of images, and we will use it to show that the properties of self-similarity at different resolutions, and the related notion of fractal dimension, exist in fractal coding of images. Moreover, applications of these properties will be given.

The chapter is organized as follows: Section 5.1 reviews the formulation of partitioned iterated function systems (PIFS) coding. Notations and illustrative examples are also given. Section 5.2 presents the main result. It contains a theorem, relating the different resolutions of a signal to its PIFS code, and gives a hierarchical interpretation of this theorem. Section 5.3 develops a matrix form of the PIFS code; a formulation which is needed for the applications sections that follow. In Sections 5.4–5.6 various applications are described: Section 5.4 describes a fast decoding method. Section 5.5 defines the important notion of the PIFS embedded function, and applies it to achieve super-resolution. Section 5.6 revisits the definition of the embedded function, this time to examine different sampling methods. Conclusions are presented in Section 5.7. Finally, the proofs of the three main theorems appear in Addenda A, B, and C.

5.1   Formulation of PIFS Coding/Decoding

In this section we discuss and show an example of a different formulation of PIFS coding and decoding. Also, we will refer to 1-dimensional signals, and we will call them either vectors or blocks. Extensions to two-dimensions are immediate in most cases. The following notation will be used: vectors are in boldface letters (like **a**) and matrices are in bold uppercase (like **A**).

Encoding

The task of finding the PIFS code of a vector $\boldsymbol{\mu}_o$ is the task of finding a contractive transformation W, such that its fixed point is as close as possible to $\boldsymbol{\mu}_o$. Formally, this task can be described as follows.

Consider the complete metric space $(\mathbb{R}^N, d^\infty)$, where:

1. $\mathbb{R}^N$ denotes the N-dimensional Cartesian product of the real numbers. Each point in $\mathbb{R}^N$ is a column-vector of size N of real numbers. Thus

$$\mathbf{x} \in \mathbb{R}^N \qquad \text{implies} \qquad \mathbf{x} = \begin{pmatrix} x_1 \\ x_2 \\ \vdots \\ x_N \end{pmatrix}. \tag{5.1}$$

2. d^∞ is the metric defined by:

$$\mathbf{x}, \mathbf{y} \in \mathbb{R}^N \qquad d^\infty(\mathbf{x}, \mathbf{y}) = \max_{i=1,\dots,N} |x_i - y_i|. \tag{5.2}$$

The vector to be encoded is $\boldsymbol{\mu}_o \in \mathbb{R}^N$. We seek a transformation W, such that the following three requirements are fulfilled:

1. W maps the space into itself:

$$\begin{aligned} W &: \quad \mathbb{R}^N \quad \rightarrow \quad \mathbb{R}^N \\ & \quad\;\; \mathbf{v} \quad \mapsto \quad \mathbf{u} = W(\mathbf{v}). \end{aligned} \tag{5.3}$$

2. W is a contractive transformation:

$$\exists\, s \in [0, 1) \; | \quad \forall\, \mathbf{x}, \mathbf{y} \in \mathbb{R}^N \qquad d^\infty(W(\mathbf{x}), W(\mathbf{y})) \le s\, d^\infty(\mathbf{x}, \mathbf{y}). \tag{5.4}$$

These first two requirements define the set of allowed transformations, $W \in \mathcal{W}$.

3. Being a contraction in a complete metric space, W has a unique fixed point $\mathbf{f}_W \in \mathbb{R}^N$ such that $\mathbf{f}_W = W(\mathbf{f}_W)$ (Contraction Mapping Theorem 2.3). The third requirement is that W minimizes the distance between its fixed point $\mathbf{f}_W$ and $\boldsymbol{\mu}_o$:

$$d^\infty(\boldsymbol{\mu}_o, \mathbf{f}_W) = \min_{v \in \mathcal{W}} d^\infty(\boldsymbol{\mu}_o, \mathbf{f}_V). \tag{5.5}$$

That is,

$$W = \arg \min_{v \in \mathcal{W}} d^\infty(\boldsymbol{\mu}_o, \mathbf{f}_V). \tag{5.6}$$

Finding such a transformation W can be a very complex problem, since it involves a minimization over many transformations. An approach to making this problem solvable is to restrict the number of allowed transformations ([45], [44]). The W found this way may be suboptimal, but this is a compromise that one has to make in order to solve the minimization problem.

Thus, we will restrict the allowed transformations W to be systems of M_R functions w_i. Each w_i is further restricted to be of the form:

$$w_i \; : \quad \mathbb{R}^D \quad \to \quad \mathbb{R}^B$$
$$\mathbf{d}_{m_i} \quad \mapsto \quad \mathbf{r}_i = w_i(\mathbf{d}_{m_i}) = a_i \varphi(\mathbf{d}_{m_i}) + b_i \mathbf{1}_B, \tag{5.7}$$

where:

$\mathbf{d}$ - is a block of D consecutive elements extracted from $\mathbf{v}$. It is called the *domain block*. $\mathbf{d}_{m_i}$ is thus the m_i-th domain block in an enumerated list of all such blocks in $\mathbf{v}$. The use of the subscript m_i stresses the fact that the domain block $\mathbf{d}_{m_i}$ is mapped to $\mathbf{r}_i$. For now, no specific mechanism for extracting the blocks will be discussed.

$\mathbf{r}$ - is a block of $B < D$ consecutive elements of $\mathbf{v}$. It is called the *range block*, and $\mathbf{r}_i$ is thus the i-th range block. $\mathbf{r}_i$ belongs to the image of W.

φ - is a *spatial contraction* function which maps blocks of size D to blocks of size B .

a_i - is a scalar *scaling* factor,

$$a_i \in \mathbb{R}, \qquad |a_i| < 1 . \tag{5.8}$$

b_i - is a scalar *offset* value, $b_i \in \mathbb{R}$.

$\mathbf{1}_B$ - is a vector of size B of all 1's.

The three parameters (a_i, b_i, m_i) are called the *transformation parameters*.

By loosely using the union notation to describe concatenation of blocks, we can write :

$$\mathbf{u} \;=\; W(\mathbf{v})$$
$$W(\mathbf{v}) \;=\; \bigcup_{i=1}^{M_R} w_i(\mathbf{d}_{m_i}), \qquad \mathbf{d}_{m_i} \in \mathbf{v}. \tag{5.9}$$

The length of $\mathbf{u}$, which is the result of concatenating M_R range blocks of size B each, is therefore:

$$N = M_R \cdot B. \tag{5.10}$$

Moreover, the concatenation of range blocks can also be written (using $\mathbf{r}_i(j)$ to mean the j-th coordinate of $\mathbf{r}_i$) as:

$$\mathbf{u}((i - 1) \cdot B + j) = \mathbf{r}_i(j); \quad i = 1, \ldots, M_R, \qquad j = 1, \ldots, B. \tag{5.11}$$

Now the mechanism of computing $\mathbf{u} = W(\mathbf{v})$, when all the parameters describing W are known, can be described as shown in Table 5.1:

Table 5.1: Algorithm for Computing the Transformation $\mathbf{u} = W(\mathbf{v})$.

---

1. For $i = 1$ to M_R:

    (a) Extract the $\mathbf{d}_{m_i}$ block from the vector $\mathbf{v}$.

    (b) Compute
    $$\mathbf{r}_i = w_i(\mathbf{d}_{m_i}) = a_i \varphi(\mathbf{d}_{m_i}) + b_i \mathbf{1}_B . \qquad (5.12)$$

2. *Concatenate* the range blocks thus obtained, $\mathbf{r}_i$, $i = 1, \ldots, M_R$, in the natural order, to get the new vector $\mathbf{u}$. The length of the vector $\mathbf{u}$ is $N = M_R \cdot B$.

The transformation W described above is called a *blockwise* transformation, the reason being evident from the computational algorithm.

So far the discussion of the w_i's was quite general. In order to make the discussion both more practical and lucid at this stage, we will make further restrictions and assumptions about the different parameters, shown in Table 5.2.

The description of the PIFS code of a vector, namely, the parameters that define W, can now be summarized. The PIFS code is shown in Table 5.3.

All other relevant parameters needed for decoding, such as $D = 2B$, $D_h = B$, $N = M_R B$, and others, are derived from the PIFS code using the previous assumptions.

The process of encoding, namely, the process of finding the M_R triplets of transformation parameters (a_i, b_i, m_i), is shown in Table 5.4. As stated, we seek to minimize $d^\infty(\boldsymbol{\mu}_o, \mathbf{f})$, where $\boldsymbol{\mu}_o$ is the original vector, and $\mathbf{f}$ is the fixed point of the sought transformation.

Since, by the Collage Theorem,

$$d^\infty(\boldsymbol{\mu}_o, \mathbf{f}) \leq \frac{1}{1-s} d^\infty(\boldsymbol{\mu}_o, W(\boldsymbol{\mu}_o)), \qquad (5.13)$$

one actually tries to minimize the upper bound on the right of Equation (5.13), by minimizing $d^\infty(\boldsymbol{\mu}_o, W(\boldsymbol{\mu}_o))$ instead of $d^\infty(\boldsymbol{\mu}_o, \mathbf{f})$. (Note that since s is the contraction factor of W, the factor $\frac{1}{1-s}$ depends on W. This factor, however, is not taken into account.) Though this method will not necessarily lead to a minimum of $d^\infty(\boldsymbol{\mu}_o, \mathbf{f})$, it is at present the most practical way of doing the coding. See [64]§4.4 for a statistical motivation for the minimization goal. Since W is a blockwise transformation, this minimization can be done in stages, as described below.

The minimization process to be described consists of finding a W that satisfies $\boldsymbol{\mu}_o \cong W(\boldsymbol{\mu}_o)$. Thus, $\boldsymbol{\mu}_o$ is approximately the fixed point of W. Since W uniquely defines its fixed point, storing W (by storing the parameters that define it) defines a lossy code for $\boldsymbol{\mu}_o$. Note that in this case both the operated-on vector and its image by the transformation W are assumed to be $\boldsymbol{\mu}_o$. Therefore, both $\mathbf{d}_{m_i} \in \boldsymbol{\mu}_o$ and $\mathbf{r}_i \in \boldsymbol{\mu}_o$.

This formulation is now demonstrated with a numerical example. Figures 5.1(a)-(b) present a vector $\boldsymbol{\mu}_o$ and its PIFS code. The PIFS is given in a table form. By performing the transformation described by the PIFS on $\boldsymbol{\mu}_o$, one can verify that in this example the vector $\boldsymbol{\mu}_o$ is a fixed

Table 5.2: Parameters, restrictions, and assumptions.

1. N - The size of the vector $\boldsymbol{\mu}_o$ to be encoded is an integer power of 2.

2. $B = 2^l$ - The size of a range block. B is therefore also some integer power of 2.

3. $D = 2B$ - The size of a domain block is twice the size of a range block.

4. $D_h = B$ - The value of D_h is defined to be the shift between consecutive domain blocks. Thus, the number of domain blocks is $M_D \equiv (\frac{N-D}{D_h} + 1)$, and each domain block is given by

$$\mathbf{d}_{m_i}(j) = \mathbf{v}((m_i - 1)D_h + j), \qquad (5.14)$$

$$m_i = 1, 2, \ldots, M_D; \quad j = 1, 2, \ldots, D.$$

Note that the domain blocks are overlapping, since $D_h < D$.

5. $\varphi(\cdot)$ - The spatial contraction function is defined to be:

$$\varphi(\mathbf{d}_{m_i})(j) \equiv \frac{1}{2}(\mathbf{d}_{m_i}(2j) + \mathbf{d}_{m_i}(2j - 1)), \qquad (5.15)$$

for $j = 1, 2, \ldots, B$. That is, $\varphi(\cdot)$ contracts blocks of size $D = 2B$ into blocks of size B by averaging pairs of adjacent elements in $\mathbf{d}_{m_i}$.

Table 5.3: PIFS code.

1. B - The size of the range blocks.

2. M_R - The number of range blocks.

3. M_R triplets of the transformation-parameters (a_i, b_i, m_i).

Table 5.4: PIFS Coding of μ_o .

---

1. Store B in the code file.

2. Store M_R in the code file, where $M_R = N/B$, and N is the length of μ_o.

3. Partition μ_o into M_R range blocks, as described in Equation (5.11),

$$\mathbf{r}_i(j) = \mu_o((i-1) \cdot B + j), \qquad\qquad (5.16)$$

$$i = 1, \dots, M_R, \qquad j = 1, \dots, B.$$

4. Extract from μ_o the $M_D = (\frac{N-D}{D_h} + 1)$ domain blocks, according to Equation (5.14)

$$\mathbf{d}_l(j) = \mu_o((l-1)D_h + j), \qquad\qquad (5.17)$$

$$l = 1, 2, \dots, M_D, \qquad j = 1, 2, \dots, D.$$

5. For $i = 1$ to M_R

    (a) Find the best parameters (a_i, b_i, m_i), such that

    $$d^\infty(\mathbf{r}_i, a_i\varphi(\mathbf{d}_{m_i}) + b_i\mathbf{1}_B) \qquad\qquad (5.18)$$

    is minimized.

    (b) Store the parameters (a_i, b_i, m_i) in the code file.

---

point of the transformation (namely, $\mu_o = W(\mu_o)$), and thus the coding in this case is lossless. Indeed, as shown in Figure 5.1(c), the result of computing the first code-line, namely, w_1, on μ_o, produces exactly $\mathbf{r}_1$.

Decoding

The process of decoding is straightforward, since it involves the finding of a fixed point of a contractive transformation W. This can be done by repeatedly iterating W on *any* initial vector until a desired proximity to the fixed point is reached [4].

In Table 5.5, the decoding of the PIFS code of Figure 5.1b is demonstrated, starting from an initial vector of all 0's.

Looking at the example in Figure 5.1, an important fact about the decoding (as well as the PIFS definition) can be observed. In the example, the PIFS code, with the prescribed $B = B_1 = 4$, resulted in a transformation $W^1 : \mathbb{R}^{16} \mapsto \mathbb{R}^{16}$ with fixed point $\mathbf{f}^1 \in \mathbb{R}^{16}$. Suppose, however, that the value of B is changed to some other value than the one prescribed in the PIFS

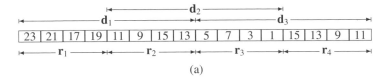

(a)

Range-block index i	Scale a_i	Domain-block index m_i	Offset b_i
1	0.5	1	12
2	0.5	3	8
3	0.5	2	0
4	0.5	1	4

$B = 4$, $M_R = 4$

(b)

$$
\begin{aligned}
\mathbf{r}_1 &= 0.5 \cdot \varphi(\mathbf{d}_1) + 12 = \\
&= 0.5 \cdot \varphi([23, 21, 17, 19, 11, 9, 15, 13]) + 12 \\
&= 0.5 \cdot [22, 18, 10, 14] + 12 \\
&= [23, 21, 17, 19].
\end{aligned}
$$

(c)

Figure 5.1: (a) An original vector $\boldsymbol{\mu}_o$. (b) The PIFS code of $\boldsymbol{\mu}_o$. (c) An example of computing $\mathbf{r}_1$ using the first code line in (b) and $\mathbf{d}_1 \in \boldsymbol{\mu}_o$ given in (a).

code, e.g., $B = \frac{1}{2}B_1 = 2$. We have thus created a new transformation, denoted here by $W^{\frac{1}{2}}$, which is clearly a contractive transformation in $\mathbb{R}^8$. The decoding process of $W^{\frac{1}{2}}$ will therefore yield a fixed-point vector $\mathbf{f}^{\frac{1}{2}}$ of length $N = N_{\frac{1}{2}} = 8$, which is half the length of $\mathbf{f}^1$. Thus, we conclude that the PIFS can be decoded in different spaces, yielding a diffcrent fixed point in each space. Through the remainder of the chapter, we will keep using the notation W^q (and $\mathbf{f}^q$) to denote the transformation (and fixed point) which results from the PIFS code when using $B = qB_1$; where $B = B_1$ is the size of the original PIFS code range block corresponding to W^1 and $\mathbf{f}^1$. In Figures 5.2(a)-(c), three fixed points of the same PIFS code (using different B's) are described, each one being in a different space – $\mathbb{R}^{16}$, $\mathbb{R}^8$, and $\mathbb{R}^4$, respectively.

The exact relation between these different fixed points, and its interpretation, is actually the main subject of this chapter and is examined in detail next.

Table 5.5: Decoding by iterations.

iter	vector															
0	0	0	0	0	0	0	0	0	0	0	0	0	0	0	0	0
1	12	12	12	12	8	8	8	8	0	0	0	0	4	4	4	4
2	18	18	16	16	8	8	10	10	4	4	0	0	10	10	8	8
3	21	20	16	17	10	8	13	12	4	5	2	0	13	12	8	9

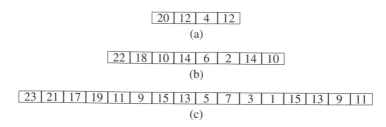

Figure 5.2: Decoding of the PIFS code of Figure 5.1 with (a) $B = 1$, (b) $B = 2$, and (c) $B = 4$.

5.2 Hierarchical Interpretation

As described in the previous section, each value of B leads to a different transformation with a different fixed point. The following theorem describes the relation between two different fixed points that arise when B is halved.

Theorem 5.1 (Zoom) *Given a PIFS code that leads to W^1 with $B = B_1$, and to $W^{\frac{1}{2}}$ with $B = B_1/2$:*

1. **Zoom-out:** *Let $\mathbf{f}^1$ be the fixed point of W^1. Then the fixed point $\mathbf{f}^{\frac{1}{2}}$ of $W^{\frac{1}{2}}$ is given by:*

$$\mathbf{f}^{\frac{1}{2}}(j) = \frac{1}{2}\left\{\mathbf{f}^1(2j) + \mathbf{f}^1(2j - 1)\right\}, \qquad j = 1, \ldots, \frac{N_1}{2}, \tag{5.19}$$

where $N_1 \equiv M_R \cdot B_1$.

2. **Zoom-in:** *Let $\mathbf{f}^{\frac{1}{2}}$ be the fixed point of $W^{\frac{1}{2}}$. Then the fixed point $\mathbf{f}^1$ of W^1 is given by:*

$$\mathbf{f}^1((i - 1)B_1 + j) = a_i \mathbf{f}^{\frac{1}{2}}((m_i - 1)D_h^{\frac{1}{2}} + j) + b_i, \tag{5.20}$$

$$i = 1, \ldots, M_R, \qquad j = 1, \ldots, B_1,$$

where $D_h^{\frac{1}{2}} \equiv \frac{D_h^{1}}{2} = \frac{B_1}{2}$.

The proof of the theorem is given in Addendum A at the end of the chapter.

An interpretation of the theorem is as follows: In order to compute each element of $\mathbf{f}^{\frac{1}{2}}$ from a given $\mathbf{f}^1$, one has to take the average of two adjacent elements in $\mathbf{f}^1$, as described in Equation (5.19). On the other hand, in order to compute $\mathbf{f}^1$ from a given $\mathbf{f}^{\frac{1}{2}}$, one follows Equation (5.20), which is similar to computing W^1 itself (compare with Equation (5.12)).

Theorem 5.1 establishes a relation between the pair $\mathbf{f}^1$ and $\mathbf{f}^{\frac{1}{2}}$. The same relation is carried over to the pair $\mathbf{f}^{\frac{1}{2}}$ and $\mathbf{f}^{\frac{1}{4}}$,

$$\mathbf{f}^{\frac{1}{4}}(j) = \frac{1}{2}\left\{\mathbf{f}^{\frac{1}{2}}(2j) + \mathbf{f}^{\frac{1}{2}}(2j - 1)\right\}, \tag{5.21}$$

$$\mathbf{f}^{\frac{1}{2}}((i - 1)B_{\frac{1}{2}} + j) = a_i \mathbf{f}^{\frac{1}{4}}((m_i - 1)D_h^{\frac{1}{4}} + j) + b_i. \tag{5.22}$$

The relation also holds for the pair $\mathbf{f}^{\frac{1}{4}}$ and $\mathbf{f}^{\frac{1}{8}}$, and so on. The collection of fixed points can thus be described in terms of a hierarchical structure, as was shown in Figure 5.2. This structure is a *pyramid of the PIFS fixed points*, with $\mathbf{f}^{1/2^p}$ comprising the p-th level of the pyramid. Thus, $\mathbf{f}^{1/2^p}$ has $N_1/2^p$ elements, and $\mathbf{f}^1$ corresponds to $p = 0$ and is of length N_1. The level with the coarsest resolution (Figure 5.2c) is called the *top level*.

The relations between two adjacent levels in the pyramid can be summarized as follows:

1. In order to go *up* the pyramid, from level p to level $p + 1$, the operation is as follows (zoom-out):

$$\mathbf{f}^{1/2^{p+1}}(j) = \frac{1}{2}\left\{\mathbf{f}^{1/2^p}(2j) + \mathbf{f}^{1/2^p}(2j - 1)\right\}, \tag{5.23}$$

$$j = 1, \ldots, N_1/2^{p+1},$$

   and is identical to Equation (5.19) for $p = 0$.

2. In order to go *down* the pyramid, from level $p + 1$ to level p, the operation is as follows (zoom-in):

$$\mathbf{f}^{1/2^p}((i - 1)B_{1/2^p} + j) = a_i\mathbf{f}^{1/2^{p+1}}((m_i - 1)D_h^{1/2^{p+1}} + j) + b_i, \tag{5.24}$$

$$i = 1, \ldots, M_R, \qquad j = 1, \ldots, B_{1/2^p},$$

   where $D_h^{1/2^{p+1}} \equiv D_h^1/2^{p+1} = B_1/2^{p+1}$. Here, Equation (5.20) is obtained for $p = 0$.

An intuitive understanding of the process can be gained by noting that the domain blocks of $\mathbf{f}^1$, after contraction, are actually blocks which are contained in $\mathbf{f}^{\frac{1}{2}}$. Formally, this can be shown by writing down the expression for the l-th element in the contraction of the m_i-th domain block of $\mathbf{f}^1$ (Equations (5.14)-(5.15)):

$$\varphi(\mathbf{d}_{m_i}{}^1)(l) = \frac{1}{2}\left\{\mathbf{f}^1((m_i - 1)D_{h1} + 2l - 1) + \mathbf{f}^1((m_i - 1)D_{h1} + 2l)\right\} \tag{5.25}$$

$$l \in 1, \ldots, B_1.$$

According to Equation (5.19), the right-hand side is *exactly* $\mathbf{f}^{\frac{1}{2}}((m_i - 1)D_{h1}/2 + l)$, and if we further denote, as before, $D_h^{\frac{1}{2}} \equiv D_{h1}/2$ and $\mathbf{d}_{m_i}{}^{\frac{1}{2}}$ as the m_i-th domain block of $\mathbf{f}^{\frac{1}{2}}$, we conclude that:

$$\varphi(\mathbf{d}_{m_i}{}^1)(l) = \mathbf{d}_{m_i}{}^{\frac{1}{2}}(l), \qquad l \in 1, \ldots, B_1. \tag{5.26}$$

The question arises: what is the smallest size of the top level such that the above relations between two adjacent levels still hold? This is answered by the following corollary.

Corollary 5.1 *Let $\mathbf{f}^1 \in \mathbb{R}^N$ be a fixed point of a given PIFS, and let $B = B_1 = 2^l$, $D = D_1 = 2B_1$, and $D_h = B_1$. Then, the number of levels in the pyramid of PIFS fixed points is*

$$\log_2(B_1) + 1 = l + 1, \tag{5.27}$$

leading to a top-level size of $N/2^l(= M_R)$.

Proof: Ascending one level in the hierarchy means halving the size of the range block B. Since this size must be at least 1 in order that the PIFS can be applied, the corollary follows. ∎

Table 5.6 summarizes the notation.

In some cases, the top level is directly derivable from the code. For example, if the coding procedure is modified so that domains are orthogonalized with respect to DC (monotone) blocks, as in Chapter 8, then the top level is the DC value associated with each range block. Further relation between this method and the method of Chapter 8 can be found in [65].

Table 5.6: Pyramid of fixed points: Notation summary.

Level-num. p	Range-block size B	Num. of elements N	fixed point $\mathbf{f}$
0	B_1	$N_1 = M_R \cdot B_1$	$\mathbf{f}^1$
1	$B_{1/2} = B_1/2$	$N_{1/2} = N_1/2$	$\mathbf{f}^{1/2}$
$\vdots$	$\vdots$	$\vdots$	$\vdots$
$\log_2 B_1 = l$ (top level)	$B_{1/B_1} = 1$	$N_{1/B_1} = M_R$	$\mathbf{f}^{1/B_1}$

5.3 Matrix Description of the PIFS Transformation

In this section, a matrix description for the PIFS transformation W described above is given. We assume that a PIFS code is given. This code includes all the necessary information for computing W, as described in Section 5.1.

The transformation $\mathbf{u} = W(\mathbf{v})$ is composed of $\{w_i\}_{i=1}^{M_R}$, where each w_i operates on a domain block. The j-th element of the i-th range block of $\mathbf{u}$ is, according to Equation (5.11):

$$\mathbf{r}_i(j) = \mathbf{u}((i-1) \cdot B + j) ; \quad i = 1, \ldots, M_R ; \quad j = 1, \ldots, B. \tag{5.28}$$

This element is, according to Equation (5.12), the result of transforming the appropriate domain block, as given by the PIFS code:

$$\mathbf{r}_i(j) = a_i \cdot \varphi(\mathbf{d}_{m_i})(j) + b_i . \tag{5.29}$$

Substituting the definition of φ given in Equation (5.15) and using Equation (5.14) gives

$$\mathbf{r}_i(j) = \frac{a_i}{2} (\mathbf{v}((m_i - 1)D_h + 2j) + \mathbf{v}((m_i - 1)D_h + 2j - 1)) + b_i . \tag{5.30}$$

Writing the last equations in a matrix form for all the elements in $\mathbf{u}$ yields:

$$\mathbf{u} = \mathbf{F}\mathbf{v} + \mathbf{b}, \tag{5.31}$$

where $\mathbf{F}$ and $\mathbf{b}$ are described below:

- **b** - An $N \times 1$ offset vector. It is a block vector, composed of (N/B) basic blocks of length B,

$$\mathbf{b} = \begin{bmatrix} \mathbf{b}_1 \\ \mathbf{b}_2 \\ \vdots \\ \mathbf{b}_{N/B} \end{bmatrix}, \tag{5.32}$$

where,

$$\mathbf{b}_i = \begin{bmatrix} b_i \\ \vdots \\ b_i \end{bmatrix}, \qquad i = 1, 2, \ldots, \frac{N}{B} \tag{5.33}$$

is a $B \times 1$ vector.

- **F** - An $N \times N$ transfer matrix, given by :

$$\mathbf{F} = \mathbf{A} \cdot \frac{1}{2} \cdot \mathbf{D}. \tag{5.34}$$

In this decomposition of **F** we have:

- **D** - A domain-location matrix. The product $\frac{1}{2}\mathbf{D} \cdot \mathbf{v}$ plays the role of $\varphi(\mathbf{d}_{m_i})$, and therefore has a dual role:

  1. Extract $\mathbf{d}_{m_i}$ from **v**.

  2. Perform the spatial contraction operation on $\mathbf{d}_{m_i}$.

  D is a block matrix with a structure that is best demonstrated by the example below. In this example $\mathbf{d}_3$ is transformed to $\mathbf{r}_1$, $\mathbf{d}_1$ to $\mathbf{r}_2$, and $\mathbf{d}_4$ to $\mathbf{r}_3$

$$\mathbf{D} = \begin{bmatrix} \mathbf{0}_B & \mathbf{0}_B & \boxed{\mathbf{D}_{B \times D}} & \mathbf{0}_B & \mathbf{0}_B & \mathbf{0}_B & \mathbf{0}_B & \cdots \\ \boxed{\mathbf{D}_{B \times D}} & \mathbf{0}_B & \mathbf{0}_B & \mathbf{0}_B & \mathbf{0}_B & \mathbf{0}_B & \mathbf{0}_B & \cdots \\ \mathbf{0}_B & \mathbf{0}_B & \mathbf{0}_B & \boxed{\mathbf{D}_{B \times D}} & \mathbf{0}_B & \mathbf{0}_B & \mathbf{0}_B & \cdots \\ \vdots & & \vdots & & \vdots & & \vdots & \ddots \end{bmatrix}_{N \times N} \tag{5.35}$$

The matrix $\mathbf{0}_B$ denotes an all-zero matrix of size $B \times B$. The location of the $B \times D$ blocks $\mathbf{D}_{B \times D}$ is determined by the indices m_i, whereas the spatial contraction is performed by its elements, as explained below:

  1. If domain block m_i is transformed to range block i, then a block matrix $\mathbf{D}_{B \times D}$ would be positioned with top left element at

$$((i - 1)B + 1, (m_i - 1)D_h + 1)). \tag{5.36}$$

2. The spatial contraction mapping is performed by $\mathbf{D}_{B \times D}$, which has the following structure:

$$\mathbf{D}_{B \times D} = \begin{bmatrix} 1 & 1 & 0 & 0 & 0 & 0 & 0 & \cdots \\ 0 & 0 & 1 & 1 & 0 & 0 & 0 & \cdots \\ 0 & 0 & 0 & 0 & 1 & 1 & 0 & \cdots \\ \vdots & & & & & \ddots & & \vdots \\ 0 & 0 & & & & 0 & 1 & 1 \end{bmatrix}_{B \times D} \tag{5.37}$$

Namely, $\frac{1}{2}\mathbf{D}_{B \times D}$ maps blocks of size D to blocks of size B by *averaging* every two adjacent elements.

- **A** - A scaling matrix. It is a block-diagonal matrix:

$$\mathbf{A} = \begin{bmatrix} \mathbf{A}_1 & \mathbf{0}_B & \mathbf{0}_B & \cdots \\ \mathbf{0}_B & \mathbf{A}_2 & \mathbf{0}_B & \cdots \\ \mathbf{0}_B & \mathbf{0}_B & \mathbf{A}_3 & \cdots \\ \vdots & \vdots & \vdots & \ddots \end{bmatrix}_{N \times N} \tag{5.38}$$

where

$$\mathbf{A}_i = a_i \cdot \mathbf{I}_{B \times B}, \qquad i = 1, 2, \ldots, \frac{N}{B}. \tag{5.39}$$

The matrix $\mathbf{I}_{B \times B}$ denotes the $B \times B$ identity matrix. If, for a certain code, $a_i = a$ for every i, then $\mathbf{A} = a\mathbf{I}_{N \times N}$.

The size of each of the matrices $\mathbf{F}, \mathbf{A}, \mathbf{D}$, as well as the size of the vector $\mathbf{b}$, are seen to depend directly on B. Indeed, as we have already seen before, changing the value of B results in a different W and therefore leads also to different matrices, though their *basic structure* remains the same. For example, the matrix form of the PIFS code described in Figure 5.1(b) with $B = 2$ is described in Figure 5.3.

5.4   Fast Decoding

A fast decoding method, which we call *hierarchical decoding*, follows directly from the hierarchical interpretation of the PIFS code. In this method, one begins by computing the *top level*. This can be done either by computing it directly, when possible (for example, with the method of Chapter 8), or by iterating the PIFS with $B = 1$, that is, by applying W to an initial vector of size M_R until a fixed point is reached (or closely approximated). Then, one follows the deterministic algorithm Equation (5.24) to advance to a higher resolution. The process of advancing to a higher resolution is repeated until the desired vector size is achieved. This method is compared below with the conventional *iterative decoding* method [48], where the iterations are done on a *full-scale* image. The computational savings obtained in using the hierarchical decoding method stems mainly from the fact that the iterations are done *only* in order to find the top level, which is a small-size vector.

$$\mathbf{A} = \tfrac{1}{2} \cdot \mathbf{I}_{8\times 8}, \qquad \mathbf{b} = [12, 12, 8, 8, 0, 0, 4, 4]^T$$

$$\mathbf{D} = \begin{bmatrix} 1 & 1 & 0 & 0 & 0 & 0 & 0 & 0 \\ 0 & 0 & 1 & 1 & 0 & 0 & 0 & 0 \\ 0 & 0 & 0 & 0 & 1 & 1 & 0 & 0 \\ 0 & 0 & 0 & 0 & 0 & 0 & 1 & 1 \\ 0 & 0 & 1 & 1 & 0 & 0 & 0 & 0 \\ 0 & 0 & 0 & 0 & 1 & 1 & 0 & 0 \\ 1 & 1 & 0 & 0 & 0 & 0 & 0 & 0 \\ 0 & 0 & 1 & 1 & 0 & 0 & 0 & 0 \end{bmatrix} \begin{matrix} \mathbf{d}_1 & \to & \mathbf{r}_1 \\ \\ \mathbf{d}_3 & \to & \mathbf{r}_2 \\ \\ \mathbf{d}_2 & \to & \mathbf{r}_3 \\ \\ \mathbf{d}_1 & \to & \mathbf{r}_4 \end{matrix}$$

Figure 5.3: The matrices **A** and **D**, and the vector **b** describing the PIFS code of Figure 5.1, for $B = 2$.

Computational Cost

A detailed discussion of the computational cost, for both the 1-dimensional (1-D) and 2-dimensional (2-D) cases, is given in [2]. Here we concentrate on the simple 1-D case, assuming a floating point processor (which means that addition and multiplication operations take the same time to perform).

The following notation and definitions are used:

t_{op} - the computation time of a summation or a multiplication operation.

t_{tot} - the total computation time.

I - the number of iterations.

In arriving at the following results, multiplications by $\tfrac{1}{2}$ or $\tfrac{1}{4}$ were not counted (counting them shows even a greater savings in using the hierarchical decoding, because they occur mainly in the iterative method, when applying φ).

With $N = N_1$ denoting the input vector length and $B = B_1$ the range block size, we have:

- *Iterative decoding* - Referring to Equations (5.12) and (5.15), for a single iteration, the computation time is:

$$N_1 \cdot [(1 + 1)\text{sums} + 1\text{mult.s}] = N_1 \cdot 3t_{op} . \tag{5.40}$$

Thus, the total computation time is:

$$t_{tot}^i = I \cdot N_1 \cdot 3t_{op} . \tag{5.41}$$

- *Hierarchical decoding* - Recall that at the top level there are M_R elements (by Corollary 5.1). Hence, according to the result in Equation (5.41), we know the computation time needed to compute the top level. According to Corollary 5.1, there are $\log_2(B_1)$

levels in the pyramid, excluding the top level, with $N_1/2^p$ elements in the p-th level ($p = \log_2 B_1$ at the top level). Referring to Equation (5.24), we can compute the cost of transforming from the top level to $\mathbf{f}^1$:

$$\sum_{p=1}^{\log_2(B_1)} \left(2^p \frac{N_1}{B_1} \right) \cdot [1\text{sums} + 1\text{mult.s}] = (B_1 - 1)\frac{N_1}{B_1} 2 \cdot 2t_{op} \approx N_1 \cdot 4t_{op} \, . \tag{5.42}$$

The total computation time is therefore

$$t_{tot}^h \approx N_1 \cdot \left(\frac{3I}{B_1} + 4 \right) \cdot t_{op} \, . \tag{5.43}$$

Thus, the ratio of computation times is :

$$Q \equiv \frac{t_{tot}^h}{t_{tot}^i} = \frac{(\frac{3I}{B_1} + 4)}{3I} = \left(\frac{1}{B_1} + \frac{4}{3I} \right) . \tag{5.44}$$

It is seen that the larger the values of I and B_1, the more advantageous the hierarchical decoding method is. For example, for a typical case in which $I = 8$, $B_1 = 8$, one gets $Q = 0.3$. In [2] it is shown that for the 2-D case (assuming $B_1{}^2 \gg I$),

$$Q_{2-D} = \frac{8}{15 \cdot I} \, . \tag{5.45}$$

Thus, the savings can reach more than an order of magnitude.

5.5   Super-resolution

The subject of resolution is inherently related to the discretization of a function of a continuous variable. The process of discretization is called sampling. In the following definition we define a specific method of sampling.

Definition 5.1 *Given a function* $G(x) \in L^\infty [0, 1]$, *define* $G_r(i)$ *by*

$$G_r(i) \equiv r \int_{\frac{(i-1)}{r}}^{\frac{i}{r}} G(x) \, dx, \qquad i = 1, \ldots, r \, , \tag{5.46}$$

which is the function $G(x)$ *at resolution* r. *We say that* $G_{r_1}(i)$ *is finer (i.e., having higher resolution) than* $G_{r_2}(i)$ *(which is* coarser*) if* $r_1 > r_2$.

The following theorem introduces the new notion of the PIFS embedded function and relates it to the PIFS fixed point.

Theorem 5.2 (PIFS Embedded Function) *Given a PIFS code, there exists a unique function* $G(x) \in L^\infty [0, 1]$ *such that a vector* $\mathbf{v}_N \in \mathbb{R}^N$ *is a fixed point of the PIFS iff it is equal to the function* $G(x)$ *at resolution* $r = N, \forall N$, *i.e.,*

$$\mathbf{v}_N(j) = G_N(j), \qquad j = 1, 2, \ldots, N. \tag{5.47}$$

The function $G(x)$ *is called the* PIFS embedded function.

The theorem is proved in Addendum B.

In Figure 5.4, an intuitive demonstration of the PIFS embedded function theorem is given. The PIFS is the one described previously in Figure 5.1. Its fixed points, for $B = 1$, $B = 2$, and $B = 4$ are shown in Figure 5.4 as functions of a continuous variable $x \in [0, 1]$. For example, the fixed point using $B = 1$ is the vector $[20, 12, 4, 12]$, which has $N = 4$ elements. Thus, the fixed point is drawn as a piecewise constant function:

$$f(x) = \begin{cases} 20 & x \in [0, \frac{1}{4}) \\ 12 & x \in [\frac{1}{4}, \frac{1}{2}) \\ \vdots & \vdots \end{cases} \tag{5.48}$$

The embedded function $G(x)$ is also shown. One easily sees that the fixed points "approach" the PIFS embedded function. Moreover, it is seen that the value of each function, in each of its intervals, equals the mean of the PIFS embedded function over the appropriate interval as described by Equations (5.46) and (5.47).

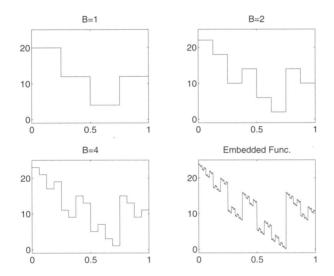

Figure 5.4: Fixed points for $B = 1$, $B = 2$, $B = 4$, and the corresponding PIFS embedded function.

Super-resolution deals with finding a higher resolution of a given discrete signal [73]. For example, suppose that a vector $\mathbf{v}^1$ with elements $\mathbf{v}^1(i) = h_N(i)$, $i = 1, 2, \ldots, N$ is given, where $h_N(i)$ is the unknown signal $h(x) \in L^\infty[0, 1]$ at resolution N. The goal is to find a vector $\mathbf{f}^2$ of length $2N$, which approximates the signal $h(x)$ at resolution $2N$, namely, $h_{2N}(i)$, $i = 1, 2, \ldots, 2N$. The process of transforming from a given resolution to a higher one is also called zoom-in (or simply *zooming*).

The hierarchical representation suggests a simple method for finding higher-resolution representations from a given resolution. Given $\mathbf{v}^1$, we can find its PIFS code. This PIFS code will be, in general, a lossy one. Thus, its fixed point $\mathbf{f}^1$ will only be an approximation of $\mathbf{v}^1$. The PIFS code enables us to build a hierarchical structure, which we called the pyramid of fixed points. The PIFS code also gives us an algorithm for relating two adjacent levels in the pyramid (zoom theorem; Equations (5.19)-(5.20)). Thus, after finding the PIFS code and $\mathbf{f}^1$, all that is needed in order to get a vector $\mathbf{f}^2$ of length $2N$ is to apply the zoom-in algorithm in Equation (5.24) to $\mathbf{f}^1$. The vector $\mathbf{f}^2$ is an integral part of the pyramid of fixed points and is the fixed point of the PIFS code when using $B = 2B_1$. This vector can be used as an approximation of the higher-resolution representation, namely, an approximation of h_{2N}.

The Fractal Dimension of the PIFS Embedded Function

We have already introduced the PIFS embedded function. It is of interest to note that its *fractal dimension* can be bounded directly from the matrix representation of the PIFS code, as described by the following theorem (see also [5]).

Theorem 5.3 (Dimension Bound of the PIFS Embedded Function) : *Given a PIFS code W, let* $\mathbf{F}, \mathbf{A}, \mathbf{D}$, *and* $\mathbf{b}$ *denote its matrix representation, using* $B = 1$ *(see Section 5.3), such that*

$$W(\mathbf{v}) = \mathbf{F}\mathbf{v} + \mathbf{b} = \left(\mathbf{A}\frac{1}{2}\mathbf{D} \right)\mathbf{v} + \mathbf{b}. \tag{5.49}$$

Let $G(x)$ *denote the related PIFS embedded function. The fractal dimension of* $G(x)$ *is* $\mathcal{D}$ *where*

$$1 \le \mathcal{D} \le 1 + \log_2(\lambda). \tag{5.50}$$

and where λ *is the largest real eigenvalue of the matrix* $(\mathbf{A}_{abs} \cdot \mathbf{D})$, *and* $\mathbf{A}_{abs}$ *denotes a matrix whose elements are the absolute values of the corresponding elements in* $\mathbf{A}$.

The proof of the theorem is given in Addendum C.

The importance of the fractal dimension stems from the fact that it tells us about the nature of the super-resolution vectors (see [4] for a detailed discussion concerning fractal interpolation functions). By introducing certain constraints in the coding procedure (for example, on $\frac{D}{B}$ and/or $|a_i|$), one can change the fractal dimension of the resultant PIFS embedded function. Thus, the above-described super-resolution method results in fractal curves (vectors) with a fractal dimension which can be computed from the PIFS code and can be affected by the constraints on transformation parameters.

An application of the super-resolution method is demonstrated in Figure 5.5, in which one sees the "fractal nature" of the interpolation. A vector of size $N_1 = 256$ serves as the original vector, namely, $\mathbf{v}^1$. This vector was used to determine a PIFS code having a fixed point $\mathbf{f}^1$. The fractal dimension of the PIFS embedded function was found (Theorem 5.3) to be 1.16. This PIFS code was then used to find a vector $\mathbf{f}^4$ of length $4 \cdot N_1 = 1024$. The first 32 elements of $\mathbf{f}^4$ are shown ("+" combined with a dash-dot line) on the same graph with the linear-interpolation of the first 8 elements of $\mathbf{v}^1$ ("$\circ$" combined with a dotted line).

The following features of the interpolation are observed:

1. The *mean* of each 4-elements of $\mathbf{f}^4$ is approximately equal to the appropriate element of $\mathbf{v}^1$ (a strict equality would hold if the coding was lossless, i.e., $\mathbf{f}^1 = \mathbf{v}^1$).

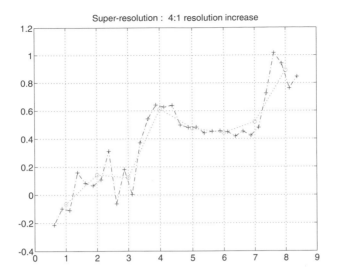

Figure 5.5: Super-resolution via PIFS code ("+" and a dash-dot line) versus linear interpolation ("○" and a dotted line).

2. The interpolation function, introduced by the super-resolution method, *does not* necessarily pass through the given points of $\mathbf{v}^1$. Furthermore, if we try to compute $\mathbf{f}^8$, it will not necessarily pass through the points of $\mathbf{f}^4$.

3. Linear interpolation tends to smooth the curve, while the PIFS code interpolation preserves the fractal dimension of the curve, which ensures the richness of details even at high resolutions. This feature is most evident when dealing with textures: while linear interpolation typically results in a blurred texture, the PIFS code interpolation preserves the appearance of the texture.

Rational Zoom Factors

After describing the method for achieving super-resolution, the following question arises:

According to the described method, one can only get resolutions which are N_1 times a power of 2, i.e., N_1, $2N_1$, $4N_1$, ... Can one also achieve different resolutions as well, such as $\frac{3}{2}N_1$, for example?

This question is addressed below.

In the description of the decoding process of a PIFS code (see Section 5.1), the possibility of replacing the prescribed $B = B_1$ by a new value was suggested. In the construction of the pyramid of fixed points, if super-resolution is desired, the new value of B is taken as $B_2 = 2B_1$. This yields a fixed point of length $N_2 = 2N_1 = 2B_1 M_R$. Suppose, however, that we choose to take $B = B_1 + 1$, which we define as $B_{(B_1+1)/B_1}$. This, in turn, yields a fixed point of length

$$N_{(B_1+1)/B_1} \equiv B_{(B_1+1)/B_1} \cdot M_R = N_1 + M_R . \tag{5.51}$$

Similarly, taking $B = B_{(B_1+2)/B_1} = B_1 + 2$ leads to

$$N_{(B_1+2)/B_1} \equiv B_{(B_1+2)/B_1} \cdot M_R = N_1 + 2M_R . \tag{5.52}$$

Pursuing the same reasoning, we see that we get a whole range of resolutions, quantized in steps of M_R. These resolution levels yield a pyramidal structure, which we call a *rational pyramid*. Each level of the rational pyramid is a fixed point of the PIFS, and the earlier pyramid of fixed points is contained in this rational pyramid.

For example, Figure 5.6(a)-(c) demonstrates the decoding of the PIFS code given in Figure 5.1, with $B = 4$, $B = 3$, and $B = 2$, respectively. In this case, $M_R = 4$.

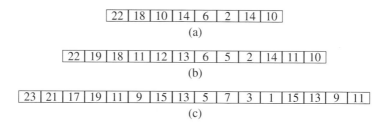

| 22 | 18 | 10 | 14 | 6 | 2 | 14 | 10 |

(a)

| 22 | 19 | 18 | 11 | 12 | 13 | 6 | 5 | 2 | 14 | 11 | 10 |

(b)

| 23 | 21 | 17 | 19 | 11 | 9 | 15 | 13 | 5 | 7 | 3 | 1 | 15 | 13 | 9 | 11 |

(c)

Figure 5.6: Decoding with (a)$B = 2$, (b) $B = 3$ (approximated to the nearest integer value), and (c) $B = 4$.

It is easily shown, by the use of Theorem 5.2, that given the PIFS code and some level of the rational pyramid, one can compute any other level of the rational pyramid, though not necessarily directly. For example, given the PIFS code and $\mathbf{f}^1$ that corresponds to $B = B_1 = 3$, what is the algorithm for computing $\mathbf{f}^{\frac{4}{3}}$ which corresponds to $B = 4$? The method is to first compute $\mathbf{f}^2$ from $\mathbf{f}^1$ and the given PIFS code, according to Equation (5.24). Here, $\mathbf{f}^2$ corresponds to $B = 6$. Then, compute $\mathbf{f}^4$ in the same manner, with $\mathbf{f}^4$ corresponding here to $B = 12$. Each element of $\mathbf{f}^{\frac{4}{3}}$ is now directly computable by averaging of every three consecutive elements from $\mathbf{f}^4$ (this follows directly from Theorem 5.2). The method is summarized as follows (see also Equation (5.46)):

1. Compute $\mathbf{f}^4$ from $\mathbf{f}^1$ and from the PIFS code, by twice applying Equation (5.24) (note that p can be negative).

2. Compute $\mathbf{f}^{\frac{4}{3}}$ by:

$$\mathbf{f}^{\frac{4}{3}}(k) = \frac{1}{3}(\mathbf{f}^4(3k) + \mathbf{f}^4(3k - 1) + \mathbf{f}^4(3k - 2)), \tag{5.53}$$

$k = (1, 2, \ldots, M_R \cdot (B_1 + 1))$, where $M_R \cdot (B_1 + 1) = M_R \cdot 4$.

5.6 Different Sampling Methods

Let $G(x) \in L^{\infty}[0, 1]$ denote a given embedded function of some PIFS code, and let $g_N(i)$, $i = 1, \ldots, N$, denotes its sampling by integration at resolution N, according to Equation (5.46), i.e.,

$$g_N(i) \equiv N \int_{(i-1)\frac{1}{N}}^{i\frac{1}{N}} G(x)\, dx, \qquad i = 1, \ldots, N. \tag{5.54}$$

As stated in Theorem 5.2, $g_N(i)$ is a fixed point of the corresponding PIFS, with a contraction function φ as defined in Equation (5.15):

$$\varphi(\mathbf{d}_l)(j) = \frac{1}{2}(\mathbf{d}_l(2j) + \mathbf{d}_l(2j - 1)). \tag{5.55}$$

Finding the PIFS code for g_N with this φ would result in the correct PIFS (the one with embedded function $G(x)$), and the coding would be lossless. There could be cases where more than one PIFS code has the same fixed-point vector at resolution N, but these codes correspond to different embedded functions. Such cases are not of interest here, so we will not consider such situations in remaining discussion.

Suppose, however, that now $G(x)$ is sampled in a different manner, denoted *point sampling*, to create $g_N^s(i)$;

$$g_N^s(i) \equiv G\left((i - 1)\frac{1}{N}\right), \qquad i = 1, \ldots, N. \tag{5.56}$$

All the previous results will be shown to hold here too if we define:

$$\varphi^s(\mathbf{d}_l)(j) \equiv \mathbf{d}_l(2j - 1), \tag{5.57}$$

meaning that φ^s is now also a point sampling operation. An earlier work on the subject can be found in [53], where a method to compute discrete values of a fractal function is discussed.

Now, let us restate few of the previous theorems and definitions, for this case of point sampling (the proofs are omitted since they closely follow the previous ones):

Theorem 5.4 (Point-Sampling Zoom) *Given a PIFS code with $\varphi^s(\mathbf{d}_l)(j) \equiv \mathbf{d}_l(2j - 1)$, which leads to W^1 with $B = B_1$, and to $W^{\frac{1}{2}}$ with $B = B_1/2$. Let the fixed points of these transformations be $\mathbf{f}^1$ and $\mathbf{f}^{\frac{1}{2}}$, respectively, then:*

1. **Zoom-out:**

$$\mathbf{f}^{\frac{1}{2}}(j) = \mathbf{f}^1(2j - 1), \qquad j = 1, \ldots, M_R(B_1/2). \tag{5.58}$$

2. **Zoom-in:**

$$\mathbf{f}^1((i - 1)B_1 + j) = a_i \mathbf{f}^{\frac{1}{2}}((m_i - 1)D_h^{\frac{1}{2}} + j) + b_i, \tag{5.59}$$

$$i = 1, \ldots, M_R, \qquad j = 1, \ldots, B_1. \tag{5.60}$$

where $D_h^{\frac{1}{2}} \equiv \frac{D_h^1}{2} = \frac{B_1}{2}$.

Definition 5.2 *Let* $G(x) \in L^\infty [0, 1]$. *Define* $G_r^s(i)$ *by*

$$G_r^s(i) \equiv G\left((i-1)\frac{1}{r}\right), \qquad i = 1, \ldots, r. \tag{5.61}$$

$G_r^s(i)$ *denotes the* function $G(x)$ *at point sampling resolution r. We say that* $G_{r_1}^s(i)$ *is finer (i.e., having higher resolution) than* $G_{r_2}^s(i)$ *(which is* coarser) *if* $r_1 > r_2$.

Theorem 5.5 (PIFS Embedded Function) [1] *Given a PIFS code with* $\varphi^s(\mathbf{d}_l)(j) \equiv \mathbf{d}_l(2j - 1)$, *there exists a unique function* $G(x) \in L^\infty [0, 1]$ *such that a vector* $\mathbf{v}_N^s \in \mathbb{R}^N$ *is a fixed point of the PIFS iff it is equal to the function* $G(x)$ *at point sampling resolution* $r = N$, *i.e.,*

$$\mathbf{v}_N^s(j) = G_N^s(j), \qquad j = 1, 2, \ldots, N. \tag{5.62}$$

The function $G(x)$ *is called the* PIFS embedded function.

The importance of formulating and treating different sampling methods lies in the fact that the coding problem can usually be stated as follows: One has a model for a continuous signal, be it a voice signal, an image, etc. The model states that the continuous signal is of fractal type. This signal is then sampled, and the work of finding the code is performed on the discrete signal.

As we have shown, the proper method for coding, namely, the method which enables finding the correct code, depends on the sampling method. So, given a model and the method of sampling, one can use the corresponding coding method.

It is worth noting that the embedded function for *both* sampling methods is the same. This should not be surprising, since as N grows, $g_N^s(i)$ and $g_N(i)$ approach each other, as seen from the definitions in Equations (5.56) and (5.54).

5.7   Conclusions

This chapter describes a hierarchical interpretation of the PIFS coding problem. Decoding the PIFS code at different resolutions results in a *pyramid of fixed points*. The ideas presented suggest many directions for future research. Some of them are:

- *Fractal image-interpolation* - This subject was briefly described in Figure 5.5.

- *Tighter Collage Theorem bound* - The Collage Theorem 2.1 (see also [4]) uses the distance between the image and its collage to bound the distance between the image and the related fixed point. However, the pyramidal structure suggests that the distance between the image and its collage in *different resolutions* should also be considered. Namely, instead of using

$$d(\boldsymbol{\mu}_o^1, \mathbf{f}^1) \le \frac{1}{1-s} d(\boldsymbol{\mu}_o^1, W^1(\boldsymbol{\mu}_o^1)),$$

  one can use

$$d(\boldsymbol{\mu}_o^1, \mathbf{f}^1) \le d(\boldsymbol{\mu}_o^1, W^1(\boldsymbol{\mu}_o^1)) + s \cdot d(\boldsymbol{\mu}_o^{\frac{1}{2}}, W^{\frac{1}{2}}(\boldsymbol{\mu}_o^{\frac{1}{2}})) + \cdots.$$

  For further details, see [65].

---

[1]The reason for using the same name for the two sampling methods will be justified below.

- *Fractal image segmentation* - The embedded function relates a certain fractal dimension to the image. However, there may be cases when more then one fractal dimension can be related to the image. This may be done by investigating more thouroughly the matrix structre of the transformation. Thus, the image can be segmented into regions with different fractal dimensions.

We believe that the combination of the evolving area of fractal image coding and the established use of pyramids in signal processing [13] will provide a useful common ground for future activity in this area.

Addendum A
Proof of Theorem 5.1 (Zoom)

In this addendum, we will prove Theorem 5.1 (zoom) stated in Section 5.2. The proof is divided into two parts, according to the two parts of the theorem.

A.1 Proof of Theorem Zoom-out

In order to prove that $\mathbf{f}^{\frac{1}{2}}$, as given by Equation (5.19), is indeed the fixed point of $W^{\frac{1}{2}}$, all we need to show is that it satisfies

$$\mathbf{f}^{\frac{1}{2}}(l) = W^{\frac{1}{2}}(\mathbf{f}^{\frac{1}{2}})(l), \qquad l = 1, \ldots, M_R B_1/2. \tag{5.63}$$

In the following, it is convenient to express l as follows:

$$l = (i - 1) \cdot B_1/2 + k, \qquad i = 1, \ldots, M_R; \quad k = 1, \ldots, B_1/2. \tag{5.64}$$

This expression for l emphasizes that the l-th element of $\mathbf{f}^{\frac{1}{2}}$ is actually the k-th element of the i-th range block of $\mathbf{f}^{\frac{1}{2}}$. Also, let the superscript 1 or $\frac{1}{2}$ denotes a symbol as belonging to $\mathbf{f}^1$ or $\mathbf{f}^{\frac{1}{2}}$, respectively. Thus, for example, we let $D_h{}^{\frac{1}{2}} = B_1/2$ (in accordance to our basic assumptions in Section 5.1); it denotes the shift between two adjacent domain blocks in $\mathbf{f}^{\frac{1}{2}}$. We start by substituting $\mathbf{f}^{\frac{1}{2}}$ into the right-hand side of Equation (5.63), and taking the relevant w_i,

$$
\begin{aligned}
W^{\frac{1}{2}}(\mathbf{f}^{\frac{1}{2}})(l) &= W^{\frac{1}{2}}(\mathbf{f}^{\frac{1}{2}})((i-1)\frac{B_1}{2} + k) \\
&= a_i \cdot \varphi(\mathbf{d}_{m_i}^{\frac{1}{2}})(k) + b_i \ .
\end{aligned} \tag{5.65}
$$

Substituting for $\varphi(\cdot)$ from Equation (5.15), Equation (5.65) becomes:

$$
\begin{aligned}
a_i \cdot \frac{1}{2} &\left\{ \mathbf{f}^{\frac{1}{2}}((m_i - 1)D_h{}^{\frac{1}{2}} + 2k) + \mathbf{f}^{\frac{1}{2}}((m_i - 1)D_h{}^{\frac{1}{2}} + 2k - 1) \right\} + b_i \\
&= \frac{1}{2} \cdot \left\{ [a_i \mathbf{f}^{\frac{1}{2}}((m_i - 1)D_h{}^{\frac{1}{2}} + 2k) + b_i] \right. \\
&\left. + [a_i \mathbf{f}^{\frac{1}{2}}((m_i - 1)D_h{}^{\frac{1}{2}} + 2k - 1) + b_i] \right\} \ .
\end{aligned} \tag{5.66}
$$

Now, let us further explore the first term in the last equation:

$$a_i \cdot \mathbf{f}^{\frac{1}{2}}((m_i - 1)D_h^{\frac{1}{2}} + 2k) + b_i$$

$$= a_i \cdot \frac{1}{2} \left\{ \mathbf{f}^1((m_i - 1) \cdot 2D_h^{\frac{1}{2}} + 4k) + \mathbf{f}^1((m_i - 1) \cdot 2D_h^{\frac{1}{2}} + 4k - 1) \right\} + b_i$$

$$= a_i \cdot \frac{1}{2} \left\{ \mathbf{f}^1((m_i - 1)D_h^1 + 4k) + \mathbf{f}^1((m_i - 1)D_h^1 + 4k - 1) \right\} + b_i$$

$$= a_i \cdot \varphi(\mathbf{d}_{m_i}^1)(2k) + b_i . \tag{5.67}$$

but since $\mathbf{f}^1$ is the fixed point of W^1, it follows that the last equation is equal to

$$\mathbf{f}^1((i - 1)B_1 + 2k). \tag{5.68}$$

Treating the second term in Equation (5.66) the same way, leads to:

$$a_i \cdot \mathbf{f}^{\frac{1}{2}}((m_i - 1)D_h^{\frac{1}{2}} + 2k - 1) + b_i = \mathbf{f}^1((i - 1)B_1 + 2k - 1). \tag{5.69}$$

Thus, Equation (5.66) can be written as:

$$\frac{1}{2} \left\{ \mathbf{f}^1((i - 1)B_1 + 2k) + \mathbf{f}^1((i - 1)B_1 + 2k - 1) \right\} \tag{5.70}$$

which, by the theorem statement (5.19), is just $\mathbf{f}^{\frac{1}{2}}((i - 1)\frac{B_1}{2} + k)$ which, in turn, is the left side of Equation (5.63).

A.2   Proof of Theorem Zoom-in

The procedure here is similar to the previous one, so we skip most of the details. It is needed to show that $\mathbf{f}^1$ obtained by Equation (5.20) is a fixed point of W^1:

$$W^1(\mathbf{f}^1)((i - 1)B_1 + j) = a_i \cdot \varphi(\mathbf{d}_{m_i}^1)(j) + b_i$$

$$= a_i \cdot \frac{1}{2} \left\{ \mathbf{f}^1((m_i - 1)D_h^1 + 2j) \right.$$

$$\left. + \mathbf{f}^1((m_i - 1)D_h^1 + 2j - 1) \right\} + b_i . \tag{5.71}$$

By Equation (5.19), which we have just proved, the last equation reduces to

$$a_i \mathbf{f}^{\frac{1}{2}} \left((m_i - 1)\frac{D_h^1}{2} + j \right) + b_i = a_i \mathbf{f}^{\frac{1}{2}}((m_i - 1)D_h^{\frac{1}{2}} + j) + b_i$$

$$= \mathbf{f}^1((i - 1)B_1 + j). \tag{5.72}$$

Addendum B
Proof of Theorem 5.2 (PIFS Embedded Function)

The proof is constructive. It is based on finding the PIFS embedded function $G(x)$ and then showing that $G_N(x)$ is indeed a fixed point of the PIFS at resolution N.

Suppose that the PIFS is composed of M_R transformations. Let (a_i, b_i, m_i) denote the transformation parameters of each range block $\mathbf{r}_i$, $i = 1, \ldots, M_R$,

For operating with the PIFS code on a function of a continuous argument $x \in [0, 1]$, we look at the following analogies between the current continuous variable case (functions) and the previously discussed discrete case (vectors):

discrete	$\mathbf{v}_N(i)$	$i \in (1, 2, \ldots N)$	B	$D = 2B$	$D_h = B$
continuous	$f(x)$	$x \in [0, 1]$	$\frac{B}{r}$	$2 \cdot \frac{B}{r}$	$\frac{B}{r}$

discrete	i-th range block : $\mathbf{v}_N(j)$,	$j = (i - 1)B + 1, \ldots, iB$
continuous	i-th range block : $f(x)$,	$x \in [\frac{(i-1)B}{r}, \frac{iB}{r})$

Such a configuration is demonstrated in Figure 5.7, where the i-th range block is

$$x \in [a, b), \qquad a = (i - 1)\frac{B}{r} \text{ and } b = i\frac{B}{r}, \tag{5.73}$$

and the m_i-th domain block is

$$x \in [c, d), \qquad c = (m_i - 1)\frac{B}{r} \text{ and } d = (m_i - 1)\frac{B}{r} + 2 \cdot \frac{B}{r}. \tag{5.74}$$

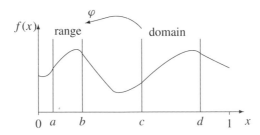

Figure 5.7: Domain to range transformation.

Define the following transformation $w_i : L^\infty[0, 1] \to L^\infty[0, 1]$:

$$w_i(f(x)) = \begin{cases} a_i f(2(x - a) + c) + b_i & a \le x < b \\ 0 & \text{otherwise}. \end{cases} \tag{5.75}$$

Then, since $|a_i| < 1$, w_i is contractive. Next, define the transformation

$$W = \bigcup_{i=1}^{M_R} w_i, \tag{5.76}$$

which is a contractive PIFS [45], and thus has a unique fixed point defined to be the function $G(x)$.

Now it is left to show that $G_N(x)$ is indeed a fixed point of the PIFS code. We take

$$\mathbf{v}_N(i) = G_N(i) \tag{5.77}$$

and look for consistency with the fixed-point criterion, i.e., verify that $\mathbf{v}_N = W(\mathbf{v}_N)$. Take, for example, the j-th element in the i-th range block. Let (a_i, b_i, m_i) denote the appropriate transformation parameters, so the fixed-point equation for this element is:

$$
\begin{aligned}
\mathbf{v}_N((i-1)B + j) &= a_i \frac{1}{2}(\mathbf{v}_N((m_i - 1)B + 2j) + \\
&\quad \mathbf{v}_N((m_i - 1)B + 2j - 1)) + b_i, \\
&\quad j = 1, 2, \ldots, B.
\end{aligned}
\tag{5.78}
$$

This time we deal with elements rather then whole blocks, so that the following notation is used (see also Figure 5.8):

- Boundaries of the j-th element in the i-th range block are

$$\tilde{a} = ((i-1)B + j - 1)\frac{1}{r} \text{ and } \tilde{b} = ((i-1)B + j)\frac{1}{r}. \tag{5.79}$$

- Boundaries of the $(2j-1)$-th and the $(2j)$-th elements in the m_i domain block are

$$\tilde{c} = ((m_i - 1)B + 2j - 2)\frac{1}{r}; \tag{5.80}$$

$$\tilde{e} = \frac{\tilde{c} + \tilde{d}}{2} = ((m_i - 1)B + 2j - 1)\frac{1}{r}; \tag{5.81}$$

$$\tilde{d} = ((m_i - 1)B + 2j)\frac{1}{r}. \tag{5.82}$$

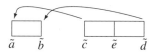

Figure 5.8: Elements of transformations.

Substituting $G_N(i)$ for $\mathbf{v}_N(i)$ in the right side of Equation (5.78) yields:

$$
\begin{aligned}
\mathbf{v}_N((i-1)B + j) &= a_i \frac{1}{2} r (\int_{\tilde{c}}^{\tilde{e}} G(x)\, dx + \int_{\tilde{e}}^{\tilde{d}} G(x)\, dx) + b_i \\
&= a_i \frac{1}{2} r \int_{\tilde{c}}^{\tilde{d}} G(x)\, dx + b_i.
\end{aligned}
\tag{5.83}
$$

The left side of Equation (5.78) yields

$$\mathbf{v}_N((i-1)B + j) = r \int_{\tilde{a}}^{\tilde{b}} G(x)\, dx. \tag{5.84}$$

Since $G(x) = w_i(G(x))$, and since the range $x \in [\tilde{c}, \tilde{d})$ is mapped to $x \in [\tilde{a}, \tilde{b})$, we obtain:

$$r \int_{\tilde{a}}^{\tilde{b}} [a_i G(2(x - \tilde{a}) + \tilde{c}) + b_i] \, dx$$

$$= a_i r \int_{\tilde{a}}^{\tilde{b}} G(2(x - \tilde{a}) + \tilde{c}) \, dx + r \int_{\tilde{a}}^{\tilde{b}} b_i \, dx$$

$$= a_i \frac{1}{2} r \int_{\tilde{c}}^{\tilde{d}} G(x) \, dx + b_i . \tag{5.85}$$

Thus, since Equations (5.83) and (5.85) agree, we've established the consistency of the fixed point equation for $\mathbf{v}_N$, under the theorem assumption that $\mathbf{v}_N(i) = G_N(i)$.

Addendum C
Proof of Theorem 5.3 (Fractal Dimension of the PIFS Embedded Function)

A related formal proof can be found in [3] and [34]. We will follow the informal proof, similar to the one given in [4] about the fractal dimension of a fractal interpolation function.

Let $\epsilon > 0$, and let $G(x)$ be superimposed on a grid of square boxes of side length ϵ. Let $N(\epsilon)$ denote the number of grid boxes of side length ϵ which intersect $G(x)$. Since $G(x)$ has fractal dimension $\mathcal{D}$, it follows that the following relation exits:

$$N(\epsilon) \sim \epsilon^{-\mathcal{D}} \quad \text{as} \quad \epsilon \to 0. \tag{5.86}$$

Our goal is to evaluate the value of $\mathcal{D}$.

Denote the number of boxes intersecting $G(x)$, for $x \in (\frac{(i-1)B}{r}, \frac{iB}{r})$, by

$$N_i(\epsilon), \qquad i = 1, \dots, M_R. \tag{5.87}$$

Namely, $N_i(\epsilon)$ is the number of intersecting boxes in the i-th range block.

Since $\epsilon \to 0$, the contribution of boxes at the edges of range blocks can be ignored. Therefore we can write

$$N(\epsilon) \cong \sum_{i=1}^{M_R} N_i(\epsilon). \tag{5.88}$$

Now, let us investigate more thoroughly the value of $N_i(\epsilon)$. Let m_i denote the index of the domain block mapped to the i-th range block by the PIFS. This domain block is composed of two adjacent range blocks with indices denoted as m_i^1 and m_i^2.

After transforming the m_i-th domain block onto the i-th range block, each column of grid squares is mapped into a column of grid-rectangles, with height $|a_i|\epsilon$ and width $\frac{B}{D}\epsilon$. Let us define $q \equiv \frac{B}{D}$, and hence, according to our assumptions, $q = \frac{1}{2}$. Therefore, if a column of width ϵ in the domain block intersects $G(x)$ in L boxes, then after transformation the column will be of width $q\epsilon$, and the number of boxes of size $q\epsilon$ intersecting $G(x)$ in the transformed column is therefore at most, but possibly less than, $|a_i| L/q$ boxes.

Summing on all the domain columns yields

$$N_i(q\epsilon) \leq \frac{|a_i|}{q}(N_{m_i^1}(\epsilon) + N_{m_i^2}(\epsilon)). \tag{5.89}$$

Denoting by c_i the proportionality factor, i.e,

$$N_i(\epsilon) = c_i \cdot \epsilon^{-\mathcal{D}} \quad \text{as} \quad \epsilon \to 0, \tag{5.90}$$

we can write

$$c_i \cdot (q\epsilon)^{-\mathcal{D}} \leq \frac{|a_i|}{q}(c_{m_i^1}\epsilon^{-\mathcal{D}} + c_{m_i^2}\epsilon^{-\mathcal{D}}). \tag{5.91}$$

With the same reasoning for all others range blocks, we get

$$c_i \cdot (q)^{-\mathcal{D}+1} \leq |a_i| (c_{m_i^1} + c_{m_i^2}), \qquad i = 1, \ldots, M_R. \tag{5.92}$$

Recalling the definition of the matrix $\mathbf{D}$ in Equation (5.35), the last equation, for all the blocks, can be written as:

$$\mathbf{c} \cdot (q)^{-\mathcal{D}+1} \leq (\mathbf{A}_{abs} \cdot \mathbf{D}) \cdot \mathbf{c}, \qquad \mathbf{c} = [c_1, c_2, \ldots, c_{M_R}]^T. \tag{5.93}$$

It can be shown that $(q)^{-\mathcal{D}+1}$ is bounded by the spectral radius λ of $(\mathbf{A}_{abs} \cdot \mathbf{D})$, and hence

$$(q)^{-\mathcal{D}+1} \leq \lambda, \tag{5.94}$$

and hence,

$$\mathcal{D} \leq 1 + \log_{\frac{1}{q}}(\lambda). \tag{5.95}$$

Since we assume $q = \frac{B}{D} = \frac{1}{2}$, the theorem follows immediately.

It is worth noting that:

- Equation (5.93) holds even when $q \neq \frac{1}{2}$, and so does Equation (5.95). For example, it can be shown immediately that in the case of uniform fractal interpolation of functions [4], where the whole function is mapped to each interval, the following results:

$$D = N, \qquad q = \frac{D}{B} = M_R, \tag{5.96}$$

$$\mathcal{D} = 1 + \log_{M_R}\left(\sum_{i=1}^{M_R} |a_i|\right) \tag{5.97}$$

  which agrees with the result there.

- The matrix $q\mathbf{D}$ is a stochastic matrix. That is, the sum of the elements in each row equals 1. Thus, all its eigenvalues are equal or smaller then 1 (in magnitude), and at least one eigenvalue equals 1. For example, if

$$\mathbf{A} = a\mathbf{I}_{M_R \times M_R} \tag{5.98}$$

one can write

$$\mathbf{A}_{abs} \cdot \mathbf{D} = |a|\, \mathbf{I}_{M_R \times M_R} \cdot \frac{1}{q} q \mathbf{D} = \frac{|a|}{q}(q\mathbf{D}) \qquad (5.99)$$

since $|a|/q$ is a scalar factor, and $q\mathbf{D}$ is stochastic, the largest eigenvalue is $|a|/q$. Thus, Equation (5.95) results in

$$\mathcal{D} = 2 + \log_{\frac{1}{q}} |a|\,. \qquad (5.100)$$

If $q = \frac{1}{2}$, then

$$\mathcal{D} = 2 + \log_2 |a|\,, \qquad (5.101)$$

which approaches $\mathcal{D} = 2$ as $|a|$ approaches 1, and $\mathcal{D} = 1$ if $|a| \leq \frac{1}{2}$ (by Equation (5.50)).

Chapter 6

Fractal Encoding with HV Partitions

Y. Fisher and S. Menlove

In this chapter we describe a way to partition images that gives significantly better results. The method, which we call fractal encoding with HV partitioning, also includes several optimizations in the storage of the transformations, the coding, and the decompression algorithm. The partitioning method used to generate the ranges is based on rectangles. A rectangular portion of the image is subdivided, either horizontally or vertically, to yield two new rectangles. The main advantage of this method over the quadtree method is clear: the ranges are selected adaptively,[1] with partitions that can correspond to image edges (at least horizontal and vertical ones). This strength also leads to the main drawback: since the domain size is a multiple of the range size, the large number of range sizes leads to a large number of domains and thus to a large number of domain-range comparisons. Also, there are many ways to select the location of the partition, and we will only describe our final implementation. Our search through all the possible implementations was neither systematic nor complete, and it is clear that further work could yield improvements. Nevertheless, this method gives results better than any similar method of which we are aware.

6.1 The Encoding Method

As with the quadtree method discussed in Chapter 3, an image is encoded by repeating two basic steps: recursive partitioning and domain search. The partitioning establishes the nonoverlapping ranges, and the domain search determines the domain that will map onto a particular

---

[1] See also Appendix C.

range. Each pixel in the original image will be assigned to exactly one range (through partitioning), but a pixel can exist in multiple domains, regardless of its range assignment. We require that the domains be strictly larger than the ranges (by a factor of 2 or 3), and that the affine transformation of the pixel values be contractive.

For example, let's consider in detail one iteration (the first) in this process. At the outset, the entire image is defined as one potential range. This range is meaningless, however; there is no domain that can map to it. In order to map to a range, a domain must exist with dimensions that are at least twice as large as the range's, both horizontally and vertically. This potential range is thus partitioned into two rectangles, each constituting a new potential range. This process is repeated. When the potential ranges are sufficiently small, they are compared with domains (as described in Chapter 3 and below, for example). Potential ranges are subdivided further if the domain-range comparison yields a computed difference below some predetermined threshold. We now consider these steps in more detail.

The HV Partitioning Method

In order to partition the image, a pixel value average is calculated for each pixel row and column in the range to be divided. These averages are then used to compute successive differences between the averages. Next, a linear biasing function is applied to each of these differences, multiplying them by their distance from the nearest side of the rectangular range. In other words, if the range contains the pixel values r_{ij} for $0 \leq i < N$ and $0 \leq j < M$, then the horizontal sums $\sum_i r_{i,j}$ and vertical sums $\sum_j r_{i,j}$ are computed, subtracted, and multiplied by the biases $\min(j, M - j - 1)/(M - 1)$ and $\min(i, N - i - 1)/(N - 1)$, respectively. This gives the biased horizontal differences,

$$h_j = \frac{\min(j, M - j - 1)}{M - 1} \left(\sum_i r_{i,j} - \sum_i r_{i,j+1} \right),$$

and biased vertical differences,

$$v_i = \frac{\min(i, N - i - 1)}{N - 1} \left(\sum_j r_{i,j} - \sum_j r_{i+1,j} \right).$$

The first partition is established in the location of maximum value, horizontal or vertical. That is, we seek the values of j and i such that $|h_j|$ and $|v_i|$ are maximal, and we make the partition at horizontal position j or vertical position i, depending on which difference is larger. This yields two rectangles which tend to partition the given range along strong vertical or horizontal edges while avoiding partitions that result in narrow rectangles.

The Domain Search

The domain search is almost identical with the domain search in the quadtree method (Chapter 3). Once a rectangle has been divided, a domain is sought for the largest currently uncovered range. The domains are selected from the image with their centers located on a lattice with spacing equal to one half the domain size. The chosen domain must have sides that are larger than the range by a factor of 2 or 3 (though each side can have a different factor). The pixels in the domain are then either subsampled or averaged in groups of 2×2 (even if the ratio of domain to range side is 3). These domain pixels are regressed against the range pixel values to find the best scaling and offset values in the least-squares sense. This also computes the matching criterion of the domain and range, discussed below.

The Domain-Range Comparison

Unlike the quadtree method, it is not the rms difference that is compared with a predetermined threshold to determine when partitioning takes place. Instead, the square difference of the pixel values is used. If the square difference is smaller than the predetermined threshold, the transformation is accepted. Otherwise, the range is partitioned into two new ranges. Partitioning continues until a sufficiently low squared-difference criterion is attained or until the ranges become smaller than some predetermined size. This particular optimization yields a significant improvement in the perceived image fidelity.

Optimizing the Encoding

We included several optimizations to speed the encoding time:

- Quadrant classification;

- Encoding by range size;

- Domain-range ratio restriction;

We discuss each of these below.

Quadrant Classification

The quadrant classification method we use is similar to the method discussed in Chapter 3 §3.1. A rectangular portion of the image (which is either a domain or a range) is divided into quadrants, and each quadrant's pixel values are averaged. Then, the quadrants are ranked in order of magnitude, brightest to darkest. Unlike the quadtree method, in this case we get six different canonical orientations, which have the brightest quadrant in the upper left corner (this is because the domain-range mapping roughly maintains the aspect ratio of the rectangles). As before, only domains and ranges that fall into the same canonical group are compared, and only with the symmetry operations (rotation and flip) that fix or invert (for the negative scaling case) the ranking of the quadrants. That is, the domain is rotated so that its brightest quadrant maps to the brightest range quadrant, its second brightest quadrant maps to the second brightest range quadrant, and so forth. If no rotation results in this match, the domain is not compared with the range at all. Additionally, a negative-scale search can compare ranges and domains with inverse-magnitude matches.

Since small differences in the average pixel values of the quadrants can lead to different classifications, the next best classification was also searched, depending on the desired speed of the algorithm. The next best classification was taken to be the one resulting from swapping the two most similar quadrant averages. Combinations of these classification techniques yield a finer gradient of time versus performance results.

Encoding by Range Size

The biggest available range is always chosen for comparison with the domains (and then possibly for partitioning into two new ranges). Here, "biggest" and "size" are measured by the maximal smaller dimension. Thus, a potential range of size 10×8 will be compared to domains before a potential range of size 20×7, even though the first contains fewer pixels. This way, ranges are mapped in decreasing order, which results in sequential range size repetition. That is, two ranges of the same size will always be defined one after the other. Same-size ranges will potentially be compared to the same set of possible domains, and since they are evaluated sequentially, domain classification is not duplicated. Once a new range size is established, however, domain information is erased and the new possible domains must be classified. Since the ranges are mapped in order of decreasing size, the same domain sizes are never classified twice.

Domain-Range Ratio Restriction

Ratio restriction limits the domain-to-range ratios that are evaluated. With no optimization, ratios of 2×2, 3×3, 3×2, and 2×3 are compared for an optimal fit. More ratios increase the search time and increase the resulting fidelity due to a better domain-range match. Restricting the ratios results in faster encoding and worse fidelity. Adding new ratios (we tried using a factor of 4 as well as 2 and 3) led to worse compression since the new, potentially better domain-range matches cannot compensate for the increased transform size.

6.2 Efficient Storage

As in the method discussed in Chapter 3, the transformations are not stored as matrices that define affine transformations. The range position is stored implicitly by specifying the partition type, horizontal or vertical, and its position, an offset from the upper or left side. This offset requires fewer bits to store as the partition gets refined. The range partitions are stored recursively from the top left to the bottom right, so the position information can be derived from the partition direction and offset.

   The scaling and offset values are quantized and Huffman coded using a fixed code based on a collection of test images. There are 32 scale values, which require slightly more than 4 bits each to store after Huffman coding. The 128 possible offset values require roughly 6 bits each to encode.

   Typically, there are many small ranges. The size of the smallest range that was partitioned for a given compression is stored along with the transformation data. Then, all potential ranges smaller than this size that occur during the decoding are known to be actual ranges that are not partitioned. Thus, the bit that specifies whether or not a potential range is partitioned can be excluded for all these ranges.

   The domain position is specified by specifying an index into a list of all possible domains. Only the domains searched (depending on the optimization) are listed. If more than one domain-range ratio was used during the encoding, the ratio of a particular domain-range pair is given implicitly by the domain index. For example, if the range size and image size were such that the image could produce 100 domains twice as big as the range, and the domain index was 101, the domain would be the first possible match with domain-range ratio of three. Since there are significantly more domains twice as big as the range (as opposed to three times as big), this storage method occasionally saves one bit (over simply using one bit to explicitly encode the ratio). The indexing is organized so that domains with ratio 2×2 are specified, then 3×3 domains. The same method is used to differentiate between 2×3 and 3×2 domains.

The rotation information is specified by using two bits per transform to encode the four possible rotations compared (note that there are eight possible rotations for the quadtree method). Finally, if the scale value is zero, the rotation and domain data are omitted, since they are superfluous; all domains give the same result when the pixel scaling value is zero.

6.3 Decoding

We implemented a decoding method that is more efficient than the standard iteration to the fixed point. There were two central optimizations: pixel referencing and low-dimensional fixed-point approximation.

Pixel Referencing

Rather than scanning through the domain-range transformations and mapping the domain pixels to the range pixels, a pointer array is computed ahead of time. This array assigns a domain pixel in the image to each range pixel in the image. In decoding, this saves the overhead (e.g. determining a rotation and scanning through a sub-image) of referencing through a transformation structure.

Since we averaged groups of 2×2 domain pixels onto each range pixel, there is a slight complication. In fact, the value in each array entry was the location of the upper left pixel of that entry's referent domain group, the four pixels from the domain whose average is mapped onto the range pixel. The domain group is made up of four pixels regardless of the range-domain ratio, so that the ratio itself need not be stored in the structure. (Alternatively, the domain group can be subsampled to achieve a single range value; however, this method yields inferior results.)

To save memory, the scaling and offset values for each pixel are stored only once per range and are referenced through another (transform index) array.

The decoding proceeds as before, but using pixel referencing array as follows: A cell in the pointer array references a group of four pixels in the bitmap, which are then averaged. The scaling and offset used on this group are derived from the transform index array, and the scale and offset are applied to the averaged group and mapped into the image. This process repeats for every pixel in the bitmap, and repeats several times for the image as a whole.

Lower-Dimensional Fixed-Point Approximation

This optimization significantly reduces decoding time. If we consider an $N \times M$ image as an element of $\mathbb{R}^{N \times M}$, then we can approximate the fixed point in a lower-dimensional space. For example, we can find the fixed point in $\mathbb{R}^{N/2 \times M/2}$ by scaling the transformations spatially by $\frac{1}{2}$ and iterating on an $N/2 \times M/2$ image. We used the image at $\frac{1}{16}, \frac{1}{8}, \frac{1}{4}$, and $\frac{1}{2}$ of the final image size. The number of operations is significantly reduced, and the low-dimensional approximation required only two iterations at the final image size to be indistinguishable from the final fixed point. For a more sophisticated description of a similar idea, see Chapter 5.

More specifically, the iterations of the decoding process give a progressively better approximation to the image. In order to save time, these iterations were performed on a small scale. A major feature of fractal decompression is that the image can be decompressed to any size, simply by scaling the range and domain of each individual transform by the appropriate amount. This is used in the early iterations of a decoding by decoding to a small size. So, for example, a 256×256 image can be decoded to a 32×32 image, then to a 64×64 image, then to a 128×128 image, and finally to a 256×256 image. This process of decoding with ascending sizes greatly reduces the number of calculations that must be done at the early iterations. In addition, early iterations can be done using domain subsampling rather than averaging, which further reduces the computation time. Using ascending decoding sizes, only one or two iterations must be done at full size, using averaging for optimal results. All the results presented here used two decoding iterations at full size, which was sufficient for all the test images with which we experimented.

Postprocessing

Depending on the criterion used to determine the allowable error for a domain-range match and also depending on the image itself, the boundaries of the ranges are sometimes visible in the reconstituted image. This can give it a blocky appearance, especially in regions which were covered by large ranges. This problem is effectively addressed by a smoothing technique applied to the image as the very last step in decoding. Smoothing basically consists of averaging the pixel values on the range boundaries. There are several factors that come into play during this process: how the average should be weighted, how many pixels deep into the range averaging should take place, and how big a range should be in order to warrant smoothing. If too much smoothing is done (by smoothing small ranges, not weighing the averages, or smoothing too deeply in the range) the smoothed image looks blurry. If too little smoothing is done the blocky artifacts are not eliminated. We discuss our (heuristic) approach below. This method provided good results across a wide range of images.

A typical range has four edges which can be smoothed (this is not the case for ranges on an edge of the image). It is desirable to smooth the top and bottom of a tall, skinny range while not altering the long edges. If the long edge were smoothed, it would be possible to replace a great majority of pixels in the range with averaged values, and this would result in a hazy image. A good rule of thumb is to smooth the border pixels in a range if it is six or more pixels long in the dimension being considered. If the range is ten or more pixels long, it is good to smooth both the pixels on the boundary and their immediate internal neighbors. The weights applied to these averages can also be altered to control the blocky/blurry tradeoff.

We experimented with different averaging weights and settled on a ratio of 2:1. That is, if the pixel values a and b occur at the boundary of two different ranges, then a is replaced by $\frac{2a+b}{3}$ and b is replaced by $\frac{a+2b}{3}$ (see Figure 6.1). For large ranges in which we averaged both the boundary pixels and their immediate internal neighbors, we used weights of 3:2:1 (see Figure 6.2).

For four pixels on a given border, two border pixels and two internal neighboring pixels, this would seem to require six additions and four divisions. In fact, it can be done with four additions and one division. If the pixel values are a, b, c, and d, as in Figure 6.2, then we wish to find $a' = \frac{3a+2b+c}{6}$, $b' = \frac{2b+c}{3}$, $c' = \frac{2c+b}{3}$, and $d' = \frac{3d+2c+b}{6}$. In order to do this, we smooth ranges from upper left to lower right. First set pixel b to b'. Next, set a' to be the simple average of a

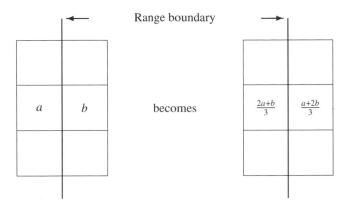

Figure 6.1: The pixels with values *a* and *b* in different ranges are smoothed with averaging weights of ratio 2:1.

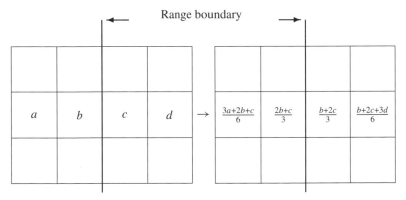

Figure 6.2: The pixels with values *a* and *d* that are neighbors to boundary pixels in distinct ranges are smoothed with averaging weights of ratio 3:2:1.

and b',

$$a = \frac{a+b'}{2} = \frac{a+\frac{2b+c}{3}}{2} = \frac{3a+2b+c}{6}.$$

Next, set c' to be the simple average of c and b' and d' to be the simple average of d and c'. Since simple averages are divisions by two, they can be implemented as an arithmetic shift, so the whole procedure requires only one division. This process is repeated on all border pixels of ranges with sufficient size.

6.4   Results

In this section, we present results at a variety of compression ratios and optimization levels. Since the encoding times for fractal encoding methods are high and of importance, we include them (as well as the decoding times) for various sample compressions. We conclude by showing the effect of smoothing on the final image. See Appendix D for a comparison of these results with others, including JPEG. Times are given in CPU seconds on a Silicon Graphics Personal IRIS 4D/35.

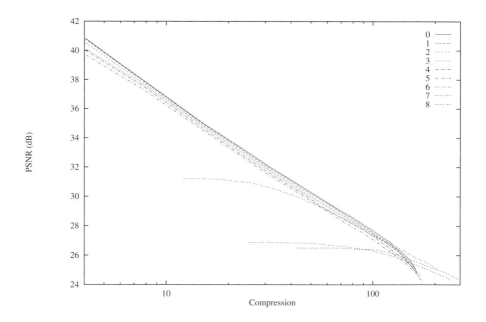

Figure 6.3: A plot of PSNR versus compression for various optimization levels of the 512×512 image of Lenna.

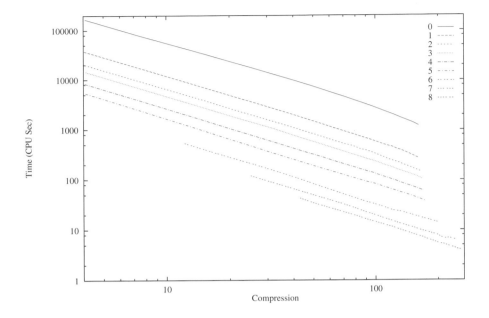

Figure 6.4: A plot of encoding time versus compression for various optimization levels of the 512 × 512 image of Lenna.

Optimization Levels

Figure 6.3 is a plot of the peak signal-to-noise ratio (PSNR) versus compression ratio for a 512 × 512 pixel image of Lenna. The graph shows several lines corresponding to the various optimizations. These optimization levels consist of the following searches:

Level 0. Compare each range with each possible domain in the image, using every possible symmetry operation. The ratio of the lengths of the sides of the domain and range blocks are 2 × 2, 2 × 3, 3 × 2, and 3 × 3.

Level 1. As above, but compare only the domains in the two classes (in the quadrant-average classification scheme) that best match the range class. Each domain is compared using two orientations, corresponding to positive and negative scale values, respectively.

Level 2. As above, but compare only domains that are factors of 2 × 2 and 3 × 3 bigger than the range.

Level 3. As above, but compare only domains that are factors of 2 × 2 bigger than the range.

Level 4. As above, but compare just one domain class.

Level 5. As above, but only positive scale values are used. This means that each range is compared with a candidate domain in only one orientation.

Level 6. Decimate the image by two (to one-quarter its original size) and encode it using optimization level 5.

Level 7. Decimate the image by a three and encode it using optimization level 5.

Level 8. Decimate the image by a four and encode it using optimization level 5.

Figures 6.4 and 6.8 show encoding time versus compression ratio and PSNR for these levels of optimization, respectively. The encoding times vary greatly, ranging from 167,292 seconds (almost 2 days!) to just a few seconds. The high optimization levels yield very rapid compression at high compression ratios, but the fidelity leaves something to be desired. We include these latter cases for comparison and for their academic interest.

Figures 6.5–6.7 show high- to low-fidelity encoding results. Since the meanings of "high" and "low" depend on the application, perhaps it is better to classify these as low artifacts to high artifacts results. These results are summarized in Table 6.1; all were postprocessed.

Table 6.1: Results for Figures 6.5–6.7.

Figure	Comp. Ratio	PSNR (dB)	Enc. Time	Level
6.5	14.7	34.62	979.9 sec	5
6.6	39.4	30.99	1122.1 sec	2
6.7	80.2	28.15	42.5 sec	6

Discussion

The relationship between the PSNR, compression ratio, and encoding time is remarkably linear (in the log plots). As expected, encodings with higher compression ratios require less time to encode and have poorer fidelity; there are fewer transformations to be found, and hence their necessarily larger ranges contribute more error due to poorer fits with domain. Using optimization levels 6–8 involves decimating the image, which means that the scheme simply cannot reach high-fidelity values. However, Figure 6.3 shows that while decimation is inefficient at low compression ratios, at high compression ratios it works better than using all the image data. In fact, below about 30 dB (or above 50:1 compression), decimating each 2×2 pixel group in the image into one pixel and encoding the decimated image (optimization level 6) yields results that are comparable (in PSNR and compression) to using the whole image. Further decimation is only useful at much higher compression ratios. (Whether there are applications that can use images at the resulting miserable signal to noise ratios is a separate point). Optimization levels 0–5 each contribute a slight degradation in image quality (measured by PSNR), with an overall difference of less than 1 dB. A typical cross section for this data range is shown in Table 6.2, which contains the compression ratios, PSNR and encoding times for levels 0–5.

The variation in encoding time is large. At high fidelity, the encoding times are very high; at moderate quality they are just a few minutes. The effect of classification is greatest between no optimization (level 0) and the first optimization level, which cuts encoding time by a factor of roughly 4.8. Compared to no optimization, level 5 cuts encoding time by a factor of about 35, with a cost of with a cost of $\frac{1}{2}$ to 1 dB, depending on the compression ratio. This is a

Figure 6.5: The 512 × 512 Lenna image compressed by a factor of 14.7 with 34.62 dB PSNR.

Figure 6.6: The 512×512 Lenna image compressed by a factor of 39.4 with 30.99 dB PSNR.

Figure 6.7: The 512 × 512 Lenna image by a factor of 80.2 with 28.1 dB PSNR.

Table 6.2: Sample data from encodings of 512 × 512 Lenna.

Level	Comp. Ratio	PSNR (dB)	Enc. Time (sec)
0	31.61	32.00	13065.4
1	31.18	31.97	2724.8
2	31.18	31.85	1502.3
3	31.39	31.73	1074.6
4	30.66	31.68	600.1
5	30.16	31.59	366.6

significant benefit. At compression ratios above 50:1, there is a further savings in time (with inconsequential image degradation) if the image is decimated before compression. For this range of compression, level 6 cuts encoding time by a factor of about 97, compared to no optimization.

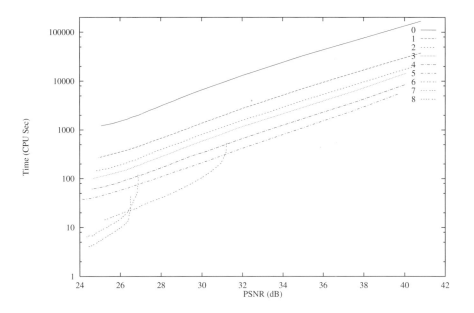

Figure 6.8: A plot of encoding time versus PSNR for various optimization levels of the 512×512 image of Lenna.

Decoding

In this section we discuss decoding time. Figure 6.9 contains a plot of the decoding time for various compression ratios of both the 512 × 512 and 256 × 256 images of Lenna. These times do not include disk access to save the image, but they do include disk access to read the compressed image data. Postprocessing did not affect the times. The time values graphed for different compressions are measured with a resolution of 0.1 sec.

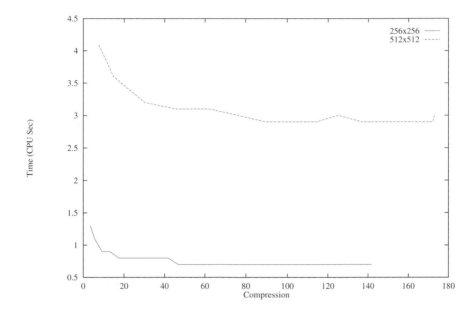

Figure 6.9: A plot of decoding time versus compression ratio for the 512×512 and 256×256 image of Lenna.

Discussion

The decompression times scale roughly with the output image size. However, the decompression times are almost independent of the original image size and compression ratio. For example, when a 21:1 compression of the 256×256 Lenna image is decompressed at sizes 256×256 and 512×512, the decompression times are 0.7 seconds and 3.1 seconds, respectively. Only when the compression ratio is low, corresponding to a large number of transformations, does the initial overhead of scanning through the transformations to build the pixel pointer array begin to contribute to the decoding time.

Smoothing

Figure 6.10 shows sample images, with and without postprocessing. In the moderate fidelity encodings (top pair), postprocessing causes some blurring and some smoothing; the smoothing reduces the PSNR. In the low-fidelity encodings (middle pair), the decrease in PSNR along

Table 6.3: Data showing the effect of postprocessing to smooth encodings of the 256×256 Lenna and Collie images in Figure 6.10.

Figure	Smoothed	Comp. Ratio	PSNR (dB)	Enc. Time
6.10a	Yes	11.0	31.38	156.3
6.10b	No	11.0	31.54	156.3
6.10c	Yes	34.4	26.67	35.0
6.10d	No	34.4	26.62	35.0
6.10e	Yes	47.9	28.13	23.3
6.10f	No	47.9	27.95	23.3

the true edges is canceled by the smoothing of artificial block edges; however, while the postprocessing is helpful, it does not completely overcome the artifacts. In the very low fidelity encodings (bottom pair), the PSNR is improved slightly. The data for these encodings are summarized in Table 6.3.

Discussion

Decompression times were identical (that is, within the resolution of the measurement) for both the smoothed (postprocessed) and unsmoothed images. This is not surprising, since the smoothing just takes place on the boundaries of the sufficiently large ranges, which means that the number of affected pixels is small. In some cases, smoothing actually increases the signal-to-noise ratio slightly. This tends to happen at high compression in which the PSNR is quite poor to begin with. In high-fidelity encodings, the smoothing does cause some blurring and reduction of PSNR, but not much, since such images contain many small ranges, and these are not smoothed.

6.5 More Discussion

A reasonable question is: why does fractal encoding with HV partitions method work better than the quadtree method? The average number of bits per transformation for this method is approximately 33, compared with 27 bits per transform for the quadtree method. The improvement in encoding time and fidelity is only partially a result of the Huffman coding of the pixel transformation coefficients.

Aside from the adaptive nature of the method, which is an inherent improvement, the scheme has the ability to use larger ranges. When a potential range is partitioned, the resulting potential ranges are (on average) half the size of the original range, as opposed to a fourth in the quadtree case. This method, therefore, can and does use fewer ranges. Finally, the relatively larger domain pool also contributes to the improved results.

6.6 Other Work

There is another approach using HV partitioning, which we only discuss briefly here. In this scheme (see [25] and [27]), the domains are selected from the partition itself, as opposed to a

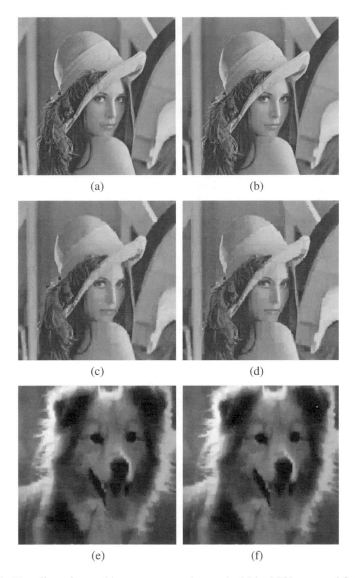

(a)　　　　　　　　　　　(b)

(c)　　　　　　　　　　　(d)

(e)　　　　　　　　　　　(f)

Figure 6.10: The effect of smoothing–postprocessing on the 256×256 Lenna and Collie images at various fidelities (encoded at optimization level 4). Data for these images appear in Table 6.3.

scan of the image. The orientation and position of the partition are chosen so that the partition tends to occur at positions in the image that are "meaningful" in the following sense. When a rectangle contains either horizontal or vertical edges, the split should occur along the strongest such edge; when a rectangle contains other edges, the split should occur in such a way that the edge runs "diagonally" through one of the generated rectangles, with the other rectangle having no edge (see Figure 6.11).

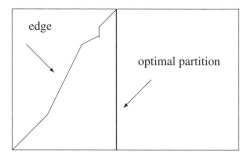

Figure 6.11: A rectangle is split vertically so that an edge runs diagonally through one of the sub-rectangles

Two ideas motivate this scheme. First, the image partitions will tend to be self-similar at different scale lengths. This facilitates efficient matching of domains and ranges. Second, the number of domains is very small compared with a domain pool generated by scanning the image, so encoding times are low. Results from this scheme, however, are considerably worse than the results presented here.

Finally, there are other adaptive partition schemes possible. These include hexagonal partitions, triangular partitions, and polygonal partitions. See Appendix C for a discussion of some of these.

Acknowledgments

Spencer Menlove would like to thank Joanna Gardner for her editing help.

Chapter 7

A Discrete Framework for Fractal Signal Modeling

L. Lundheim

In this chapter we present a description, used in the following three chapters, of fractal signal modeling in a discrete domain setting using linear algebra. We derive Collage Theorems, and we show how the contractivity factors of a certain class of affine operators are derived. A least squares approach for finding the optimal collage is outlined for a broad class of non-linear operators, including the affine case.

The central idea of fractal data compression is to exploit the redundancy of an information source by use of a fractal signal model. Such a model is valid if the signals generated by the source exhibit some kind of self-similarity, or, more generally, piecewise self-transformability. This means that small pieces of the signal can be well approximated by transformed versions of some other larger pieces of the signal itself.

To be able to describe, analyze, and use such models, we need a mathematical framework to define *signals, pieces of signals,* and allowable *transformations of such pieces.* The framework should be chosen to make the description well suited to capture the nature of the signals in question, to analyze the models and algorithms used in practical compression schemes, and to facilitate the implementation of the algorithms in software or hardware.

In the literature, three basic frameworks have been used in the investigation of fractal signal modeling and coding, which in varying degree meet the above requirements:

1. **Spaces of measures** (reducing to spaces of compact subsets of the plane in the case of representation of two-level images) [4, 45].

2. **Spaces of continuous functions** defined on $\mathbb{R}$ or $\mathbb{R}^2$ [63].

3. **Discrete frameworks** where each signal is modeled as a point in $\mathbb{R}^M$ [53, 57, 58, 59, 64, 66].

The measure-theoretical approach has the advantage that it elegantly describes what is actually happening when an image is created, e.g., on a photographic film. It is also useful for accurately describing the attractor of iterated function systems with probabilities assigned to the individual mappings. However, the underlying mathematics is rather abstract, and does not directly reflect what is going on in a practical implementation working on sampled signals or images.

A description based on continuous functions avoids some of the more difficult mathematics, but lacks also a direct connection to the sampled domain.

The discrete framework reflects the fact that images and signals to be compressed are provided in a discrete representation, i.e., in a sampled format, and that all algorithms are to be performed by digital circuits (e.g., by a general-purpose computer.) Hence, it is natural that the mathematical framework also reflects these aspects. And, as we shall shortly see, the resulting analysis may be performed using fairly elementary linear algebra.

7.1   Sampled Signals, Pieces, and Piecewise Self-transformability

A sampled signal, be that an image, audio, or other signal, may be thought of as a collection of numbers, where each number represents a grey-tone value, a sound pressure, or another physical quantity, depending on the origin of the signal.

Each number, or sample, denotes the signal value at a particular time or at a specified spatial location. Thus, to represent useful information, the samples must be presented in a particular temporal or spatial order. For audio signals, this order is sequential, resulting in a 1-dimensional signal, whereas for an image signal, one needs a 2-dimensional ordering corresponding to two spatial coordinates. Video sequences require an ordering, or indexing, of the samples in three dimensions.

Irrespective of the signals being 1-, 2-, or 3-dimensional, we restrict our discussion to signals of finite length. That is, we seek to model a discrete signal consisting of a finite number M of real numbers. A 1-dimensional signal will then be thought of as a sequence $\{x_i ; i = 1, \ldots, M\}$ indexed by one integer i, an image by an array $\{x_{ij} ; i = 1, \ldots, M_1, j = 1, \ldots, M_2\}$, and a video sequence by the indexed set $\{x_{ijk} ; i = 1, \ldots, M_1, j = 1, \ldots, M_2, k = 1, \ldots, M_3\}$.

For the following discussion we want to treat a finite length discrete signal as an element of $\mathbb{R}^M$. For a 1-dimensional signal, this is done simply by considering the sequence of samples as a column vector $\mathbf{x} = [x_1, x_2, \ldots, x_M]^T \in \mathbb{R}^M$, where the T superscript represents the transpose. For signals of higher dimensions, the procedure is as follows: When the dimensions M_1, M_2, or M_1, M_2, M_3 are decided for a particular class of signals, it is straightforward to define a *domain coordinate mapping* from a tuple of *domain coordinates* such as (i, j) or (i, j, k) to a single index i, arranging all the samples of a 2- or 3-dimensional sequence as a 1-dimensional one. Thus, a finite-length sampled signal may be thought of as vector $\mathbf{x} = [x_1, x_2, \ldots x_M]^T \in \mathbb{R}^M$ where $M = M_1 M_2$ for a 2-dimensional signal and $M = M_1 M_2 M_3$ for a 3-dimensional one.

Pieces and Blocks

Having identified a sampled signal with a vector (or point) in $\mathbb{R}^M$, we are now ready to define precisely what is meant by a *piece* of a signal.

In a discrete domain setting, a piece of a signal is simply a collection of samples which in some sense may be said to be adjacent to one another. For an image, this could mean that the samples all belong to a region of a certain shape (rectangular, triangular, etc.), size, and location. The size of such a piece may be defined by the number of samples it consists of. Depending on the chosen mapping from domain coordinates to vector indices, the samples constituting a piece of size M_r will have uniquely determined positions in the vector representation, specified by a subset of sample indices $I = \{i_1, i_2, \ldots, i_{M_r}\} \subset \{1, 2, \ldots, M\}$. Such a collection of indices corresponding to the samples of a *piece* will be called an *index block*. We will call the column vector consisting of the actual sample values corresponding to the indices of I a *block* and denote it $\mathbf{x}_I = [x_{i_1}, x_{i_2}, \ldots, x_{i_{M_r}}]^T$. We say that $\mathbf{x}_I$ is the contents of $\mathbf{x}$ *over* the index block I.

An example is shown in Figure 7.1, where a triangular piece of an image is mapped onto a block by a zig-zag domain coordinate mapping.

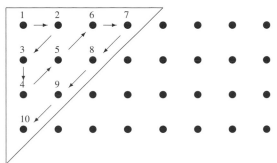

Figure 7.1: Example of domain coordinate mapping for triangular piece.

Piecewise Mappings

A central concept in fractal modeling schemes is a mapping that transforms one piece of a signal to another (usually of a smaller size). We will call such a transform a *piecewise mapping*, and we will call the piece being transformed *the domain block* and the result of the transformation *the range block*. In our setting a piecewise mapping may be defined as a function $T_n : \mathbb{R}^M \to \mathbb{R}^M$, for which we have chosen a *domain index block* I and a *range index block* J. The mapping is characterized by two properties:

1. $T_n \mathbf{x}$ is dependent only of the values of $\mathbf{x}$ belonging to (having indices in) the domain index block I.

  2. $T_n\mathbf{x}$ is a vector which is zero for all samples except those belonging to the range index block J.

In this chapter we will only study piecewise mappings which can be decomposed in the following way. The transformation of domain block $\mathbf{x}_I$ into a range block over J is performed in four stages:

$$T_n\mathbf{x} = \mathbf{P}_J K (\mathbf{DF}_I\mathbf{x}), \tag{7.1}$$

where

- $\mathbf{F}_I : \mathbb{R}^M \to \mathbb{R}^{M_d}$ is a "fetch operator" which selects the M_d sample values over a domain index block I, i.e., $\mathbf{F}_I\mathbf{x} = \mathbf{x}_I$.

- $\mathbf{D} : \mathbb{R}^{M_d} \to \mathbb{R}^{M_r}$ is a "decimation" operator which "shrinks" the chosen block; that is, it reduces the number of samples in such a way that essential properties of the signal piece in question are preserved.

- $K : \mathbb{R}^{M_r} \to \mathbb{R}^{M_r}$ is the *kernel* of the mapping which performs the non-trivial part (more about this later).

- $\mathbf{P}_J : \mathbb{R}^{M_r} \to \mathbb{R}^M$ is the "put" operator which places the output form K over the proper range index block J.

The fetch and put operators may be described by matrices consisting of ones and zeros only.

$$
\mathbf{F}_I = \left[\mathbf{0} \;
\overbrace{\begin{matrix} 1 & 0 & & 0 \\ 0 & 1 & & 0 \\ & & \ddots & \\ 0 & 0 & & 1 \end{matrix}}^{I}
\; \mathbf{0} \right] \quad ; \quad
\mathbf{P}_J = \left.\left[\begin{matrix} & & \mathbf{0} & \\ 1 & 0 & & 0 \\ 0 & 1 & & 0 \\ & & \ddots & \\ 0 & 0 & & 1 \\ & & \mathbf{0} & \end{matrix} \right]\right\} J \; .
$$

The shrink operation may be performed in several ways, the two most common being "pure" decimation (subsampling) and "decimation by averaging." Both strategies require that the ratio of the domain block length M_d and the range block length M_r be an integer number r, the *decimation ratio*. More general schemes involving interpolation may be used when this is not the case (see [59]).

In the case of subsampling, a block of length $M_r = \frac{1}{r}M_d$ is obtained by picking M_r of the M_d samples from $\mathbf{x}_I$. For 1-dimensional signals this means selecting each r-th sample, while for a 2-dimensional signal with rectangular domain blocks, the picking of samples should correspond to a regular, two-dimensional subsampling as illustrated in Figure 7.2 a.

Again, in the case of subsampling, $\mathbf{D}$ is adequately described by a matrix consisting of zeros and ones (at most, one zero in each row), the actual arrangement depending on the domain coordinate mapping. Shrinking by (weighted) averaging is performed by letting each sample in $\mathbf{Dx}_I$ be a weighted average of r samples of $\mathbf{x}_I$ (see Figure 7.2 b). In this case, $\mathbf{D}$ is a matrix with r non-zero entries in each row, each corresponding to the weight each sample is given in the average.

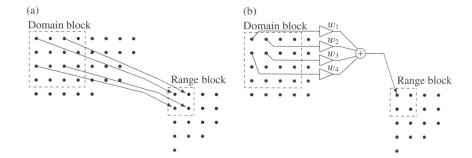

Figure 7.2: Two ways of shrinking a domain block: (a) Subsampling; (b) Averaging. In both cases the decimation ratio is $r = 4$.

In the rest of this chapter we shall assume that both domain and range index blocks consist of a *consecutive sequence* of indexes from the set $1, \ldots, M$. No generality is lost by this, since, if the blocks are actually distributed over a non-consecutive set of indices, this may be compensated for by multiplying $\mathbf{F}$ and $\mathbf{P}$ by suitable permutation matrices.

Thus, the decimation operator will obtain the following shape:

$$
\mathbf{D} = \begin{bmatrix}
w_1 & w_2 & \cdots & w_r & & & & & \\
 & & & w_1 & w_2 & \cdots & w_r & & \\
 & & & & & & & \ddots & \\
 & & & & & w_1 & w_2 & \cdots & w_r
\end{bmatrix}. \tag{7.2}
$$

The sequence of weights $\mathbf{w} = [w_1, w_2, \ldots, w_r]^T$ will be called the *decimation filter*. When subsampling is used, we have a decimation filter $\mathbf{w} = [1, 0, \ldots, 0]^T$, whereas decimation by averaging, giving equal weight to each sample, yields $\mathbf{w} = [\frac{1}{r}, \frac{1}{r}, \ldots, \frac{1}{r}]^T$.

So far, we have only treated how to fetch a domain block, shrink it, and put the transformed contents back in the position of a range block. The actual transformation, the kernel K, will be treated later on. As we shall see in Section 7.4 it is essential that K be a nonlinear operator.

Piecewise Transforms

Note that even if we think of T_n as being an operator working on pieces of a signal, we have chosen to define it as an operator on the entire signal $\mathbf{x}$, setting all samples to zero except the ones belonging to one particular range block. The reason for doing this is to facilitate the next step, which is the definition of a *piecewise transform*.

Before doing so, we need to define two new terms. An essential step in fractal coding is the subdivision of a signal into range blocks. When this subdivision consists of nonoverlapping blocks of the same size M_r, we have a *uniform partition* of the signal. In our vector representation this corresponds to a partition of the indices $1, \ldots, M$ into $N = M/M_r$ subsets $\{J_n\}_{n=1}^N = \mathcal{J}$. This collection will be called the *range partition* of the fractal model.

The range partition consists of what will become the "small pieces" when self-transformability is defined.

In addition to the small pieces, we need a collection of "large pieces" from which the small ones may be derived. These are the domain blocks mentioned above. For a certain source model a certain collection of domain blocks will be at our disposal. The corresponding collection of domain index blocks will be called the *library* and denoted $\mathcal{I} = \{I_1, I_2, \ldots, I_Q\}$.

When a range partition $\mathcal{J}$ and a library $\mathcal{I}$ is defined, a piecewise transform may be defined as follows: For each of the range index blocks $J_n \in \mathcal{J}$ a domain index block $I_{q(n)}$ is chosen from $\mathcal{I}$, and a kernel transform K_n is chosen from a predefined family $\mathcal{K}$. Then a piecewise mapping T_n is defined by $T_n\mathbf{x} = \mathbf{P}_{J_n} K(\mathbf{DF}_{I_{q(n)}}\mathbf{x})$. By summing, the piecewise mappings define a piecewise transform. That is, the piecewise transform $T : \mathbb{R}^M \to \mathbb{R}^M$ is defined by

$$T\mathbf{x} = \sum_{n=1}^{N} T_n\mathbf{x}. \tag{7.3}$$

7.2 Self-transformable Objects and Fractal Coding

In the previous section, we demonstrated how to describe an operator that transforms large pieces of a signal into smaller ones. If an object has the property of being left unchanged by such an operation we may say that it is piecewise self-transformable or self-similar. This is a well-known property, possessed in an exact sense by many fractal objects [4, 60] and approximately by many naturally occurring signals, such as images.

If T is a piecewise transform of the kind described in Section 7.1, a signal $\mathbf{x} \in \mathbb{R}^M$ is *piecewise self-transformable under T* if $T\mathbf{x} = \mathbf{x}$. In mathematical terms, $\mathbf{x}$ is a *fixed point* of T. If $\mathbf{x}_T$ is the only fixed point of T, a specification of T would indirectly yield an exact description of $\mathbf{x}_T$, provided we have a way of finding fixed points of arbitrary operators. This is the essence of fractal modeling of images and other signals.

Formally, fractal signal modeling may be described as follows:

1. We consider signals of fixed length M, generated from an information source. Let $S \subset \mathbb{R}^M$ consist of all signals the particular source can possibly generate.

2. We define a family $\mathcal{T} = \{T_{\mathbf{a}} : \mathbb{R}^M \to \mathbb{R}^M \,|\, \mathbf{a} \in \mathcal{A}\}$ of piecewise transforms, where each $T_{\mathbf{a}}$ is constructed according to Equations (7.1) and (7.3). Each transform of the family is uniquely identified by a parameter vector $\mathbf{a}$ belonging to a parameter space $\mathcal{A}$. Furthermore, all operators of $\mathcal{T}$ should be of a kind having unique fixed points which are easily computed once the operator, or, equivalently, the corresponding parameter vector is known. Each of the operators of $\mathcal{T}$ now has a fixed point with self-transformability properties defined by the actual operators. (Many of these will have fractal properties).

3. Now, provided that the source S consists of signals approximately sharing the self-transformabilities defined by $\mathcal{T}$, the collection of fixed points

$$\hat{S} = \{\mathbf{x}_T \,|\, T\mathbf{x}_T = \mathbf{x}_T \text{ for some } T \in \mathcal{T}\}$$

constitutes a *fractal source model* for S.

4. Given a signal $\mathbf{x} \in S$, a *fractal signal model* $\mathbf{x}_T \in \hat{S}$ may be found such that $\mathbf{x}_T$ approximates $\mathbf{x}$ in some sense. The process of finding such a model we will call *fractal signal modeling*. This is the essential step in a fractal data compression scheme.

As will be shown below, there is one property of operators guaranteeing the uniqueness and the computability of a fixed point, namely, that the operator be *eventually contractive*. Then the fixed point may be found iteratively, and is usually called an *attractor* of the operator.

7.3 Eventual Contractivity and Collage Theorems

Contractive Operators

In order to evaluate the quality of our approximations, to analyze our methods, and, generally, to express what is going on, we need some means to measure the distance or difference between two signals. Having decided to work in $\mathbb{R}^M$ it is natural to use *norms* for this purpose. The norm of a vector is a real, non-negative number measuring the *length* of a vector and is denoted $\|\mathbf{x}\|$. Thus, the distance between signals $\mathbf{x}$ and $\mathbf{y}$ can be evaluated by $\|\mathbf{x} - \mathbf{y}\|$. Various such norms exist, the most well known being the Euclidean or quadratic norm $\|\mathbf{x}\|_2 = \sqrt{\sum_{i=1}^M x_i^2}$.

Generally, an operator on $\mathbb{R}^M$ will change the length, or norm, of the vector it operates upon. Often there is some bound on how much the norm of an arbitrary vector may be increased by a certain operator. Then the operator is said to be *Lipschitz*, and the bound is expressed by the *Lipschitz factor* defined by the real number $s = \sup_{\mathbf{x}} \frac{\|T\mathbf{x}\|}{\|\mathbf{x}\|}$, also called the norm of T and denoted $\|T\|$. If the Lipschitz factor of an operator T is strictly less than 1, the operator is called *contractive*; that is, it will essentially reduce the norm of any vector. Then s is called the *contractivity (factor)* of T. More generally, if the K-th iterate T^K, $K \geq 1$, of an operator is contractive, then T is said to be *eventually contractive (at the K-th iterate)*.

For such operators the following theorem[1] holds:

Theorem 7.1 (Contraction Mapping Theorem) *If $T : \mathbb{R}^M \to \mathbb{R}^M$ is an eventually contractive operator, then T has a unique fixed point $\mathbf{x}_T$ which is given by $\mathbf{x}_T = \lim_{n \to \infty} T^n \mathbf{x}$, where $\mathbf{x}$ is an arbitrary point in $\mathbb{R}^M$.*

A proof may be found in Chapter 2.

Thus, we see that eventual contractivity both secures the necessary uniqueness of the fixed point required for fractal modeling, *and* provides an iterative method for finding it. Equally important is the fact that it helps us in actually finding a good fractal approximation for a given signal, as we shall shortly see.

Collage Theorems

Given a signal $\mathbf{x}$ and a family of piecewise similar operators, the problem is to find the operator having an attractor $\mathbf{x}_T$ closest to the signal, i.e., minimizing the distance $\|\mathbf{x} - \mathbf{x}_T\|$.

---

[1] This is a finite-dimensional version of the more general *Banach's Fixed-Point Theorem*, originally valid for general complete metric spaces (see, for example, [51]).

Finding this minimum for a non-trivial operator family is in general a very difficult task, and several methods for finding sub-optimal, but computable solutions have been proposed. Most of these are based on some kind of *Collage Theorem*.

In the visual arts, a collage is a collection of pieces cut out from various fabrics and pasted together in a pattern, often to resemble something. The counterpart in our context consists of using a piecewise transform T on a signal $\mathbf{x}$ to "cut out" or "fetch" the contents of chosen domain blocks, transforming them and "pasting" them up as range blocks to constitute a "collage" $\mathbf{x}_c = T\mathbf{x}$. The idea of a Collage Theorem in search for an attractor $\mathbf{x}_T$ close to $\mathbf{x}$ is to utilize the fact that under some conditions a collage close to the original signal implies an attractor approximating the original well.

More precisely:

Theorem 7.2 *Let $T : \mathbb{R}^M \to \mathbb{R}^M$ be an eventually contractive operator at the K-th iterate, with fixed point $\mathbf{x}_T$. For a given signal $\mathbf{x}$ and positive real number ϵ, if $\|\mathbf{x} - \mathbf{x}_c\| \le \epsilon$, then*

$$\|\mathbf{x} - \mathbf{x}_T\| \le \frac{1 + \sum_{k=1}^{K-1} s_k}{1 - s_K}\epsilon, \tag{7.4}$$

where $s_k, k = 1, 2, \ldots, K$ denote the Lipschitz factors of the iterates T^k.

The proof may be found in Chapter 2. In this result the Lipschitz factors of the $K - 1$ iterates of T are used. From Equation (7.4) the following result, which depends only on s_1 and s_K, is easily derived:

$$\|\mathbf{x} - \mathbf{x}_T\| \le \frac{1 - s_1^K}{(1 - s_1)(1 - s_K)}\epsilon. \tag{7.5}$$

In the special case of T being (strictly) contractive, Equation (7.4) is simplified to

$$\|\mathbf{x} - \mathbf{x}_T\| \le \frac{1}{1 - s_1}\epsilon. \tag{7.6}$$

All the Equations (7.4)–(7.6) may be referred to as Collage Theorems whose practical consequence is that if we find an operator whose collage, when applied to a signal $\mathbf{x}$, is close to the signal itself, the fixed point will also be close to $\mathbf{x}$ provided that the right-hand side expressions are not too large.

The idea of using a Collage Theorem for fractal coding was first presented by Barnsley, Ervin, Hardin, and Lancaster [6] for strictly contractive operators, giving a result corresponding to Equation (7.6). The more general result expressed by Equation (7.5) was introduced independently by Fisher, Jacobs, and Boss [26] and Lundheim [59]. For special cases of operators, improved versions have been found by Øien [64].

7.4   Affine Transforms

We have now demonstrated how a piecewise self-transformable object in $\mathbb{R}^M$ can be described as the fixed point or attractor of an eventually contractive piecewise transform. We have seen that this transform is the sum of a number of piecewise mappings, one for each range block. We will now return to the inner workings of these mappings. More precisely, we will consider the choice of the kernel K (see Equation (7.1)). Apart from the fact that T be eventually contractive, one additional requirement is necessary for a useful source model, namely, that T be nonlinear. This follows from the fact that $\mathbf{0}$ is always a fixed point for a linear transform, and, since we require that T have a unique fixed point, $\mathbf{0}$ must be the only one if T were linear. Thus, a linear transform can not be used to model a non-zero signal in the manner outlined in Section 7.2.

From the construction of T by Equations (7.1)–(7.3), it is easily realized that T is linear if and only if the kernels of all the piecewise mappings are linear. Thus, our next consideration will be the selection of a family $\mathcal{K} = \{K : \mathbb{R}^{M_r} \to \mathbb{R}^{M_r}\}$ of nonlinear kernel transforms.

Among all the possible nonlinear transforms, we will choose the class which – in a sense – may be said to be the simplest one, namely, the class of affine transforms. If L' is a linear transform on $\mathbb{R}^{M_r}$ (i.e., an $M_r \times M_r$ matrix) and $\mathbf{t}'$ an arbitrary vector in $\mathbb{R}^{M_r}$, then $K : \mathbb{R}^{M_r} \to \mathbb{R}^{M_r}$ given by $K(\mathbf{x}) = L'\mathbf{x} + \mathbf{t}'$ is an *affine transform* on $\mathbb{R}^{M_r}$. It is easily seen that if $\mathcal{K}$ consists of affine transforms on $\mathbb{R}^{M_r}$, then

$$\mathcal{T} = \left\{ T = \sum_{n=1}^{N} T_n \mid T_n(\mathbf{x}) = \mathbf{P}_{J_n} K_n(\mathbf{DF}_{I_{q(n)}}\mathbf{x}); \ K_n \in \mathcal{K} \right\}$$

will in turn consist of affine transforms on $\mathbb{R}^M$.

As an example of a class of such transforms, we will again choose the simplest one, where the linear part of each of the kernels consists of only a multiplication with a constant β_2 (amplitude scaling), and the constant part is a vector of equal valued samples $\mathbf{t} = \beta_1[1, 1, \ldots, 1]^T$ (a DC term). This kernel family corresponds to the one used by Jacquin [45].

Then a member of $\mathcal{T}$ would correspond to the operation $T\mathbf{x} = L\mathbf{x} + \mathbf{t}$ where *the linear term* L is given by the matrix

$$L = \sum_{n=1}^{N} \beta_2^{(n)} \mathbf{P}_{J_n} \mathbf{DF}_{I_{q(n)}} = \begin{bmatrix} & & & & \boxed{\beta_2^{(1)}\mathbf{D}} \\ & \boxed{\beta_2^{(2)}\mathbf{D}} & & & \\ & & & \ddots & \\ \boxed{\beta_2^{(N)}\mathbf{D}} & & & & \end{bmatrix} \tag{7.7}$$

consisting of zeros except for the outlined sub-matrices, one for each range block. The block structure arises from the convention of consecutive blocks (see page 141). The horizontal position of a non-zero sub-matrix $\beta_2^{(n)}\mathbf{D}$ corresponds to the position of the assigned domain index block $I_{q(n)}$, while the vertical position is determined by the range index block J_n. The *translation term* obtains the shape

$$\mathbf{t} = [\beta_1^{(1)}, \beta_1^{(1)}, \ldots, \beta_1^{(1)}, \beta_1^{(2)}, \beta_1^{(2)}, \ldots, \beta_1^{(2)}, \ldots, \beta_1^{(N)}, \beta_1^{(N)}, \ldots, \beta_1^{(N)}]^T.$$

7.5 Computation of Contractivity Factors

We have seen that the Lipschitz factor of an operator is important both for the Collage Theorems and for the convergence of an iterative decoding algorithm. For analysis purposes, and even as part of practical algorithms, it is therefore desirable to be able to compute this factor. We will now show how this can be done for an affine operator family with uniform range partition, nonoverlapping domain blocks, and a linear term of the amplitude scaling type described above. We also assume that the library consist of Q adjacent domain blocks, each of length M/Q. The results are derived for the l_2 norm only.

First observe that the Lipschitz factor s for an affine operator is equal to the Lipschitz factor of its linear term $\mathbf{L}$, i.e., the matrix norm of the linear part, $s = \|\mathbf{L}\|$. In the l_2 case, this norm is given by the square root of the largest eigenvalue of $\mathbf{L}^T\mathbf{L}$ (see, e.g., [76]). We will therefore study this latter matrix in more detail.

From the initial assumptions about the linear part, we find that $\mathbf{L}^T\mathbf{L}$ will have a block diagonal structure

$$
\mathbf{L}^T\mathbf{L} =
\begin{bmatrix}
\mathbf{L}_1^* & & & \\
& \mathbf{L}_2^* & & \\
& & \ddots & \\
& & & \mathbf{L}_Q^*
\end{bmatrix}.
$$

This matrix consists of one sub-matrix $\mathbf{L}_q^*$ for each of the domain blocks $q = 1, 2, \ldots, Q$ of the library. By studying the block nature pictured in Equation (7.7), it is easily realized that each sub-matrix is made up by a sum over *all the range blocks which use that particular domain block*, resulting in

$$
\mathbf{L}_q^* = \sum_{I_{q(n)}=I_q} (\beta_2^{(n)}\mathbf{D})^T(\beta_2^{(n)}\mathbf{D}) = \sum_{I_{q(n)}=I_q} (\beta_2^{(n)})^2\mathbf{D}^T\mathbf{D}.
$$

By inspection of Equation (7.2), we see that each of the sub-matrices will again have a block diagonal structure, since

$$
\mathbf{D}^T\mathbf{D} =
\begin{bmatrix}
\mathbf{w}\mathbf{w}^T & & & \\
& \mathbf{w}\mathbf{w}^T & & \\
& & \ddots & \\
& & & \mathbf{w}\mathbf{w}^T
\end{bmatrix}.
$$

Now, the eigenvalues of $\mathbf{L}^T\mathbf{L}$ are the roots of the determinant $|\mathbf{L}^T\mathbf{L} - \lambda\mathbf{I}|$. Since the determinant of a block diagonal matrix equals the product of the determinants of the sub-matrices, we have

$$
|\mathbf{L}^T\mathbf{L} - \lambda\mathbf{I}| = \prod_{q=1}^{Q} |\beta_q^*\mathbf{w}\mathbf{w}^T - \lambda\mathbf{I}|^{\frac{M}{Qr}}, \tag{7.8}
$$

where we have used the notation

$$
\beta_q^* = \sum_{I_{q(n)}=I_q} (\beta_2^{(n)})^2.
$$

Finally, we need an expression for the determinant $\left|\beta_q^* \mathbf{w}\mathbf{w}^T - \lambda \mathbf{I}\right|$ which will depend on the chosen decimation filter $\mathbf{w}$. We derive the expression for the two special cases of subsampling and averaging only.

For the subsampling case $\mathbf{w} = [1, 0, \ldots, 0]^T$, yielding

$$\left|\beta_q^* \mathbf{w}\mathbf{w}^T - \lambda \mathbf{I}\right| = \begin{vmatrix} \beta_q^* - \lambda & & & \\ & \beta_q^* - \lambda & & \\ & & \ddots & \\ & & & \beta_q^* - \lambda \end{vmatrix} = (\beta_q^* - \lambda)^r$$

which, inserted into Equation (7.8), gives

$$\left|\mathbf{L}^T\mathbf{L} - \lambda \mathbf{I}\right| = \prod_{q=1}^{Q} (\beta_q^* - \lambda)^{\frac{M}{Q}} ;$$

that is, $\mathbf{L}^T\mathbf{L}$ has Q eigenvalues $\lambda_q = \beta_q^*$, each with multiplicity $\frac{M}{Q}$.

For the averaging case, all the decimation filter weights are equal to $\frac{1}{r}$, and we get

$$\left|\beta_q^* \mathbf{w}\mathbf{w}^T - \lambda \mathbf{I}\right| = \begin{vmatrix} \frac{\beta_q^*}{r^2} - \lambda & \frac{\beta_q^*}{r^2} & \frac{\beta_q^*}{r^2} & \cdots & \frac{\beta_q^*}{r^2} \\ \frac{\beta_q^*}{r^2} & \frac{\beta_q^*}{r^2} - \lambda & \frac{\beta_q^*}{r^2} & \cdots & \frac{\beta_q^*}{r^2} \\ \frac{\beta_q^*}{r^2} & \frac{\beta_q^*}{r^2} & \ddots & & \frac{\beta_q^*}{r^2} \\ \vdots & \vdots & & & \vdots \\ \frac{\beta_q^*}{r^2} & \frac{\beta_q^*}{r^2} & \frac{\beta_q^*}{r^2} & \cdots & \frac{\beta_q^*}{r^2} - \lambda \end{vmatrix} .$$

It is easily proved (see Addendum 7.7) that the determinant is equal to $(-1)^r \lambda^{r-1}(\lambda - \frac{\beta_q^*}{r})$, which inserted into Equation (7.8) yields

$$\left|\mathbf{L}^T\mathbf{L} - \lambda \mathbf{I}\right| = \prod_{q=1}^{Q} (-1)^{\frac{M}{Q}} \lambda^{\frac{r-1}{Qr}M} \left(\lambda - \frac{\beta_q^*}{r}\right) ; \tag{7.9}$$

that is, $\mathbf{L}^T\mathbf{L}$ has Q (possibly) non-zero eigenvalues $\lambda_q = \frac{\beta_q^*}{r}$.

The preceding results may now be used to find the Lipschitz factor of an affine, piecewise similar operator as follows:

Decimation by subsampling:

$$s = \sqrt{\max_{n=1:N} \sum_{I_{q(n)}=I_q} (\beta_2^{(n)})^2}$$

Decimation by averaging:

$$s = \sqrt{\frac{1}{r} \max_{n=1:N} \sum_{I_{q(n)}=I_q} (\beta_2^{(n)})^2}$$

To find out whether an operator of the described type is contractive or not, one must for each of the domain blocks compute the sum of the squared amplitude scaling factors for each of the range blocks which use the domain block in question. Then one finds the maximum of all these sums. The square root of the result is the Lipschitz factor in the subsampling case. If averaging is used, the result is divided by the decimation ratio before taking the square root.

Note that even if the stated results are shown to be valid only for the l_2 norm, they may be used as contractivity tests for *any* desired norm on $\mathbb{R}^M$, since all norms on finite-dimensional vector spaces are equivalent.

Since the Lipschitz factor is reduced by the ratio $1/r$ in the averaging case as compared to subsampling, the result may explain the fact that using subsampling often results in very slow convergence [59], whereas decimation by averaging usually converges in a moderate number of steps.

7.6   A Least-squares Method

The framework introduced so far may now be used to formulate a practical signal modeling procedure as follows:

- For each range index block J_n:

  - Select the domain index block $I_{q(n)} \in \mathcal{I}$ and kernel $K \in \mathcal{K}$ such that $\left\| \mathbf{x}_{J_n} - K(\mathbf{DF}_{I_{q(n)}}\mathbf{x}_{J_n}) \right\|$ is minimized.

If we are able to complete this procedure, we have found the operator $T = \sum_{n=1}^{N} T_n$ giving the collage $\mathbf{x}_c$ closest possible to $\mathbf{x}$. By the Collage Theorem, this should also yield a fixed point $\mathbf{x}_T$ approximating $\mathbf{x}$, provided the factor in front of ϵ in one of the Equations (7.4)–(7.6) is not too large.

For each range block, the procedure consists of two tasks: find the right domain block, i.e., establish the function $q(n)$, and the right kernel. The former problem is dealt with in Chapter 8, while the latter will be considered in the following.

To begin with, we will make as few assumptions about the kernel family $\mathcal{K}$ as possible. For instance, we shall not demand that it consist of affine transforms, only that the transforms be nonlinear. However, the family as a whole shall fulfill another kind of linearity condition, in the sense that if K_1 and K_2 both are members of the family, then the one defined by $K\mathbf{x} = \beta_1 K_1\mathbf{x} + \beta_2 K_2\mathbf{x}$ should also be a member, when β_1 and β_2 are arbitrary real numbers. This implies (by an elementary property of linear spaces) that $\mathcal{K}$ must have a *basis* $\{K_1, K_2, \ldots, K_P\}$ such that *any* member of $\mathcal{K}$ may be written as a linear combination $K = \beta_1 K_1 + \beta_2 K_2 + \cdots + \beta_P K_P$. Thus, we have found a way to parametrize the kernel family $\mathcal{K}$.

Now, given a signal $\mathbf{x}$ to be modeled, for each range block we must find the optimal domain block, and the optimal set of coefficients $\beta_1, \ldots \beta_P$. Assuming a domain index block I is chosen, and omitting the indices on $\mathbf{P}$ and $\mathbf{F}$ for readability, the collage over a given range index block J_n may be expressed by

$$T_n\mathbf{x} \quad = \quad \mathbf{P}K(\mathbf{DFx})$$

$$= \mathbf{P} \sum_{p=1}^{P} \beta_p K_p (\mathbf{DFx})$$

$$= \mathbf{PHb}$$

where we have introduced the matrix

$$\mathbf{H} = [K_1(\mathbf{DFx}) \; K_2(\mathbf{DFx}) \; \cdots \; K_P(\mathbf{DFx})] \tag{7.10}$$

and $\mathbf{b}$ is the parameter vector $[\beta_1 \; \beta_2 \cdots \; \beta_P]^T$.

Thus, for each range block, we have to solve the minimization problem of finding the vector $\mathbf{b}$ minimizing the distance $\|\mathbf{x}_{J_n} - \mathbf{Hb}\|$.

Assuming that a quadratic norm is used, we have from elementary linear algebra (see [76]) that the solution is given by solving the normal equation

$$\mathbf{H}^T \mathbf{Hb} = \mathbf{H}^T \mathbf{x}_{J_n}. \tag{7.11}$$

Note that in this section we have not assumed that the kernel family consist of affine operators. This means that the outlined method for finding the (collage-) optimal kernel is valid for all kinds of nonlinear operators, as long as the kernel family is allowed to encompass all linear combinations of the nonlinear transforms in question.

If the function $q(n)$, assigning a domain block to each range block, is given, the optimal parameter vector $\mathbf{b}$ is now found by solving Equation (7.11). To find the best assignment pattern of domain blocks to range blocks requires a search of the library. This method will give an operator producing the optimal collage. However, it is not guaranteed that the operator will be (eventually) contractive. This must be checked separately by computing the Lipschitz factor by the method described earlier, for example.

By an exhaustive search of the whole library, the best domain block is found for each of the range blocks by finding the optimal parameter kernel for each combination of range and domain block, and calculating the resulting distance between range block and collage by the following algorithm:

> - Initialize $q(n) := 1$ for all n.
>
> - For each domain index block I_q:
>
> - Compute the matrix $\mathbf{H}$ according to Equation (7.10)
>
> - For each range index block J_n:
>
>   - Find the optimal $\mathbf{b}$ by solving Equation (7.11).
>
>   - Compute $\|\mathbf{x}_{J_n} - \mathbf{Hb}\|$, and update $q(n)$ if the result is improved.

As an example, consider again the kernel family of affine operators:

$$K(\mathbf{x}) = \beta_2 \mathbf{x} + \beta_1 \mathbf{u},$$

where $\mathbf{u} = [1, 1, \ldots, 1]^T \in \mathbb{R}^{M_r}$, i.e., we have a kernel family of dimension $P = 2$.

Using the notation $\mathbf{x}^{(n)} = \mathbf{x}_{J_n}$ (range block n) and $\mathbf{y}^{(n)} = \mathbf{DF}_I\mathbf{x}$ (the decimated domain block), we get

$$\mathbf{H} = [\mathbf{y}^{(n)}\mathbf{u}]$$

which, inserted into Equation (7.11) gives the normal equation for range block n:

$$\langle \mathbf{y}^{(n)}, \mathbf{y}^{(n)} \rangle \beta_1^{(n)} + \langle \mathbf{y}^{(n)}, \mathbf{u} \rangle \beta_2^{(n)} = \langle \mathbf{y}^{(n)}, \mathbf{x}^{(n)} \rangle$$
$$\langle \mathbf{u}, \mathbf{y}^{(n)} \rangle \beta_1^{(n)} + \langle \mathbf{u}, \mathbf{u} \rangle \beta_2^{(n)} = \langle \mathbf{u}, \mathbf{x}^{(n)} \rangle.$$

where $\langle\ ,\ \rangle$ denotes the inner (dot) product. These equations have the solution

$$\beta_1 = \frac{\sum_{m=1}^{M_r} x_m \sum_{m=1}^{M_r} y_m^2 - \sum_{m=1}^{M_r} y_m \sum_{m=1}^{M_r} x_m y_m}{M_r \sum_{m=1}^{M_r} y_m^2 - \left(\sum_{m=1}^{M_r} y_m\right)^2}$$

$$\beta_2 = \frac{M_r \sum_{m=1}^{M_r} x_m y_m - \sum_{m=1}^{M_r} x_m \sum_{m=1}^{M_r} y_m}{M_r \sum_{m=1}^{M_r} y_m^2 - \left(\sum_{m=1}^{M_r} y_m\right)^2},$$

where superscripts $^{(n)}$ are omitted for readability.

7.7   Conclusion

By using a discrete domain description for the theory of fractal signal modeling and coding, we are immediately able to use the powerful tools of linear algebra for the design of analysis/synthesis algorithms. The present chapter presents only some examples of how this might be done. Other instances can be found in the other chapters of this book (for example Chapter 8), and in the theses by Lepsøy, Lundheim, and Øien [59, 64, 53]. In particular, note how orthogonalization is utilized in Øien's work as a sophistication of the least squares method presented above.

Almost all literature on fractal coding until now has been focused on affine transforms. Further progress in the field depends perhaps on using other kinds of non-linearities. In that case some of the tools from linear algebra will be of no use, but the discrete domain framework should still have much to offer in clarity and by bringing the theory close to actual implementations.

Acknowledgments

I would like to thank Geir Egil Øien and Yuval Fisher for proofreading the manuscript and for offering several valuable suggestions.

Addendum A
Derivation of Equation (7.9)

In the derivation of Equation (7.9) we needed an expression for the r-th order determinant with the following structure:

$$D_r(a) = \begin{vmatrix} a-\lambda & a & a & \cdots & a \\ a & a-\lambda & a & \cdots & a \\ a & a & \ddots & & a \\ \vdots & \vdots & & & \vdots \\ a & a & a & & a \\ a & a & a & \cdots & a-\lambda \end{vmatrix}.$$

By subtracting the second row from the first one, and developing the determinant by the first row, we find

$$D_r(a) = -\lambda D_{r-1}(a) - \lambda \begin{vmatrix} a & a & a & \cdots & a \\ a & a-\lambda & a & \cdots & a \\ a & a & \ddots & & a \\ \vdots & \vdots & & & \vdots \\ a & a & a & & a \\ a & a & a & \cdots & a-\lambda \end{vmatrix}. \tag{7.12}$$

Subtracting the first row from all the remaining ones, the determinant of Equation (7.12) becomes triangular and equal to the product of its diagonal, and we obtain the recurrence relation

$$D_r(a) = a\lambda^{r-1}(-1)^{r+1} - \lambda D_{r-1}(a). \tag{7.13}$$

Defining $D_0(a) = 1$, it is easily shown by induction from Equation (7.13) that

$$D_r(a) = (-1)^r \lambda^{r-1}(\lambda - ra) \tag{7.14}$$

In the derivation of Equation (7.9) we have used Equation (7.14) for $D_r(\frac{\beta_q^*}{r})$.

Chapter 8

A Class of Fractal Image Coders with Fast Decoder Convergence

G. E. Øien and S. Lepsøy

In this chapter we introduce a class of fractal image coders which have the remarkable property of giving *exact decoder convergence* in the *lowest possible number of iterations* (which is image independent). The class is related to that introduced by Jacquin [45, 46, 47, 48], employing simple affine mappings working in a blockwise manner. The resulting decoder can be implemented in a pyramid-based fashion, yielding a computationally very efficient structure. Also, a coder offering non-iterative decoding and direct attractor optimization in the encoder is included as a special case. Other benefits of the proposed coder class include more optimal quantization and an improved Collage Theorem.

The encoder/decoder structure in a fractal image coder is depicted in Figure 8.1. The encoder optimizes a mapping T such that the distance between the image x to be coded, and the *collage* Tx of x with respect to T is minimized. The decoder generates the *attractor* x_T of T by the iterative equation

$$x_T = \lim_{k \to \infty} T^k x_0 \text{ for an arbitrary } x_0. \tag{8.1}$$

The attractor will generally not be exactly equal to the collage, but the two images can be made close to each other by constraining the mapping T. This can be seen from the *Collage Theorem* 2.1, which traditionally has been the justification behind the coder structure in Figure 8.1.

In this chapter we shall limit ourselves to considering *affine* mappings, operating on the space of discrete images of a given size M pixels. The main purpose is to introduce and describe a certain limited class of such mappings that secures fast, image independent decoder convergence, without impairing the image quality relative to that obtained with the coder class

153

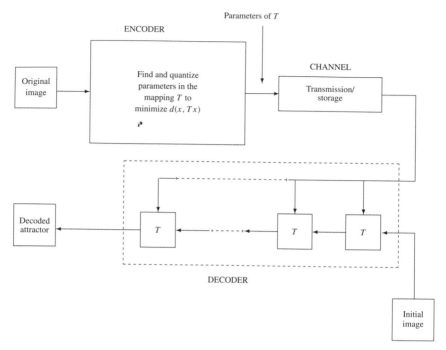

Figure 8.1: A general fractal encoder/decoder structure.

introduced by Jacquin. Initially we describe the class of mappings suggested by Jacquin, but within a discrete framework suited for the treatment of digital images.

8.1 Affine Mappings on Finite-Dimensional Signals

Let us assume that each image is partitioned into N nonoverlapping *range blocks*, each of which is to be coded separately. Range block n will be assumed to be of size $B_n \times B_n$, such that $M = \sum_{n=1}^{N} B_n^2$. Furthermore, we will assume that all such images are ordered as M-dimensional column vectors, according to some prescribed convention (which does not need to be specified further here, but may for example be defined by the *Hilbert curve* described in [7, pp. 39–41]. See also Chapter 7). An image ordered as such a vector will be denoted by a boldface letter, e.g., $\mathbf{x}$. (That is, the vector $\mathbf{x} \in \mathbb{R}^M$ corresponds to an image x with M pixels.) Applying an affine mapping T to a vector-ordered image $\mathbf{x}$ may be described by a matrix multiplication (the *linear* term) and an addition of a fixed vector (the *translation* term):

$$T\mathbf{x} = \mathbf{Lx} + \mathbf{t} = \begin{bmatrix} l_{11} & \cdots & l_{1M} \\ \vdots & \ddots & \vdots \\ l_{M1} & \cdots & l_{MM} \end{bmatrix} \cdot \begin{bmatrix} x_1 \\ \vdots \\ x_M \end{bmatrix} + \begin{bmatrix} t_1 \\ \vdots \\ t_M \end{bmatrix}. \tag{8.2}$$

For the mappings introduced by Jacquin, the linear term may be decomposed as a sum over all range blocks,

$$\mathbf{L} = \sum_{n=1}^{N} \mathbf{L}_n,$$ (8.3)

where $\mathbf{L}_n$ is given by

$$\mathbf{L}_n = \beta_2^{(n)} \mathbf{P}_n \mathbf{D}_n \mathbf{F}_n.$$ (8.4)

In this decomposition,

- $\mathbf{F}_n : \mathbb{R}^M \to \mathbb{R}^{D_n^2}$ fetches the correct *domain block*, here assumed to be of size $D_n \times D_n = D_n^2$ pixels, and may also shuffle its pixels internally in a prescribed manner

- $\mathbf{D}_n : \mathbb{R}^{D_n^2} \to \mathbb{R}^{B_n^2}$ shrinks (decimates) the domain block to range block size $B_n \times B_n = B_n^2$ pixels

- $\mathbf{P}_n : \mathbb{R}^{B_n^2} \to \mathbb{R}^M$ puts the decimated domain block in the correct range block position, and sets the rest of the image to zero

- $\beta_2^{(n)} \in \mathbb{R}$ scales the dynamic range of the decimated domain block.

The collage approximation to the n-th range block is thus a sum of a decimated, grey-tone-scaled domain block found by a systematic search through the set of all allowed decimated domain blocks for range block n (denoted $\mathcal{L}_n$), and a translation term. In his work, Jacquin constrained the translation term $\mathbf{t}$ to be a *blockwise constant* (DC) term. It has been demonstrated how this can be generalized [59, 64, 66], but here we will constrain ourselves to Jacquin's choice, which is the most important special case for image coding purposes. We may then write

$$\mathbf{t} = \sum_{n=1}^{N} \left(\beta_1^{(n)} \mathbf{P}_n \mathbf{b}_1^{(n)} \right),$$ (8.5)

where $\mathbf{b}_1^{(n)}$ is a *constant* B_n^2-vector,[1] e.g., containing only 1's. For each domain block that we consider for a given range block, we then have *an optimization problem in a 2-dimensional linear subspace of the range block space*. The decimated domain block, which we will denote as $\mathbf{b}_2$, acts as a second basis vector in the approximation of the range block in question.

ℓ^2-optimization

When optimizing the model coefficients $\{\beta_1^{(n)}\}_{n=1}^{N}$ and $\{\beta_2^{(n)}\}_{n=1}^{N}$, we shall use the ℓ^2-*metric*, defined by the expression

$$d_B(\mathbf{x}, \mathbf{y}) = \left(\sum_{j=1}^{B^2} (x_j - y_j)^2 \right)^{\frac{1}{2}} \quad \text{for all } \mathbf{x}, \mathbf{y} \in \mathbb{R}^{B^2}$$ (8.6)

as the range block space metric. This metric has several advantages:

---

[1] We will often omit the subscript and use B for the range size and D for the domain size.

- it is easily computed.

- for block sizes that are not too large,[2] it is known to correspond reasonably well with our visual perception.

- it is uniquely minimized over the elements in a linear subspace by an *orthogonal projection* onto the subspace.

The *projection theorem* [51] states that the optimal ℓ^2-approximation to a vector $\mathbf{x} \in \mathcal{X}$ within a linear subspace $\mathcal{S}$ of a linear vector space $\mathcal{X}$ is the $\mathbf{x}_p \in \mathcal{S}$ making the error vector $\mathbf{x} - \mathbf{x}_p$ *orthogonal* to the subspace. If we define the inner product on $\mathbb{R}^{B^2}$ as

$$\langle \mathbf{x}, \mathbf{y} \rangle = \sum_{j=1}^{B^2} x_j y_j \text{ for all } \mathbf{x}, \mathbf{y} \in \mathbb{R}^{B^2} \tag{8.7}$$

the ℓ^2-distance between $\mathbf{x}$ and $\mathbf{y}$ is simply

$$d_B(\mathbf{x}, \mathbf{y}) = \|\mathbf{x} - \mathbf{y}\|_2 = \langle \mathbf{x} - \mathbf{y}, \mathbf{x} - \mathbf{y} \rangle^{\frac{1}{2}} \tag{8.8}$$

where $\| \cdot \|_2$ denotes the ℓ^2-norm,

$$\|\mathbf{x}\|_2 = \langle \mathbf{x}, \mathbf{x} \rangle^{\frac{1}{2}}. \tag{8.9}$$

Here, we are to find two optimal coefficients, β_1 and β_2. Invoking the projection theorem, we find that we must solve the *normal equations*

$$\langle \mathbf{x} - \beta_1 \mathbf{b}_1 - \beta_2 \mathbf{b}_2, \mathbf{b}_l \rangle = 0 \text{ for } l = 1, 2. \tag{8.10}$$

The resulting approximation vector $\mathbf{x}_p = \beta_1 \mathbf{b}_1 + \beta_2 \mathbf{b}_2$ is called the *orthogonal projection* of $\mathbf{x}$ onto the subspace span$\{\mathbf{b}_1, \mathbf{b}_2\}$.

8.2 Conditions for Decoder Convergence

The mappings used for signal representation in fractal coding must be *contractive* – either strictly or eventually – in order for the decoding process to converge toward the correct reconstructed image. However, unconstrained ℓ^2-optimization of the collage with respect to the parameters in mappings of the type described in the previous section does not necessarily provide us with a mapping having this property.

For that class of mappings, Lundheim has derived (see Chapter 7) the following condition for strict contractivity in the ℓ^2-metric[3]:

$$\max_{l \in \mathcal{L}} \left[\sum_{n \in \mathcal{I}_l} \left(\beta_2^{(n)} \right)^2 \right] < \frac{D}{B}. \tag{8.11}$$

---

[2]Typically up to 8×8 pixels.

[3]Lundheim's proof holds for the case when all range blocks are of size $B \times B$, and the domain blocks are nonoverlapping and of size $D \times D$.

In the above equation, $\mathcal{I}_l$ is the set of indices for those range blocks that use l as a domain block, and $\mathcal{L}$ is the set of all allowed decimated domain blocks.

It is seen that securing convergence involves *constraining the sizes of the grey-tone scalings*. This can impair the image quality, since the ℓ^2-optimal values generally may not fulfill the above constraint. The *speed* of convergence is also linked to the absolute values of the grey-tone scalings. Generally, the convergence will be faster the smaller these factors are constrained to be. However, if only very small factors are allowed, this may very well have a significant negative influence on the model quality, resulting in bad collages, and hence also bad attractors.

On the other hand, if unconstrained ℓ^2-optimal model coefficients are used, the collages will be optimal, but then there is at least a theoretical possibility that the decoding algorithm may fail, i.e., that divergence occurs, or that the convergence becomes too slow to yield a practically useful decoder. Slow convergence is especially serious in applications where a simple decoding algorithm is essential, such as in image sequence coding for broadcasting purposes.

Eventual Contractivity

As has been demonstrated independently in [27], [28], and [66], constraining an affine image mapping to be a strict contraction can be a drawback as far as image quality is concerned. *Eventual* contractivity imposes weaker constraints on T – that is, greater freedom in parameter choices – than does strict contractivity. Thus, we might in principle derive slacker constraints on the grey-tone scaling coefficients when taking into consideration that eventual contractivity is all we need.

This again implies that the distance between the original image and the *collage* can be made smaller if eventually contractive mappings are allowed. As concerns the attractor, the Collage Theorem bound on its distance to the original image essentially depends on three factors, one which is a function of the Lipschitz factor s_L of the mapping T, one which is a function of the contractivity factor s_K of T^K, and one which is a function of the quality of the collage Tx. If the latter factor can be made much smaller by slightly increasing one of the first two, we might in fact obtain a stricter Collage Theorem bound.

The problem with allowing for eventually contractive mappings within the framework used so far is that it is difficult to find suitable constraints on the linear term such that eventual contractivity is secured. Also, the *speed of convergence* might again decrease when the strict contractivity constraint is relaxed, leading to a more complex decoder. However, the next section introduces a trick that allows us to obtain complete control of the decoder complexity and to significantly reduce it.

8.3 Improving Decoder Convergence

In this section we introduce a modification of the linear term, $\mathbf{L} \to \mathbf{L}_o$, designed to yield a decoding algorithm with *exact convergence in a finite number of iterations*. This is done by applying an additional operator, an *orthogonalization* operator, to each decimated domain block before it is inserted in its proper range position. We will show that this modification – under certain mild constraints – gives:

- an unchanged attractor,

- existence of an integer $K > 1$ such that $\mathbf{L}_o^K = 0$, meaning that exact convergence in a finite number of steps is always secured,[4]

- a number of decoder iterations that is only dependent on the domain and range block sizes,

- decoder convergence that is at least as fast as before (in most cases faster),

- existence of a pyramid-structured decoding algorithm with low computational complexity compared to all related algorithms previously published,

- no constraints necessary on the grey-tone scaling coefficients.

Orthogonalization of Collage Subspace Bases

Many signal processing applications involve decomposing vectors into separate components in a linear space. In such cases it is mostly desirable – for example, for reasons of computational complexity or coding efficiency – to use an *orthogonal* basis. For a given subspace and an *arbitrary* given basis for this space, an orthogonal basis may be derived by applying the well-known *Gram-Schmidt orthogonalization procedure* [51, 76].

In the fractal coding scheme we have described, the second basis vector $\mathbf{b}_2$ for each range block approximation is blockwise dependent on the image we perform the mapping on. Therefore, the bases employed in the mapping T described at the outset of this chapter are not completely orthogonal. We proceed to modify the mapping T to obtain orthogonality.

In the linear algebraic description of the mapping T, the linear term was decomposed as

$$\mathbf{L} = \sum_{n=1}^{N} \beta_2^{(n)} \mathbf{P}_n \mathbf{D}_n \mathbf{F}_n. \tag{8.12}$$

We now want to make all the decimated domain blocks orthogonal to the translation subspace basis vectors. If the decimated domain block does not lie *within* the translation subspace (in which case it should be excluded from the set of useful domain blocks, as it provides no information that cannot be expressed with the translation term alone), this can always be obtained by means of the Gram-Schmidt procedure. This corresponds to left-multiplying each $\mathbf{D}_n \mathbf{F}_n \mathbf{x}$ with an *orthogonalizing matrix*, $\mathbf{O}_n : \mathbb{R}^{B_n^2} \to \mathbb{R}^{B_n^2}$, given as

$$\mathbf{O}_n = \mathbf{I} - \mathbf{b}_1^{(n)} \mathbf{b}_1^{(n)T}, \tag{8.13}$$

where $\mathbf{I}$ is the identity matrix of dimension $B_n^2 \times B_n^2$. The modified linear term can thus be written

$$\mathbf{L}_o = \sum_{n=1}^{N} \alpha_2^{(n)} \mathbf{P}_n \mathbf{O}_n \mathbf{D}_n \mathbf{F}_n \tag{8.14}$$

---

[4]Such an $\mathbf{L}_o$ is termed *nilpotent*.

and the total mapping is expressible as

$$T_o\mathbf{x} = \mathbf{L}_o\mathbf{x} + \mathbf{t}_o, \tag{8.15}$$

where $\mathbf{t}_o$ is the modified translation term, which can be expanded as

$$\mathbf{t}_o = \sum_{n=1}^{N} \left(\alpha_1^{(n)} \cdot \mathbf{P}_n \mathbf{b}_1^{(n)} \right). \tag{8.16}$$

The matrix $\mathbf{O}_n$ can be interpreted as a *projection* onto the *orthogonal complement* of $\text{span}\{\mathbf{b}_1^{(n)}\}$. Such a projection matrix is known to possess a number of useful properties [76]:

1. It is *symmetric*, i.e.,

$$\mathbf{O}_n^T = \mathbf{O}_n. \tag{8.17}$$

2. It is *idempotent*, i.e.,

$$\mathbf{O}_n^2 = \mathbf{O}_n. \tag{8.18}$$

3. It has *unit matrix norm*, i.e.,

$$\|\mathbf{O}_n\|_2 \overset{\text{def}}{=} \max_{\|x\|_2=1} \|\mathbf{O}_n\mathbf{x}\|_2 = 1. \tag{8.19}$$

4. For all vectors $\mathbf{v} \in \mathbb{R}^{B_n^2}$,

$$\mathbf{O}_n\mathbf{v} \perp \mathbf{b}_1^{(n)}. \tag{8.20}$$

We shall now use Properties (8.17) and (8.18) to compute the coefficients $\{\alpha_1^{(n)}\}_{n=1}^N$, $\{\alpha_2^{(n)}\}_{n=1}^N$ corresponding to the orthogonalized collage subspace basis. Let us consider a given range block $\mathbf{x}$, with which there is associated an optimal decimated domain block $\mathbf{b}_2$ (we omit the index n in this derivation). Orthogonalization of $\mathbf{b}_2$ with respect to the translation basis can be written as

$$\tilde{\mathbf{b}}_2 = \mathbf{O}\mathbf{b}_2. \tag{8.21}$$

Thus, $\{\mathbf{b}_1, \tilde{\mathbf{b}}_2\}$ replaces $\{\mathbf{b}_1, \mathbf{b}_2\}$ as the basis for the collage approximation. The resulting collage block $\mathbf{c}$ – which is the same for both bases[5] – can now be expanded as

$$\mathbf{c} = \alpha_1\mathbf{b}_1 + \alpha_2\tilde{\mathbf{b}}_2 \tag{8.22}$$

The normal equations we must solve in order to find the optimal α_1 and α_2 can be written

$$\begin{bmatrix} 1 & 0 \\ 0 & \|\tilde{\mathbf{b}}_2\|^2 \end{bmatrix} \begin{bmatrix} \alpha_1 \\ \alpha_2 \end{bmatrix} = \begin{bmatrix} \langle \mathbf{b}_1, \mathbf{x} \rangle \\ \langle \tilde{\mathbf{b}}_2, \mathbf{x} \rangle \end{bmatrix}, \tag{8.23}$$

which directly yields

$$\alpha_1 = \langle \mathbf{x}, \mathbf{b}_1 \rangle \tag{8.24}$$

and

$$\alpha_2 = \frac{\langle \mathbf{x}, \tilde{\mathbf{b}}_2 \rangle}{\|\tilde{\mathbf{b}}_2\|^2} = \frac{\mathbf{x}^T \mathbf{O} \mathbf{b}_2}{\mathbf{b}_2^T \mathbf{O}^T \mathbf{O} \mathbf{b}_2} = \frac{\mathbf{x}^T \mathbf{O} \mathbf{b}_2}{\mathbf{b}_2^T \mathbf{O} \mathbf{b}_2}. \tag{8.25}$$

---

[5]This is trivial, since $\text{span}\{\mathbf{b}_1, \mathbf{O}\mathbf{b}_2\} = \text{span}\{\mathbf{b}_1, \mathbf{b}_2\}$. We have not changed the collage subspace itself, only its basis.

Expression (8.25) may be expanded as

$$\alpha_2 = \frac{\langle \mathbf{x}, \mathbf{b}_2 \rangle - \alpha_1 \langle \mathbf{b}_1, \mathbf{b}_2 \rangle}{\|\mathbf{b}_2\|^2 - \langle \mathbf{b}_1, \mathbf{b}_2 \rangle^2}, \tag{8.26}$$

which shows that there is no need to *explicitly* perform the orthogonalization of the decimated domain blocks.

The model coefficients for the case of the nonorthogonal basis in Jacquin's mappings are related to α_1, α_2 through the equations

$$\beta_2 = \alpha_2 \tag{8.27}$$

and

$$\beta_1 = \alpha_1 - \langle \mathbf{b}_1, \mathbf{b}_2 \rangle \beta_2 \tag{8.28}$$

It is seen that both β_1 and β_2 vary with both range and domain block. Hence we would have had to compute both coefficients – not just α_2, as is now the case – for each combination of range and domain block, had we not introduced orthogonalization.

Decoder Convergence with Orthogonalization

We will now state and prove the main result of this chapter – how the orthogonalization operator improves the convergence properties of the decoding algorithm. We do this by proving two important properties of the new mapping T_o, under the following constraints:

1. We assume a *quadtree range partition* which allows blocks of size $2^{b_1} \times 2^{b_1}, \ldots, 2^{b_q} \times 2^{b_q}$. The corresponding allowed domain block sizes are $2^{d_1} \times 2^{d_1}, \ldots, 2^{d_q} \times 2^{d_q}$. We assume the ordering $b_1 = b_2 + 1 = \cdots = b_q + q - 1$. The decimation factors used are $2^{d_1 - b_1}, \ldots, 2^{d_q - b_q}$ in both the x- and y-directions, with $d_i > b_i$ for all i. An example of a quadtree range partition with $q = 4$ and $b_1 = 5$ is given in Figure 8.2.

2. The range partition and domain construction is such that every domain block in the image (before the $2^{d_i - b_i}$-decimation) is made up of an integer number of range blocks. In practice this holds for the quadtree partition defined above if every allowed domain address is also a position for a range block on the top level, and if $\min_i d_i \geq \max_i b_i$.

3. The $2^{d_i - b_i}$-decimation is to be done by *pure averaging* over $2^{d_i - b_i} \times 2^{d_i - b_i}$ domain block pixels for each range pixel. If we assume that the vector ordering of pixels is such that pixels that are averaged together during the domain-to-range decimation lie as successive vector elements, we may write the decimator matrix on level i as

$$\mathbf{D}^{(i)} = 2^{2(b_i - d_i)} \cdot \begin{bmatrix} \mathbf{1} & \mathbf{0} & \cdots & \cdots & \mathbf{0} \\ \mathbf{0} & \mathbf{1} & \mathbf{0} & \cdots & \mathbf{0} \\ \vdots & \ddots & \ddots & \ddots & \vdots \\ \mathbf{0} & \cdots & \cdots & \mathbf{0} & \mathbf{1} \end{bmatrix}, \tag{8.29}$$

where $\mathbf{1}$ is a row vector of $2^{2(d_i - b_i)}$ 1's, and $\mathbf{0}$ is a row vector of $2^{2(d_i - b_i)}$ 0's.

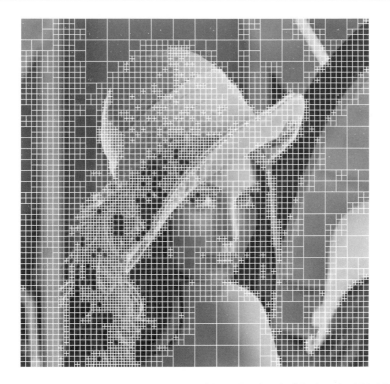

Figure 8.2: Four-level quadtree target block partition allowing for blocks of width 32, 16, 8, and 4 pixels.

It is possible that the following properties can be proved under more general constraints. However, these constraints seem to provide us with sufficient freedom in design, while making the proofs simple.

The properties of T_o are:

Property 1 *T_o and T have the same attractor $\mathbf{x}_T$.*
Proof: Let $\mathbf{x}$ be the original image to be coded. The *collage* of $\mathbf{x}$ with respect to T is by definition

$$\mathbf{x}_c = \mathbf{L}\mathbf{x} + \mathbf{t} = \mathbf{L}_o\mathbf{x} + \mathbf{t}_o \tag{8.30}$$

while the *attractor* of T is

$$\mathbf{x}_T = \mathbf{L}\mathbf{x}_T + \mathbf{t}. \tag{8.31}$$

We now define the *attractor error*

$$\mathbf{e} = \mathbf{x} - \mathbf{x}_T \tag{8.32}$$

and the *collage error*

$$\mathbf{e}_c = \mathbf{x} - \mathbf{x}_c. \tag{8.33}$$

Using Equations (8.30)– (8.33) it is easy to show that $\mathbf{e}$ and $\mathbf{e}_c$ are related by the equation

$$\mathbf{e} = \mathbf{e}_c + \mathbf{L}\mathbf{e}. \tag{8.34}$$

By expanding this equation we obtain

$$\mathbf{e} = \mathbf{e}_c + \mathbf{L}\mathbf{e}_c + \mathbf{L}^2\mathbf{e}_c + \mathbf{L}^3\mathbf{e}_c + \cdots. \tag{8.35}$$

If T_o is used instead of T then we have the attractor error

$$\mathbf{e}_o = \mathbf{e}_c + \mathbf{L}_o\mathbf{e}_c + \mathbf{L}_o^2\mathbf{e}_c + \mathbf{L}_o^3\mathbf{e}_c + \cdots. \tag{8.36}$$

Now, since the collage is ℓ^2-optimized, the content of $\mathbf{e}_c$ over each range position is orthogonal to $\mathbf{b}_1$. Thus it has zero DC component. Due to the second constraint above this also holds for each domain block, since a sum of zero sums is itself zero. Applying either of the linear terms $\mathbf{L}$ or $\mathbf{L}_o$ to an image in which all the domain blocks have a zero DC component will not introduce a nonzero DC component in the decimated domain blocks: The averaging decimator (constraint 3) does not change the DC component in the domain blocks – it just sums and then weighs all pixels by the same factor. Hence there is no DC component to remove in any of the decimated domain blocks, so the action of $\mathbf{L}_o$ and $\mathbf{L}$ *is the same* on the collage error signal. Thus the two series expansions above yield equal results in this case, i.e., $\mathbf{e} = \mathbf{e}_o$, which implies that the attractors are equal. ∎

Comment. When coefficient quantization is introduced, the above conclusion will have to be modified somewhat, but not dramatically. As long as the DC quantization is "fine enough," there is no change in practice.

Thus the introduction of orthogonalization does not affect the quality of the decoded attractor in any negative way. We proceed to prove that the attractor can be reached in very few iterations.

Property 2 $\mathbf{x}_T$ *can always be reached exactly in a finite number of iterations of T_o from an arbitrary initial image.*
Proof: The attractor $\mathbf{x}_T$ can be written

$$\mathbf{x}_T = \sum_{k=0}^{\infty} \mathbf{L}_o^k \mathbf{t}_o \tag{8.37}$$

if this series converges. Since $\mathbf{L}_o$ orthogonalizes all decimated domain blocks with respect to the DC component before placing them in their proper place, $\mathbf{L}_o\mathbf{t}_o$ has range-sized blocks of zero DC. Hence, again due to the second of the three constraints stated above, the domain blocks in this image also have zero DC. Since $\mathbf{L}_o$ also shrinks $2^{d_i} \times 2^{d_i}$ blocks to size $2^{b_i} \times 2^{b_i}$, $\mathbf{L}_o^2\mathbf{t}_o$ has zero DC blocks of size $\frac{2^{d_i}}{2^{d_i}/2^{b_i}} \times \frac{2^{d_i}}{2^{d_i}/2^{b_i}}$, and generally $\mathbf{L}_o^k\mathbf{t}_o$ consists of zero DC blocks of size $\frac{2^{b_i}}{(2^{d_i}/2^{b_i})^{k-1}} \times \frac{2^{b_i}}{(2^{d_i}/2^{b_i})^{k-1}}$ pixels. When this size becomes less than or equal to 1×1 pixel for all i, then *every pixel in the image $\mathbf{L}_o^k\mathbf{t}_o$ must be zero.* After some simple manipulations we thus obtain that $\mathbf{L}_o^k$ is zero for all k greater than or equal to

$$K = 1 + \max_{i \in \{1,\dots,q\}} \left\lceil \frac{b_i}{d_i - b_i} \right\rceil, \tag{8.38}$$

where $\lceil u \rceil$ denotes "the smallest integer larger than u."

This means that the attractor is simply given as

$$\mathbf{x}_T = \sum_{k=0}^{K-1} \mathbf{L}_o^k \mathbf{t}_o \tag{8.39}$$

with K as given by Equation (8.38). Hence, we have proved that the introduction of orthogonalization secures decoder convergence in a finite number of steps that is *only a function of the factors d_i and b_i.* ∎

Comment 1. The above conclusion holds also when coefficient quantization is introduced. This is because the elements of the orthogonalizing matrix $\mathbf{O}$ are image-independent and not part of the image code, and need not be quantized.

Table 8.1: Decoder expression for some choices of b and d.

b	d	$\mathbf{x}_T$	K
2	3	$\sum_{k=0}^{2} \mathbf{L}_o^k \mathbf{t}_o$	2
2	4	$\sum_{k=0}^{1} \mathbf{L}_o^k \mathbf{t}_o = T_o \mathbf{t}_o$	1
3	4	$\sum_{k=0}^{3} \mathbf{L}_o^k \mathbf{t}_o$	3
3	5	$\sum_{k=0}^{2} \mathbf{L}_o^k \mathbf{t}_o$	2
3	6	$\sum_{k=0}^{1} \mathbf{L}_o^k \mathbf{t}_o = T_o \mathbf{t}_o$	1
4	5	$\sum_{k=0}^{4} \mathbf{L}_o^k \mathbf{t}_o$	4

Table 8.1 shows the expression (8.39) for $\mathbf{x}_T$ for uniform range partitions ($b_i = b$ for all i) and all domain blocks of equal size ($d_i = d$ for all i). From the table we also see that if the domain blocks are large enough compared to the range blocks, the decoding in fact becomes *noniterative*; it is simply performed by applying the mapping T_o once to the image $\mathbf{t}_o$. This fact has previously been published in [54]; the above is a direct generalization of the example in that paper.

Comment 2. It is worth noting that the above convergence result is obtained *without any constraints whatsoever* on the absolute values of the grey-tone scalings. This implies that we are free to optimize the collage in an unconstrained manner, without having to worry about violation of the contractivity property. Figure 8.3 shows an example of a typical distribution of ℓ^2-optimal grey-tone scaling factors. It is seen that amplitudes up to about 8–10 may occur, which makes it clear that the freedom to use large grey-tone scalings is useful also in practice. Also, note that we may now use the unconstrained distribution of the grey-tone scaling coefficient to design an *optimal quantization strategy* for this parameter, instead of merely postulating a quantizer characteristic that fulfills the strict contractivity constraint [64]. We also mention that the introduction of the orthogonalization operator *decorrelates* the coefficients α_1, α_2, thus making the setting more optimal for scalar quantization [64].

Choice of Initial image

Iteration from an *arbitary* initial image $\mathbf{x}_0$ will produce the iterate sequence $\{\mathbf{x}_0, \mathbf{x}_1 = \mathbf{L}_o\mathbf{x}_0 + \mathbf{t}_o,$ $\mathbf{x}_2 = \mathbf{L}_o^2\mathbf{x}_0 + \mathbf{L}_o\mathbf{t}_o + \mathbf{t}_o, \ldots, \mathbf{x}_j = \sum_{k=0}^{j-1} \mathbf{L}_o^k\mathbf{t}_o + \mathbf{L}_o^j\mathbf{x}_0, \ldots\}$. The last term in the $(j + 1)$-th image

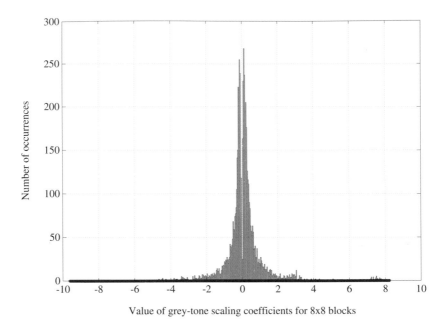

Figure 8.3: Histogram for grey-tone scaling coefficients.

$\mathbf{x}_j$ in the sequence is here zero for $j \geq K$ as given by Equation (8.38). In the special case of $\mathbf{x}_0 = \mathbf{t}_o$, however, this occurs already in the k-th image in the sequence. Hence, we need *one additional iteration* if we start from any initial image other than $\mathbf{t}_o$. Therefore, $\mathbf{t}_o$ must be deemed the best choice of initial image in a practical implementation.

Comparisons and Experiments

The iteration procedure will always converge at least as fast with orthogonalization as without. If we start with constant grey-tone blocks of size $B \times B$ and decimate them by a factor $\frac{D}{B}$ in each iteration, image details the size of one pixel cannot be created in fewer steps than we have outlined here – much less, *correct* details. In practice, iteration of T_o converges much faster than iteration of T, for most cases. Use of T may give wildly image-varying convergence rates. The two examples in Table 8.2 illustrate this. Here I_o denotes the number of iterations with T_o, and I_{no} the number of iterations with T. (Without orthogonalization, the chosen criterion for convergence in this experiment was a change in PSNR (original, iterate) that was less than 10^{-2} dB.). Both experiments were conducted with a 2-level quadtree range block partition, with $\frac{D}{B} = 2$, pure averaging in the decimation and range blocks of size 8×8 and 4×4 pixels. The initial image was the translation term in both cases, so we know in advance that I_o should be $K - 1 = 3$. Unconstrained ℓ^2-optimal coefficients were used in both cases.

We see that the number of iterations of T_o is consistent with the theory that we have just

Table 8.2: Decoder iterations when applying T_o and T to two sample images.

Image (512 × 512 pixels)	I_o	I_{no}
Lenna	3	5
Kiel Harbor	3	49

developed. For the mapping T, the number of iterations is heavily image dependent, and may be very high. In general, the tendency here seems to be that images with high activity (more small details) need more iterations than images with low activity. Jacquin and Fisher have both reported that even with contractivity-constrained grey-tone scalings, 8 to 10 iterations are needed for most images [45, 27].

Figure 8.4 depicts excerpts of the iterates of the Kiel Harbor image when T_o is applied. The figure demonstrates the fact that when $\mathbf{t}_o$ is used as an initial image (iterate 0), iterate k has correct DC over square blocks of size $\frac{B_i}{(D_i/B_i)^k} \times \frac{B_i}{(D_i/B_i)^k}$ pixels. The translation term $\mathbf{t}_o$ gives the iteration a "flying start" by providing the correct DC component over each range block, and each succeeding term adds new detail on a finer scale (high-frequency content in each block) while leaving the DC over the areas that were uniform in the previous iterate unchanged. This can be attributed to the fact that the succeeding terms are *orthogonal* to each other: $\mathbf{t}_o^T (\mathbf{L}_o^k)^T \mathbf{L}_o^j \mathbf{t}_o = 0$ for all $j \neq k$. This is easily realized by noting that areas that are uniform in one term in the sum (8.39) are DC-less in all higher terms.

Decoder Implementation and Complexity

The computational complexity of the decoder is of vital importance, because in many applications the encoding is done once and for all, while the decoding is to be repeated many times, often in man-machine interaction situations – or even in real-time applications. This is typical in the generation and use of compressed image data bases, and in processing of video sequences, respectively. In such situations, it is of utmost importance that the decoding can be done fast. Although we have seen that fractal coding can be done noniteratively in a special case, there is no guarantee that the restrictions inherent in that particular implementation (very large domain blocks compared to range blocks) will always give satisfactory coding results. Experiments indicate that for finely detailed images such as "Kiel Harbor," the noniterative solution is as good – or even better – than those obtained for the other cases, but for smoother images such as "Peppers" or "Lenna" it performs slightly worse [64]. We therefore seek a fast implementation of the decoding algorithm within the *most general* restrictions stated in this chapter.

The decoding algorithm produces iterates $\mathbf{t}_o, T_o\mathbf{t}_o, \ldots, T_o^{K-1}\mathbf{t}_o$ that consist of blocks that are *constant-valued* over $2^{b_i} \times 2^{b_i}$ pixels, $\frac{2^{b_i}}{2^{d_i}/2^{b_i}} \times \frac{2^{b_i}}{2^{d_i}/2^{b_i}}$ pixels, $\ldots$, 1×1 pixels, respectively. *But this implies that the iterates may be subsampled by factors* $\min_i 2^{b_i} \times 2^{b_i}$, $\max \left\{ \min_i \frac{2^{b_i}}{2^{d_i}/2^{b_i}} \times \frac{2^{b_i}}{2^{d_i}/2^{b_i}}, 1 \times 1 \right\}, \ldots, 1 \times 1$ *without loss of information.* Hence, the k-th iteration may be performed on an image of $\min \left\{ \max_i \frac{(2^{2d_i}/2^{2b_i})^{k-1}}{2^{2b_i}}, 1 \times 1 \right\}$ of the original size. The resulting decoder structure is depicted in Figure 8.5, for the example of an image with uniform range partition $B = 8$ and $D = 16$.

Comment. Note that the use of the name "T_o" for the mappings applied in the pyramid structure

Figure 8.4: Successive iterates for the Kiel Harbor image. PSNRs are 23.29 dB between 0th (upper left) and 1st (upper right) iterate, 29.42 dB between 1st and 2nd (lower left), and 43.53 dB between 2nd and 3rd (lower right).

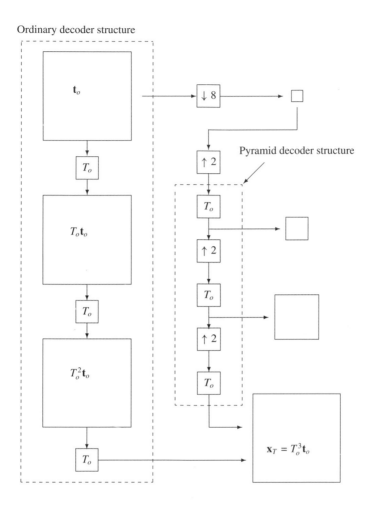

Figure 8.5: Ordinary and pyramid decoder structures.

in Figure 8.5 is not really mathematically correct, as these mappings operate on images of different sizes than the one T_o operates on. However, they perform exactly the *same type of operations* on the images on which they operate. With these reservations, there hopefully should not occur any confusion because of our lack of mathematical precision in the naming convention.

Computational Complexity

One iteration on an image of the *original* size can be performed with approximately one multiplication and three additions per pixel [53]. Hence the number of multiplications per pixel in the above "pyramid" decoder structure is

$$m_d = \sum_{k=1}^{K-1} \min \left\{ \max_i \frac{(2^{2d_i}/2^{2b_i})^{k-1}}{2^{2b_i}}, 1 \right\} \tag{8.40}$$

with K given by Equation (8.38), while the number of additions is approximately three times as high. Table 8.3 shows the resulting number of arithmetic operations for a uniform range partition, for various choices of b and d. It is seen that the decoder complexity in these examples is comparable to that of the fastest established compression techniques for all choices.

Table 8.3: Pyramid decoder complexity for some choices of b and d.

b	d	Multiplications per pixel	Additions per pixel
2	3	1.25	3.75
2	4	1	3
3	4	1.3125	3.9375
3	5	1.25	3.75
3	6	1	3

8.4 Collage Optimization Revisited

For a mapping with orthogonalization, we can obtain a new Collage Theorem. Let us again assume an ℓ^2-optimized model. Then, $\mathbf{e}_c$, and hence $\mathbf{e}$, are DC-less over each range block even *before* we start applying $\mathbf{L}_o$. Hence, for these particular images the attractor series expansion will have only $K - 1$ terms instead of K. With this in mind, we get

$$
\begin{aligned}
d_B(\mathbf{x}, \mathbf{x}_T) &= \|\mathbf{x} - \mathbf{x}_T\| \\
&\leq \|\mathbf{x} - T_o\mathbf{x}\| + \|T_o\mathbf{x} - \mathbf{x}_T\| \\
&= \|\mathbf{x} - T_o\mathbf{x}\| + \|\mathbf{e} - \mathbf{e}_c\| \\
&= \|\mathbf{e}_c\| + \left\| \sum_{k=1}^{K-2} \mathbf{L}_o^k \mathbf{e}_c \right\|
\end{aligned}
$$

$$\leq \left(\sum_{k=0}^{K-2} \|\mathbf{L}_o\|_2^k \right) \cdot d(\mathbf{x}, T_o\mathbf{x}), \qquad (8.41)$$

which is our new Collage Theorem bound. For a uniform range partition, with nonoverlapping domain blocks, this Collage Theorem bound can be explicitly calculated, as $\|\mathbf{L}_o\|_2$ in this case is given by (see Chapter 7)

$$\|\mathbf{L}_o\|_2 = \sqrt{\frac{B}{D} \cdot \max_{l \in \mathcal{L}} \sum_{n \in I_l} (\alpha_2^{(n)})^2}. \qquad (8.42)$$

Figure 8.6 depicts the bound Equation (8.41) gives on $\rho = \frac{d_B(\mathbf{x}, \mathbf{x}_T)}{d_B(\mathbf{x}, T_o\mathbf{x})}$, as a function of the matrix norm $\|\mathbf{L}_o\|_2$ for various $K > 1$. From the figure it is seen that the bound grows fast in $\|\mathbf{L}_o\|_2$ if $K > 2$. Thus, the Collage Theorem does not really justify the optimization of the collage for other mappings than those with a small Lipschitz factor and a small K, although practical results show that better results can be obtained by allowing for larger Lipschitz factors, and maybe also larger K. In [28] it was demonstrated through practical encodings that the gap between the Collage Theorem bounds and the actual distance between collage and attractor can be very large – indeed, the best coding result was there provided by a mapping with an extremely pessimistic Collage Theorem bound.

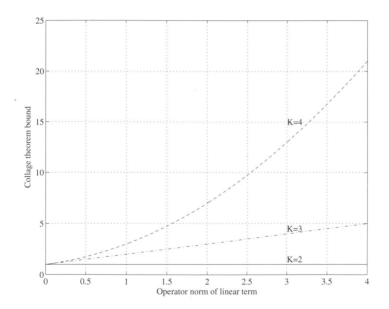

Figure 8.6: Collage theorem bound as a function of the norm of $\mathbf{L}_o$ for various K.

Hence, since the bounds given by the Collage Theorem are very pessimistic for many cases of practical interest, the theorem should be viewed mainly as an initial motivation. It is desirable to find a better justification of why the collage quality is used as an optimization criterion. In [64], a *statistically* based argument on the relationship between the attractor and the collage is given. Space does not permit reproducing this argument in its entirety here, but it basically says that the difference between the attractor and the collage in an arbitrary pixel can be interpreted as an *estimator of the expected value of the collage error* in this pixel, which is *zero* for ℓ^2-optimized code and a blockwise DC translation term. The estimator is unbiased, and has lower variance the more domain block pixels the linear term averages over to obtain each range pixel. Thus, when an averaging decimator is used, the collage-attractor difference is closer to zero the higher the domain-range decimation ratio, and the attractor converges towards the collage as we increase this ratio. Of course, this does not say anything about the quality of the collage, but it indicates that using this quality as an optimization criterion in the encoder is more valid the larger the decimation factor is. If equality between attractor and collage is the only concern, one should use a large ratio $\frac{D}{B}$.

The above argument is supported by experimental results [59, 64]. The absolute difference in PSNR between original and collage, and PSNR between original and attractor, has been measured – for different values of $\frac{D}{B}$ – on the image "Lenna." Decimation by averaging and unconstrained grey-tone scaling coefficients was used in all experiments,[6] and the range block size was held constant at 8×8 pixels. The results are collected in Figure 8.7.

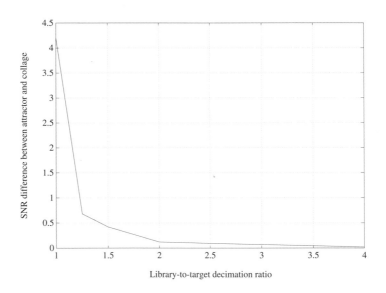

Figure 8.7: Example of the influence of $\frac{D}{B}$ on the degree of collage-attractor equality.

As is seen from this figure, the difference in quality in collage and attractor, as measured through the PSNR, monotonically and rapidly decreases as $\frac{D}{B}$ is increased. This effect, also observed for other images, supports the conclusion drawn from the analysis above. In fact the

---

[6]For the cases when $\frac{D}{B}$ is not an integer, the sub-pixel averaging scheme of Jacquin was used [45].

tendency toward equality between attractor and collage is even stronger than what we could predict by means of statistics.

Comment. When $D = B$, it is in fact easy to see that the attractor is simply $\mathbf{x}_T = \mathbf{t}_o$. In this case every domain block in $\mathbf{t}_o$ is also a block in the translation space, and is thus set to zero by $\mathbf{O}$.

Direct Attractor Optimization

There exists a special case in which both the Collage Theorem and the statistical analysis referred to above become redundant. This solution, first discovered by Lepsøy [53, 54], provides *direct attractor optimization* in the encoder. This is a simple consequence of the fact that the encoding algorithm produces a collage error which is blockwise orthogonal to the translation space, and of specific constraints made on the domain-to-range decimation factor. We illustrate this by a simple example.

Let the constraints stated earlier in this chapter be fulfilled. It can be shown that the attractor error $\mathbf{e}$ in pixel m can be written as

$$e_m = e_{c,m} + \sum_{j \in \mathcal{I}_m} \ell_{mj} e_j, \tag{8.43}$$

where $e_{c,m}$ is the collage error $\mathbf{e}_c$ in pixel m, $\mathcal{I}_m$ is the set of domain pixels used in obtaining range pixel m (the *averaging mask*), and $\{\ell_{mj}\}$ are the corresponding linear term elements (multipliers).

By substituting this expression back into itself once, we obtain

$$e_m = e_{c,m} + \sum_{j \in \mathcal{I}_m} \ell_{mj} e_{c,j} + \sum_{j \in \mathcal{I}_m} \ell_{mj} \left(\sum_{k \in \mathcal{I}_j} \ell_{jk} e_k \right). \tag{8.44}$$

Doing similar back-substitutions over and over again will produce a series that contains only $e_{c,n}$ plus linear combinations of terms of the form $\sum_{k \in \mathcal{I}_j} \ell_{jk} e_{c,k}$ for various j. Now, suppose the averaging mask for each pixel *is itself a range block* in the image. In this case, $I = D = B \times B$ and

$$\sum_{k \in \mathcal{I}_j} \ell_{jk} e_{c,k} = \langle \mathbf{l}_j, \mathbf{e}_c(\mathcal{I}_j), \rangle \tag{8.45}$$

where $\mathbf{l}_j$ is the relevant set of elements in $\mathbf{L}_o$ ordered as a B^2-vector, and $\mathbf{e}_c(\mathcal{I}_j)$ is the collage error restricted to the $B \times B$ filter mask considered. Furthermore, we have that $\ell_{jk} = \frac{\alpha_2^{(j)}}{B^2}$, since we have let the decimator be a pure averaging filter. Then,

$$\langle \mathbf{l}_j, \mathbf{e}_c(\mathcal{I}_j) \rangle = \frac{\alpha_2^{(j)}}{B^2} \cdot \sum_{k \in \mathcal{I}_j} e_{c,k}. \tag{8.46}$$

But, since we have orthogonalized the error with respect to the DC component, $\sum_{k \in \mathcal{I}_j} e_{c,k} = 0$ in this case. Hence, $e_m = e_{c,m}$ for all pixels m, which implies that the attractor and the collage are identical.

Comment 1. The above result does *not* assume a linear part with orthogonalization, but *if* we choose the orthogonal basis solution, we also obtain a *noniterative* decoder, as shown in Table 8.1. We may also point out that the Collage Theorem bound given by Equation (8.41) is in this case fulfilled with *equality*: $d_B(\mathbf{x}, \mathbf{x}_T) = d_B(\mathbf{x}, T_o\mathbf{x})$.

Comment 2. When *quantized* coefficients are used, the above proof no longer holds exactly, since it exploits the fact that the linear term in this case operates in the same way on the original image $\mathbf{x}$ and on its exact projection onto the translation subspace. After quantization, the translation component $\mathbf{t}_o$ no longer is this exact projection. But the attractor may still be written as $\mathbf{x}_T = \mathbf{L}_o\mathbf{t}_o + \mathbf{t}_o$, and since the fully quantized $\mathbf{t}_o$ is available when we start optimizing $\mathbf{L}_o$, we may perform the optimization directly on this image instead of on the original $\mathbf{x}$. If we do that, we still optimize the same image in the encoder that is synthesized in the decoder. However, the term "collage of $\mathbf{x}$" is no longer an accurate name for this image – it is instead the "collage of $\mathbf{t}_o$."

8.5 A Generalized Sufficient Condition for Fast Decoding

It should be mentioned that a more general way of expressing a sufficient condition for decoding with finitely many iterations has also been expressed. First, note that for *any* affine mapping $T\mathbf{y} = \mathbf{L}\mathbf{y} + \mathbf{t}$, the decoder expression used can be written

$$\mathbf{x}_T = \sum_{k=0}^{\infty} \mathbf{L}^k\mathbf{t} = \mathbf{t} + \mathbf{L}\mathbf{t} + \mathbf{L}^2\mathbf{t} + \cdots. \tag{8.47}$$

Therefore, a sufficient condition for decoding with finitely many iterations is simply that there exists some K such that $\mathbf{L}^K\mathbf{t} = 0$. In our case, the affine mapping may be rewritten as

$$T_o\mathbf{y} = \mathbf{L}_o\mathbf{y} + \mathbf{t}_o = \mathbf{SOB}\mathbf{y} + (\mathbf{I} - \mathbf{O})\mathbf{x}, \tag{8.48}$$

where $\mathbf{y}$ is arbitrary and $\mathbf{x}$ is the original image the mapping was optimized with respect to; $\mathbf{B} : \mathbb{R}^M \rightarrow \mathbb{R}^M$ fetches all the individual domain blocks, decimates them, and inserts them in its proper range position; $\mathbf{O} : \mathbb{R}^M \rightarrow \mathbb{R}^M$ performs orthogonalization of the individual range-sized blocks; and $\mathbf{S} : \mathbb{R}^M \rightarrow \mathbb{R}^M$ is a diagonal matrix performing blockwise scalar multiplication.

Now, suppose that there exists some integer K such that

$$\mathbf{L}_o^K = \mathbf{L}_o^K(\mathbf{I} - \mathbf{O}). \tag{8.49}$$

The operator $\mathbf{P} = \mathbf{I} - \mathbf{O}$ is simply the projection onto the translation subspace, so condition (8.49) simply means that $\mathbf{L}_o^K$ "sees" only the translation component of any signal on which it operates. This implies that for any image $\mathbf{y}$,

$$\mathbf{L}_o^K(\mathbf{I} - \mathbf{O})\mathbf{y} = \mathbf{L}_o^K\mathbf{y} \tag{8.50}$$

or, equivalently,

$$\mathbf{L}_o^K\mathbf{O}\mathbf{y} = 0 \tag{8.51}$$

Figure 8.8: Excerpt from coding (top) and original (bottom) of "Boat," for a two-level quadtree partition. Overall PSNR= 30.80 dB. Overall bit rate = 0.66 bpp.

which simply states that *the range of* **O** *lies in the nullspace of* $\mathbf{L}_o^K$. Moreover, note that $\mathbf{SOy} = \mathbf{OSy}$ for any $\mathbf{y}$, since $\mathbf{S}$ only amounts to a blockwise scalar multiplication. This implies that for any positive integer i,

$$
\begin{aligned}
\mathbf{L}_o^{K+i}\mathbf{t}_o &= \mathbf{L}_o^{i-1}\mathbf{L}_o^K\mathbf{L}_o\mathbf{x} \\
&= \mathbf{L}_o^{i-1}\mathbf{L}_o^K\mathbf{OSBx} \\
&= \mathbf{L}_o^{i-1} \cdot 0 = 0
\end{aligned}
\tag{8.52}
$$

Thus the decoder series expansion has only K terms, as we desired. The theory derived earlier in this chapter is but one way of implementing the sufficient condition $\mathbf{L}_o^K = \mathbf{L}_o^K\mathbf{P}$; other ways may also exist.

8.6 An Image Example

In Figure 8.8, a coding result for the image "Boat" is depicted. Here, $\frac{D}{B} = 2$, and a 2-level quadtree partition with 8×8 and 4×4 blocks has been used. For the grey-tone scaling, pdf-optimized scalar quantization [50, 64] with 5 bits has been used, and 8-bit uniform quantization has been used for the DC component. The complexity reduction algorithm outlined in Chapter 9 has been applied during the encoding.

Comments. The overall visual appearance of the coded image is very good. The most disturbing degradations are the "staircase" effects in some of the masts, and the partial disappearance of some thin ropes. The overall level of detail accuracy is high – for example, the lettering on the back of the boat is clearly legible after coding.

8.7 Conclusion

The contributions described in this chapter serve to further emphasize one of the strongest points of fractal coding, namely, the efficiency of the decoding scheme. This may be of vital importance if the method is to become a viable compression algorithm in the future. At present, the quality-versus-compression ratio of fractal coding does not outperform the state-of-the-art standard methods such as subband coding or transform coding. However, its efficiency of decoder implementation (as well as its multiresolution properties, which have not been discussed here, see Chapter 5) may still justify it being the method of choice in applications where the encoding is performed off line, and decoding speed is of the essence. It remains to be seen how widely used it will be, since the technique is still a young one. It is likely that its full potential has not yet been explored.

Acknowledgments

The authors would like to express their thanks to the following:

Professor Tor A. Ramstad, the Norwegian Institute of Technology, for his enthusiastic, inspirational, and knowledgeable guidance, and for his abundance of ideas.

Dr. Lars Lundheim, Trondheim College of Engineering, and Dr. Arnaud Jacquin, AT&T Bell Labs, for the inspiration they have given us through their groundbreaking work on fractal signal modeling and coding.

Dr. Yuval Fisher, University of California, San Diego, for fruitful discussions and clarifying comments, and for giving us the opportunity to participate in this book.

Zachi Baharav, Technion, for eagerness to cooperate, infectious enthusiasm, shared insights, and discussions.

All the graduate students we have collaborated with at the Norwegian Institute of Technology, and who have contributed many valuable and useful results.

The Royal Norwegian Council for Scientific Research (NTNF), for generous financial support during our Dr. Ing. studies at the Norwegian Institute of Technology.

Chapter 9

Fast Attractor Image Encoding by Adaptive Codebook Clustering

S. Lepsøy and G. E. Øien

In attractor image compression systems, the encoders are often computationally intensive. An attractor encoder involves comparisons between two sets of blocks, called *range* blocks and *domain* blocks. The most basic type of encoder compares these sets exhaustively in the sense that every possible pair of range and domain domain blocks is examined. Unless the complexity of this task can be reduced, attractor coding may be excluded from many applications. This problem is recognized, and previous publications have presented strategies to overcome it.

Jacquin [45, 46, 48] employed a scheme for classifying domain blocks and range blocks. For each range block, only the domain blocks within the corresponding class are considered. The classes and classification schemes were introduced by Ramamurthi and Gersho [72]. The classes are visually quite distinct: blocks with edges, smooth blocks, and blocks with some detail but no dominant orientation. Fisher, et al. [27] used a similar scheme with more classes, also chosen to distinguish between blocks of different appearance. These methods of block classification and fast encoding are related to *classified vector quantization* (CVQ) and *tree-structured vector quantization* (TSVQ). CVQ was proposed by Ramamurthi and Gersho as a method of image coding, aimed at preserving visually important properties like edges in blocks [72]. TSVQ was first proposed by Buzo *et al.* in the context of speech coding [14].

Block classification methods for attractor image coding have usually relied on heuristically chosen criteria. If more analytical or perhaps adaptive methods can be found, better control of the involved trade-offs (encoding time, number of classes, image quality) may be possible. This chapter presents a theory and technique for block classification, in which the definition of the classes is adapted to the original image. The adaptation attempts to maximize the attainable image quality while keeping the total complexity of adaptation, classification and subsequent encoding low.

9.1   Notation and Problem Statement

In the encoder, a block in the original image is to be approximated by a *collage* block. The collage block is a linear combination of two or several basis blocks, of which one is extracted from the original image itself and the rest are fixed, known both in the encoder and the decoder.

Let the original image be denoted by x_o and the n-th block in this image be denoted by $[x_o]_n$. The block in the original image is called a *range* block. Let the collage image be denoted by x_c and its n-th block be denoted by $[x_c]_n$. The collage block can be expressed as

$$[x_c]_n = \alpha_{1,n} b_{k(n)} + \alpha_{2,n} \xi_2 \left(+ \sum_{i=3}^{D} \alpha_{i,n} \xi_i \right).$$

The α's are scalar coefficients, $b_{k(n)}$ is the block extracted from the original image, and the ξ's are the fixed basis blocks.

The block b_k is usually produced by a linear operation that copies a *domain* block (larger than the range block) from somewhere in the image, shrinks it and shuffles that pixels around. This corresponds to the original scheme by Jacquin. There will be a set of such operations to choose from, yielding a *codebook* of blocks

$$b_k \in \{B_1 x_o, \ldots, B_K x_o\},$$

where each operator B_k produces a block.

Encoding

A range block is encoded by projecting it tentatively onto all the subspaces offered by the codebook and the fixed basis [66]. That is, it is projected onto the subspaces spanned by $\{b_k, \xi_2, \ldots, \xi_D\}$ for $k = 1, \ldots, K$. The resulting coefficients are quantized, and the codebook block that yields the smallest error is chosen. The code for one range block is thus

$$\{k(n), \alpha_{1,n}, \ldots, \alpha_{D,n}\},$$

where $k(n)$ indicates the choice of codebook block and the α's are the quantized coefficients.

Clustering

The codebook is to be divided into subsets or *clusters*. This organization of the codebook permits efficient encoding. The general procedure is:

1. **Initialization.** Subdivide the codebook into clusters $D_1, \ldots, D_M$, by computing *cluster centers* $c_1, \ldots, c_M$ and grouping the codebook blocks around the centers. (The centers have the same size as the blocks.)

2. **Encoding.** For each original block: find the center that most resembles the original; this determines a cluster. Within this cluster, find the codebook block that most resembles the original.

To use such a procedure, we need a function to measure how much blocks resemble each other, we need a function that expresses how good a set of cluster centers is, and we need a procedure for finding good cluster centers. Each of these problems is treated in this chapter.

9.2 Complexity Reduction in the Encoding Step

A set of cluster centers implies a clustering of the original blocks as well as of the codebook blocks (cluster X_m of original blocks consists of the blocks that most resemble cluster center c_m). The sizes of the clusters control the number of block comparisons that must be made during the encoding. In this section, we shall see how these sizes influence the encoding complexity.

Let us introduce the following notation:

N	-	The number of original blocks.
K	-	The number of codebook blocks.
M	-	The number of cluster centers.
X_m	-	Cluster number m of original blocks.
$\sharp X_m$	-	The number of blocks in X_m.
$\phi_m = \sharp X_m / N$	-	Size of X_m relative to the total number of original blocks.
D_m	-	Cluster number m of codebook blocks.
$\sharp D_m$	-	The number of blocks in D_m.
$\gamma_m = \sharp D_m / K$	-	Size of D_m relative to the total number of codebook blocks.

For the encoding of an image, the total number of comparisons to find optimal clusters is NM, and the total number of comparisons to find optimal codebook blocks is

$$\sum_{m=1}^{M} \sharp X_m \sharp D_m = NK \sum_{m=1}^{M} \phi_m \gamma_m. \tag{9.1}$$

Note that the relative cluster sizes satisfy

$$\sum_{m=1}^{M} \phi_m = 1; \quad \phi_m \geq 0 \quad \text{and} \quad \sum_{m=1}^{M} \gamma_m = 1; \quad \gamma_m \geq 0.$$

From Equation (9.1), the number of comparisons to find optimal codebook blocks can be made zero, by letting the vectors

$$\phi = \begin{bmatrix} \phi_1 \\ \vdots \\ \phi_M \end{bmatrix}, \gamma = \begin{bmatrix} \gamma_1 \\ \vdots \\ \gamma_M \end{bmatrix}$$

be orthogonal. This is, of course, meaningless, as it implies that there will be no codebook vectors in any chosen cluster. Therefore, we must demand that no codebook cluster is empty, i.e., that all $\gamma_m > 0$. If all clusters do contain codebook vectors but ϕ and γ are still nearly orthogonal, the effective codebook size is very small and the coding results will be accordingly bad. The vectors ϕ and γ should thus be far from orthogonal.

If these vectors are parallel, the number of block comparisons may vary between a minimum discussed below and a maximum close to the case of full codebook search. The maximum occurs when almost all vectors are grouped in one cluster. Intuitively, the relative sizes of X_m and D_m

should be nearly equal to obtain good coding results.[1] This would reflect a certain similarity in the way original and codebook blocks are distributed, as measured by the criterion function for codebook search. Let us suppose that the relative sizes are equal,

$$\phi = \gamma,$$

and see which ϕ minimizes the number of comparisons. This is equivalent to minimizing $\phi \cdot \phi = \|\phi\|_2^2$ under the constraints that ϕ is in the first quadrant and that $\|\phi\|_1 = 1$. The solution is that all clusters must be equally large

$$\phi_m = \frac{1}{M}; \quad m = 1, \ldots, M.$$

Actually, the number of block comparisons is the same even if this only holds for one of the cluster size distributions. If all $\phi_m = 1/M$, the number of comparisons is independent of γ_m because

$$\sum_{m=1}^{M} \frac{1}{M} \gamma_m = \frac{1}{M}.$$

We can now make some preliminary conclusions about three cases of cluster size distribution.

Free adjustment of ϕ and γ. If the cluster sizes can be adjusted freely to minimize total search time, almost arbitrary reductions are possible at the cost of worse coding results (collage error).

Uniform size distribution of either original block clusters or codebook block clusters. In this case, the number of block comparisons during the encoding phase becomes roughly $NM + NK/M$ instead of NK. Suppose that we have the following values, which are typical for a 256×256-image segmented into blocks of 4×4 pixels;

$$
\begin{aligned}
N &= 4096 \\
K &= 4000 \\
M &= 20.
\end{aligned}
$$

Then the number of block comparisons is reduced from about $16 \cdot 10^6$ down to about $9 \cdot 10^5$, a reduction factor of about 18.

Equal ϕ and γ. If neither the original nor the codebook blocks are grouped into clusters of uniform size, but still $\phi = \gamma$, smaller reduction factors must be expected than stated above.

This discussion has not considered the amount of computation in the initialization phase. The overall complexity of both initialization and encoding is what counts, hence the number M of clusters must be chosen after the method for initialization has been chosen.

---

[1] Suppose that the original image and the codebook contain the same number of blocks, and suppose that the collage is a perfect copy of the original image. In such a case, any clustering of the codebook would imply a clustering of the original blocks with the same sizes for X_m and D_m. Small deviations from this ideal should in most cases lead to small differences in the sizes of X_m and D_m.

9.3 How to Choose a Block

This section considers the criterion for comparing a range block to a cluster center or a codebook block, i.e., for measuring how well two blocks resemble each other. This "similarity measure" is large when two blocks are nearly parallel and small when the two blocks are nearly orthogonal. After this has been demonstrated, we make the tree-structured encoding plausible by invoking a "triangle inequality" for image blocks.

The methods discussed in this chapter rely on a condition of orthogonalized codebook blocks.[2] By this we mean that the mapping B_k that provides a codebook block $b_k = B_k x_o$ also removes components in the subspace spanned by the fixed basis blocks $\{\xi_i\}_{i=2}^D$. This is accomplished when the provided mapping is a composition of two mappings: a mapping F_k that copies a domain block from somewhere in the image, shrinks it, and shuffles pixels; and an orthogonalizing mapping O that removes components in the subspace of the fixed basis blocks.[3] Thus, $B_k = O \circ F_k$.

Under the given condition, the similarity measure is the squared inner product

$$s(x, b) = \left\langle x, \frac{b}{\|b\|} \right\rangle^2. \tag{9.2}$$

Here, x denotes the range block to be approximated ($[x_o]_n$) and b denotes a codebook block.
Proof: Suppose that the quantization of the coefficients is fine enough to be neglected, i.e., that the coefficients that weigh the basis blocks result from projecting the range block onto each basis block.[4] In the case of the codebook block, the coefficient is

$$\alpha_1 = \frac{\langle x, b_k \rangle}{\|b_k\|^2}. \tag{9.3}$$

The goal is to minimize the collage error. By the assumptions of orthogonality and projection, the collage error for one block is

$$\begin{aligned} e(k, \alpha_1, \alpha_2) &= \|x - \alpha_1 b_k - \alpha_2 \xi\|^2 \\ &= \|x\|^2 - \|\alpha_1 b_k\|^2 - \|\alpha_2 \xi\|^2. \end{aligned} \tag{9.4}$$

The codebook block can thus be chosen to make the second term large, independently of the other terms. By inserting Equation (9.3),

$$\|\alpha_1 b_k\|^2 = \left\langle x, \frac{b_k}{\|b_k\|} \right\rangle^2,$$

which was to be demonstrated. ∎

Note that

$$\left\langle x, \frac{b}{\|b\|} \right\rangle^2 = \|x\|^2 \cos^2 \angle(x, b).$$

---

[2]The condition of orthogonalization is not only fundamental to the clustering procedure; it is also beneficial in other respects. See, for example, Chapter 8.

[3]In this case, F_k performs the operations proposed by Jacquin. This only serves as an example; any mapping that produces a block could be used.

[4]We shall assume fine quantization throughout this chapter.

The similarity measure therefore concerns an *angle*; it expresses how close to parallel a codebook block is to the range block x.

We thus arrive at the procedure for search in the tree. The procedure relies on a kind of "triangle inequality" for angles between blocks.

Lemma 9.1 (A Rationale for Clustering) *Let x, b and c be vectors in an inner product space. Let $\epsilon, \delta \geq 0$ and $\delta + \epsilon \leq \pi/2$. Suppose that*

$$\cos^2 \angle(x, c) \geq \cos^2 \delta$$

and

$$\cos^2 \angle(c, b) \geq \cos^2 \epsilon.$$

Then,

$$\cos^2 \angle(x, b) \geq \cos^2(\delta + \epsilon).$$

Proof: See [53].   ■

Let $C = \{c_1, \ldots, c_M\}$ be the cluster centers, and let D_m denote the codebook cluster that descends from the center c_m. The procedure for search in such a structure is:

1. At C: select $c_m = \arg\max_{c \in C} \cos^2 \angle(x, c)$.

2. At D_m: select $b_k = \arg\max_{b \in D_m} \cos^2 \angle(x, b)$.

This procedure can be interpreted in view of Lemma 9.1. Suppose that range block x and cluster center c_m are nearly parallel. Suppose also that the center is nearly parallel to the codebook blocks in its cluster D_m. Then, the codebook blocks in D_m will also be nearly parallel to the range block. This is the rationale for the whole idea of codebook clustering.

In the remainder of the chapter, we will express the cosines by inner products:

$$\|x\|^2 \cos^2 \angle(x, c) = \left\langle x, \frac{c}{\|c\|} \right\rangle^2.$$

The factor $\|x\|^2$ does not matter, as the original block x is the same during one search for a codebook block. To avoid division with the norms of the centers, we require that they have unit norm, $\|c\| = 1$.

9.4 Initialization

So far, we have investigated the complexity of the encoding step as a function of the cluster sizes, and we have expressed a criterion for choosing a codebook block. This section introduces a method for the initialization step.

The collage errors for each block can be added to yield the collage error for a whole image. According to Equation (9.4), the (negative) contribution from the codebook blocks when clustering is used, is

$$Q(c_1, \ldots, c_M) = \sum_{m=1}^{M} \sum_{x \in X_m} \max_{b \in D_m} \left\langle x, \frac{b}{\|b\|} \right\rangle^2, \tag{9.5}$$

where X_m is a cluster of original blocks around center c_m

$$X_m = \left\{ x : \langle x, c_m \rangle^2 > \langle x, c_n \rangle^2, m \neq n \right\},$$

and D_m is a cluster of codebook blocks around center c_m

$$D_m = \left\{ b : \langle b, c_m \rangle^2 > \langle b, c_n \rangle^2, m \neq n \right\}.$$

There are two main reasons for not attempting to find cluster centers that maximize the function in Equation (9.5) (i.e., that minimize the collage error).

- A true maximization of this function will probably not lead to a faster encoding step. One way of maximizing the function is simply to put all blocks in one cluster, since encoding by an exhaustive search always yields the best image quality.

- For a given set of cluster centers, one evaluation of the function is (almost) identical to one complete encoding of an image. We believe that the centers must be computed using an iterative procedure, where the function is evaluated once for each step. Several of these steps may have to be carried out with non-optimal centers, and such a procedure may therefore be much heavier than the encoding itself. If so, little or nothing is gained.

What to Maximize

The computation of cluster centers will involve the original blocks and not the codebook blocks. This should seem plausible, as all the original blocks contribute to the expression for the collage error, whereas not all codebook blocks do. Which codebook blocks count and how many times is known only after encoding.

We shall use Lemma 9.1 to find a lower bound on the error term that Equation (9.5) expresses and arrive at a useful objective function for the clustering.

Suppose that an original block x belongs to a cluster X_m with center c_m. Suppose that x forms an angle δ_x or $\pi - \delta_x$ with the center c_m. Then

$$\langle x, c_m \rangle^2 = \|x\|^2 \cos^2 \delta_x, \tag{9.6}$$

since $\|c_m\| = 1$. Such an angle δ_x exists for all original blocks, and if we add the equations, we have

$$\sum_m \sum_{x \in X_m} \langle x, c_m \rangle^2 = \sum_m \sum_{x \in X_m} \|x\|^2 \cos^2 \delta_x. \tag{9.7}$$

Suppose that a codebook block b belongs to a cluster D_m with center c_m. Suppose that b forms an angle less than ϵ_m or larger than $\pi - \epsilon_m$ to the center c_m and that ϵ_m is the largest angle deviation from the cluster center among all the chosen blocks in the cluster. Suppose that b is the block most parallel to a given x in the cluster X_m, denoted by $b = b(x)$. Then

$$\langle \frac{b(x)}{\|b(x)\|}, c_m \rangle^2 \geq \cos^2 \epsilon_m, \tag{9.8}$$

since $\|c_m\| = 1$. We can sum over all the original blocks and get the inequality

$$\sum_m \sum_{x \in X_m} \left\langle c_m, \frac{b(x)}{\|b(x)\|} \right\rangle^2 \geq \sum_m \sum_{x \in X_m} \cos^2 \epsilon_m. \tag{9.9}$$

(As we shall invoke Lemma 9.1, we must assume that the angles obey δ_x, $\epsilon_m \geq 0$ and $\delta_x + \epsilon_m \leq \pi/2$. The second assumption is rather arbitrary, and there is no reason to believe that it will always hold. Any shortcomings of the algorithms that are proposed later in this section must partly be ascribed to this flaw.)

For any original block, Equations (9.6) and (9.8) comply with the premises in Lemma 9.1. The corresponding conclusions for all the original blocks can be added to yield the inequality

$$\sum_m \sum_{x \in X_m} \left\langle x, \frac{b(x)}{\|b(x)\|} \right\rangle^2 \geq \sum_m \sum_{x \in X_m} \|x\|^2 \cos^2 (\delta_x + \epsilon_m). \tag{9.10}$$

The left side of this inequality is exactly the sum in Equation (9.5) which expresses the negative term of the collage error. Since we cannot maximize the left side directly we shall instead try to maximize its lower bound.

The lower bound given by Equation (9.10) must be maximized indirectly by maximizing the left sides in Equations (9.7) and (9.9). Equation (9.9) involves the codebook blocks that are chosen to approximate the original blocks during the encoding. These blocks are not known during the initialization. We do certainly know that if the codebook is larger than the set of original blocks, not all of the codebook blocks can be chosen. As the codebook is not in any way adapted to the original blocks, we must expect that some codebook blocks remain unused even in a smaller codebook. Before encoding we do not know which blocks are unused, so we cannot find cluster centers that maximize the left side and minimize the angles ϵ_m. Fortunately, the sum in Equation (9.9) tends to be large if the sum in Equation (9.7) is large.

All the variables that enter Equation (9.7), the original blocks and the cluster centers, are known before encoding, so that we may try to maximize this equation. If the centers are adapted to the original blocks by maximization of Equation (9.7), that is if the centers form small angles δ_x with the corresponding original blocks, the centers will probably also be well adapted to those codebook blocks that will eventually be selected in the encoding. These codebook blocks are selected because they are nearly parallel to the original blocks, and by Lemma 9.1, they should also be nearly parallel to the centers. In short, if we succeed in making the δ_x small, we expect to make the ϵ_m small as well.

Before we state the optimization problem, we introduce another constraint on the cluster centers. The codebook blocks are orthogonalized, i.e.,

$$b_k = B_k x_o = O \circ F_k x_o.$$

Now, since $b = Ob$ and $x = (O + P)x$,[5]

$$\langle x, b \rangle = \langle Ox, Ob \rangle + \langle Px, Ob \rangle = \langle Ox, b \rangle.$$

We see that the component Px of the original block is irrelevant for the selection of a codebook block. It should therefore not be permitted to influence the selection of a cluster center. This can be done through demanding that any cluster center be orthogonal to this component, $c \perp Px$, which will be the case if $c = Oc$.

---

[5] P projects onto the space of the fixed basis. The orthogonalization operator O is idempotent, and the sum of orthogonalization and projection equals the identity operator.

We conclude that the left side of Equation (9.7) should be the objective function for the initialization of the cluster structure. The task is therefore to maximize

$$G(c_1, \ldots, c_M) = \sum_{m=1}^{M} \sum_{x \in X_m} \langle x, c_m \rangle^2, \tag{9.11}$$

under the constraints that $\|c_m\| = 1$ and $c_m = Oc_m$.

The constraint that any cluster center must obey $c = Oc$ can be fulfilled without any special attention by a simple modification of the objective function. As this constraint ensures that

$$\langle x, c \rangle = \langle Ox, c \rangle,$$

the original block x in Equation (9.11) will be replaced by orthogonalized original blocks Ox. We arrive at the objective function

$$F(c_1, \ldots, c_M) = \sum_{m=1}^{M} \sum_{x \in X_m} \langle Ox, c_m \rangle^2, \tag{9.12}$$

which is to be maximized under the constraints that $\|c_m\| = 1$. Section 9.5 shows that any cluster centers that maximize this function indeed obey $c = Oc$.

An Iterative Algorithm

We may state two necessary conditions for a cluster structure to maximize the objective function in Equation (9.12).

1. Assume that the cluster centers are given. Every original block can be assigned freely to any cluster, and the assignments that maximize each term in the objective function, will maximize the sum. Original block x is assigned to the cluster center c_i if

$$\langle Ox, c_i \rangle^2 \geq \langle Ox, c_j \rangle^2; \quad j = 1, \ldots, M.$$

2. Assume that the clusters X_m; $m = 1, \ldots, M$ are given. The cluster centers that maximize each of the inner sums in Equation (9.12)

$$c_m \text{ maximizes } \sum_{x \in X_m} \langle Ox, c_m \rangle^2,$$

will maximize the objective function.

Equivalent conditions exist for the problem of *vector quantizer design*. In the context of quantization, condition 1 describes how to find an optimal encoder for a given decoder, and the condition is called the *nearest neighbor condition*. Condition 2 concerns finding an optimal decoder for a given encoder, and it is called the *centroid condition*.

These conditions have been employed successfully for design of both scalar and vector quantizers [55, 56, 62]. The design algorithm, called the *generalized Lloyd algorithm* or the *LBG algorithm*, is iterative and produces a sequence of quantizers. The iteration is interrupted when some convergence criterion is met.

The LBG-procedure is designed for codebook training. It does not partition the given training set into clusters of exactly equal size, but for most practical sources the cluster sizes will be fairly equal.[6] In Section 9.2, it was demonstrated that a uniform cluster size distribution leads to substantial savings of complexity.

Unfortunately, the resulting quantizer depends on the initial quantizer. This is a fundamental problem that cannot be overcome. In our experiments we shall use randomly generated cluster centers to specify the initial conditions.

For our purpose, the algorithm will be as follows:

Algorithm 9.1 (Finding cluster centers)
Let an initial set of cluster centers be given.

1. For the centers, group the blocks into clusters.

2. Stop if the objective function has increased by less than a given threshold since the last iteration.

3. For the resulting clusters, compute new centers.

4. Go to 1.

Finally, the codebook blocks are grouped into clusters according to the resulting cluster centers. Steps 1 and 2 are quite simple, but step 3 requires a close examination. The next section examines step 3 for two different objective functions.

9.5 Two Methods for Computing Cluster Centers

Optimal solution of step 3 in Algorithm 9.1 above involves the maximization of l_2-norms. We shall argue that this requires too much computation, and we shall replace the l_2-norm with an l_1-norm. We can verify that this is acceptable by comparing the the resulting collage error for the two cases.

Step 3 in the initialization algorithm requires that such a norm is maximized for each of the given cluster centers. With the condition $\|c\| = 1$, this is an eigenvalue problem, as shown below.

We define a linear operator

$$H : \mathbf{R}^N \to \mathbf{R}^I,$$

given by the blocks in the cluster

$$Hc = \begin{bmatrix} \langle Ox_1, c \rangle \\ \vdots \\ \langle Ox_I, c \rangle \end{bmatrix}.$$

---

[6]If not, there would have been much to gain in entropy-coding the codewords emitted by a VQ encoder.

I is the number of blocks in the cluster, $x_1, \ldots, x_I$ are the blocks in the cluster, and N is the block size. Then the function to maximize is the squared l_2-norm of Hc;

$$\|Hc\|_2^2 = \sum_{x \in X_m} \langle Ox, c \rangle^2. \tag{9.13}$$

It is to be maximized under the constraint that $\|c\|_2 = 1$.

This amounts to computing the squared operator norm of H, which is given by the largest eigenvalue of $H^T H$ [76]. The cluster center that yields the wanted maximum is the eigenvector of $H^T H$ that corresponds to the largest eigenvalue. The equation

$$H^T Hc = \lambda_{\max} c$$

is therefore to be solved for c, which is a relatively complex task. A method recommended in [70, pp. 367–381] is reduction of $H^T H$ to tridiagonal form and application of the QL method. Below, we propose a modification of the objective function that leads to a much simpler solution. The l_2-solution given here can be used to check the success of maximizing the following modified objective function.

Before we go on to the next objective function, we show that any c that maximizes $\|Hc\|_2^2$, will obey $c = Oc$. Suppose that a candidate for a maximizing c does not obey this, i.e., that it has a non-zero component Pc. Then, $\|Oc\| < 1$, since c must have norm equal to one. Thus,

$$\|Hc\| = \|HOc\| < \left\| H \frac{Oc}{\|Oc\|} \right\|.$$

Since $\tilde{c} = Oc/\|Oc\|$ is itself a valid candidate ($\|\tilde{c}\| = 1$ and $\tilde{c} = O\tilde{c}$), we conclude that a candidate c with non-zero Pc cannot maximize $\|Hc\|$.

Let us replace the objective function (9.13) by the function

$$F_1(c_1, \ldots, c_M) = \sum_{m=1}^{M} \sum_{x \in X_m} |\langle Ox, c_m \rangle| \tag{9.14}$$

and keep the constraint that $\|c_m\|_2 = 1$.

As before, step 3 of the initialization algorithm concerns computation of the cluster centers by maximization of each of the inner sums. Using the same construction of a linear operator H as above, each of the inner sums is an l_1-norm;

$$\|Hc\|_1 = \sum_{x \in X_m} |\langle Ox, c \rangle|.$$

As for the maximization of the l_2-norm, any c that maximizes $\|Hc\|_1$ will obey $c = Oc$, by the same argument.

For a general H, maximization of $\|Hc\|_1$ under the constraint $\|c\|_2 = 1$ seems to be an open problem. Attempts at solving it are investigated by Eikseth in [22]. We propose a procedure for suggesting a maximizing c below.

Here we write $Ox = h$ for simplicity. We begin by noting that

$$\|Hc\|_1 = \max\{sHc : s = [\pm 1, \ldots, \pm 1]\}. \tag{9.15}$$

Equation (9.15) holds since

$$\|Hc\|_1 = \sum_h |\langle h, c \rangle| = \sum_h \langle h, c \rangle \cdot \mathrm{sgn}\langle h, c \rangle,$$

and

$$s^T = \vec{\mathrm{sgn}}Hc = \begin{bmatrix} \mathrm{sgn}\langle h_1, c \rangle \\ \vdots \\ \mathrm{sgn}\langle h_l, c \rangle \end{bmatrix},$$

where

$$\mathrm{sgn}\, y = \begin{cases} 1; & y \geq 0 \\ -1; & y < 0 \end{cases}.$$

Any other choice of s would make some terms in the sum negative. Thus,

$$\max_{\|c\|_2=1} \|Hc\|_1 = \max \{ sHc : \|c\|_2 = 1, s = [\pm 1, \ldots, \pm 1] \}.$$

We shall propose an iterative procedure to seek a maximum of $\|Hc\|_1$ such that the execution of Algorithm 9.1 will eventually meet its stop criterion. First, the procedure must stop after a finite (and small) number of steps. Second, the cluster center returned by the procedure must make the considered partial sum larger than did the previous cluster center accepted as input. This is necessary for the LBG procedure to yield a sequence of steadily better (never worse) codebooks with a reasonable computational load.

The principle is to construct a non-decreasing sequence of numbers

$$\{s_n H c_n\}$$

by alternating between two operations.[7] The first is to keep c fixed and to maximize for s; the second is to keep s fixed and to maximize for c.

Algorithm 9.2 (Computation of one cluster center)
Let c_0 be given.

    1. Compute

$$s_n = \vec{\mathrm{sgn}}^T H c_{n-1}$$

    2. Compute

$$c_n = \frac{H^T s_n^T}{\|H^T s_n^T\|_2}$$

    3. Repeat the two above steps until $s_n = s_{n-1}$.

The first step ensures that

$$s_n H c_{n-1} \geq s_{n-1} H c_{n-1},$$

---

[7]The procedure presented here is concerned with computing one cluster center. The index on the center c means iteration number, not cluster number.

as s_n makes all terms in the sum

$$\sum_{i=1}^{I} s_{n,i} \langle h_i, c_{n-1} \rangle$$

positive. The second step ensures that

$$s_n H c_n \geq s_n H c_{n-1},$$

as is seen from Cauchy-Schwarz inequality

$$|sHc| \leq \|H^T s^T\|_2 \|c\|_2$$

where equality holds and $\|c\|_2 = 1$ by

$$c = \pm \frac{H^T s^T}{\|H^T s^T\|_2}.$$

Thus, the sequence $\{s_n H c_n\}$ is non-decreasing. It is also bounded above by $\max \|Hc\|_1$, hence it converges. When it is applied in initializing a codebook structure, our experiments indicate that the procedure stops after less than seven iterations; it usually stops after 3 iterations.

We may note that any cluster center returned by this procedure is a linear combination of the orthogonalized blocks in one cluster. Therefore $c = Oc$.

Eikseth [22] has studied the properties of this procedure. He has found a family of examples for which the procedure stops at $sHc < \max \|Hc\|_1$. However, a common feature of these examples seems to be that the rows of H are quite far from being parallel. In our case, the rows are the image blocks in one cluster, and they are assigned to their cluster because they are less parallel to any other cluster centers than to the chosen one. We therefore expect the blocks in a cluster to be roughly parallel. Also, the starting point c_0 is the cluster center around which the rows are grouped. This may account for the success of the overall initialization procedure as it is reported in Section 9.7.

9.6 Selecting the Number of Clusters

With a given initialization procedure, we may find the number M of clusters that minimizes the total time of initialization and encoding. This section treats this problem for the simple case of one given block size, without considering the collage error.

The operation of taking an inner product is fundamental to both initialization and encoding, and those operations undoubtedly dominate the computational burden. While other operations, such as addition of blocks and multiplication of a block by a scalar contribute less, we shall take them into account by translating them into "inner product equivalents," that is expressing their complexity as fractions of the complexity of the inner product. Table 9.1 shows that block addition and scalar multiplication of a block both require half the number of arithmetic operations as does an inner product. Thus the number of inner product equivalents may be minimized.

With this approach, the computational burden can be expressed as a function of the number M of clusters, with:

Table 9.1: Complexity for some basic operations. L denotes the block width.

Operation	Multiplications	Additions
Inner product	L^2	$L^2 - 1$
Mult. of block by a scalar	L^2	0
Addition of blocks	0	L^2

N the number of original blocks;

K the number of codebook blocks;

I the expected number of iterations before convergence of the LBG algorithm;

J the expected number of iterations in finding cluster centers;

as parameters. The subdivision below refers to the steps in Algorithm 9.1.

Step 1 of the Initialization Algorithm.   This step concerns the grouping of N blocks into M clusters by comparing each block to a cluster center through an inner product. This step hence involves NM inner products.

Step 2 of the Initialization Algorithm.   This step is a summation of a few scalars computed in step 3, and is not counted.

Step 3 of the Initialization Algorithm.   This step concerns computation of new cluster centers. It is done by iterating the procedure proposed in Section 9.5. The first step in the procedure is computation of a vector of signs;

$$s_n = \vec{\text{sgn}}^T H c_{n-1}.$$

As H is an operator built up by the blocks in the cluster, this step requires as many inner products as there are blocks, $\sharp X_m$.

The second step is computation of a new cluster center;

$$c_n = \frac{H^T s_n^T}{\|H^T s_n^T\|_2}.$$

The numerator here is simply an addition of the blocks in the cluster, hence this involves $\sharp X_m - 1$ block additions which corresponds to $(\sharp X_m - 1)/2$ inner products. Computation of the denominator involves 1 inner product, and division of the numerator by the denominator is one scalar multiplication, which corresponds to $1/2$ inner product.

We assume that this procedure is iterated J times. As there are M clusters, the computation of cluster centers should then require

$$J \sum_{m=1}^{M} \left(\frac{3}{2} \sharp X_m + 1 \right) = \frac{3}{2} J N + J M$$

inner products.[8]

---

[8]There are N original blocks, hence the sum of the cluster sizes is $\sum_{m=1}^{M} \sharp X_m = N$.

Final Clustering of the Codebook Blocks. This step concerns grouping K codebook blocks into M clusters by comparing a block to a cluster center through an inner product. This step involves MK inner products.

Encoding. For simplicity we assume that the codebook clusters or the original block clusters are uniformly distributed. Then, from Section 9.2, the encoding involves $NM + NK/M$ inner products.

Total Complexity. The LBG-algorithm is assumed to require I iterations. Then the above considerations yield a sum

$$C(M) = I\left(NM + \frac{3}{2}JN + JM\right) + MK + NM + NK/M. \qquad (9.16)$$

This sum is plotted for the examples of

$$K = 2000, \quad I = 4, \quad J = 3$$

and an image size of 1024 blocks in Figure 9.1 for $M = 1, \ldots, 100$. (An image of size 256×256 pixels consists of 1024 blocks of 8×8 pixels.)

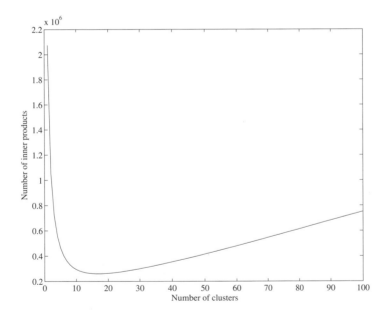

Figure 9.1: Total count of inner products as a function of the number of clusters. The image consists of 1024 blocks and the codebook consists of 2000 blocks.

With 1024 original blocks and 2000 codebook blocks as in Figure 9.1, the minimum complexity is attained with 17 clusters. Compared to the complexity of a full search (NK inner products), a cluster structure with 17 yields a complexity reduction by 8 compared to a full search encoder.

9.7 Experimental Results

The experiments have been aimed at finding the complexity reduction and the loss of image quality that result from the clustering, compared to the full search case. All experiments have been carried out using the l_1-version of the initialization. The initial cluster centers have been selected at random, but such that they fulfill the given constraints. We have used a SUN SPARC workstation.

Complexity Reduction

The 256×256 pixel Kiel test image has been encoded with a block size of 8×8, a codebook of 2000 blocks and 20 clusters. In 15 attempts, the time consumption was the following

Average:	21.7 seconds
Minimum:	21 seconds
Maximum:	23 seconds.

A full search encoding of the same image with the same block size takes 160 seconds. The complexity is thus reduced by 7.4 by the clustering procedure. The expression in Equation (9.16) predicts a reduction by 7.8 in this case.

Cluster Sizes

Section 9.2 treats the effect of the distribution of cluster sizes on the complexity of the encoding phase. The complexity of this sub-task is proportional to the inner product of two vectors. The elements of one vector ϕ are the relative sizes of the clusters of original blocks, the elements of the other vector γ are the relative sizes of the clusters of codebook blocks.

In Section 9.4, we remarked loosely that the clustering procedure would yield "fairly equal" cluster sizes. This is contrasted to the undesirable case of all blocks in one cluster; thus, the emphasis should be on the word "fairly." Figure 9.2 shows a histogram of the sizes of clusters for the original blocks. The encoded image is Peppers of size 512×512 pixels and there are 20 clusters. Even though the cluster sizes vary, the computational complexity is close to the prediction that assumed uniform sizes also in this case.

Loss of Image Quality

The initialization procedure is designed to minimize the loss of image quality (in the sense of the norm). The effect is examined here for two test images; 256×256 pixel Kiel and 512×512 pixel Peppers. Table 9.2 summarizes the results. Images are shown in Figures 9.3–9.5. The drop in image quality is visible but not annoying. We judge this loss to be comparable to the loss experienced with complexity-reduced vector quantization.

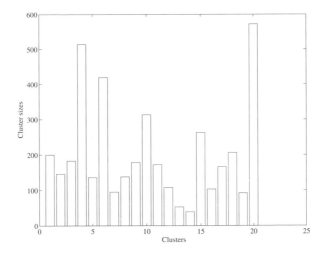

Figure 9.2: Sizes of the clusters of original blocks, when coding the Peppers image using 20 clusters. The image contains 512×512 pixels and the blocks consist of 8×8 pixels.

Table 9.2: Peak signal-to-noise ratios with and without clustering.

Image	Block size	Codebook size	Clustering	PSNR [dB]
Kiel	4×4	2000	Full search	27.43
			20 clusters	26.75
Peppers	8×8	4000	Full search	28.83
			20 clusters	28.19

Figure 9.3: The image Kiel encoded using a codebook of 2000 blocks. The image contains 256×256 pixels and the blocks consist of 4×4 pixels. The upper image is encoded with a full search, the lower image is encoded using 20 clusters.

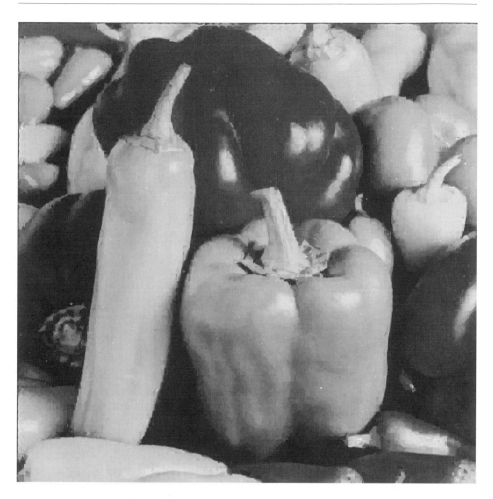

Figure 9.4: The Peppers image encoded using 20 clusters and a codebook of 4000 blocks. The image contains 512×512 pixels and the blocks consist of 8×8 pixels.

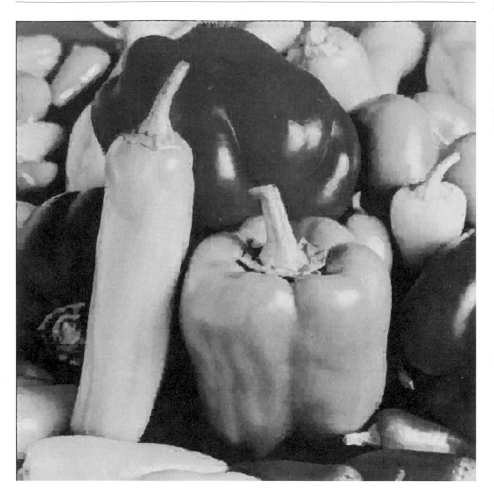

Figure 9.5: The Peppers image encoded using a full search of a codebook of 4000 blocks. The image contains 512×512 pixels and the blocks consist of 8×8 pixels.

Øien [64] has found experimentally that the image quality drops steadily as the number of clusters increases to about 25, and that it remains almost constant for numbers above 25.

9.8 Possible Improvements

Precomputed Initial Cluster Centers The initialization method presented here shares the weaknesses of the LBG algorithm. For example, the resulting cluster centers are sensitive to the initial cluster centers. Even though this problem cannot be entirely overcome, it may be made less serious.

The technique can employ a fixed, predesigned set of initial cluster centers. This set can be designed for a large set of training blocks. As it is designed once, computation time is not critical. There exist good alternatives to the LBG type of algorithm for this use. For example, Zeger, et al. [84] and Cocurullo, et al. [19] report on methods that produce better vector quantizers than LBG algorithms do.

Fixed Centers and Shrunken Blocks In our algorithm the cluster centers are adapted to the blocks in the original image. The initialization phase is rather slow because of this adaptivity. Øien [64] and Waldemar [82] have investigated methods that use fixed cluster centers that are adapted to a training set by the methods proposed in this chapter. As this method of initialization is fast, more clusters may be used. The resulting image quality is almost indistinguishable from the image quality obtained by adaptive cluster centers, while the computing time is reduced substantially.

The complexity can be reduced further by doing the search on shrunken blocks (shrunken by pixel averaging.) This method, investigated by Øien, chooses a small set of candidate blocks for search and encoding at the full resolution. This reduction of dimension seems beneficial for the LBG algorithm, and the image quality is actually improved in experiments.

With the two improvements mentioned here, the complexity is reduced by two orders of magnitude compared to encoders with exhaustive codebook search.

9.9 Conclusion

We have introduced a theory and technique for reducing the complexity in an attractor image encoder. The method uses classification of the blocks in both the codebook and the original range blocks, so that each range needs only be compared to codebook blocks in its own class.

The classes are adaptive and can be tuned to the original image or to a fixed training set. This chapter has treated the case of adaptation to the original image, and we have found an approximate expression for the complexity as a function of the number of classes.

The computational load can be reduced by a factor of eight when using image-adaptive classes. With refinements such as fixed trained classes and reduced block dimension, the method promises complexity reductions of two orders of magnitude.

The image quality is slightly affected, and this loss must be weighed against the advantage of a fast encoder.

Chapter 10

Orthogonal Basis IFS

G. Vines

The accurate coding of a range block is dependent upon there being a domain block which is self-similar in the library. As we have seen, this is normally true. However, while coding an image a range block will occasionally be encountered for which no suitable domain block can be found. Normally this situation would result in some reduction of the reconstructed signal quality in the location where the domain block resides. With more complex signals, this can happen more frequently.

Because the piecewise self-similar model is an approximation of real-world data, there is no guarantee that perfect maps can be found. As an example, in early PIFS coders the quality of the reconstructed image seems to reach an upper limit of around 32 dB PSNR for the 512×512 Lenna image. This trend was illustrated in a report comparing the so-called IFS image coding technique to the JPEG standard [18]. In this comparison, it was seen that the IFS coding technique seems to have a limit in the accuracy that an image can be coded. Lowering the compression ratio does not increase the reconstructed image quality significantly. This is due to the nature of IFS coding methods, in contrast to the JPEG standard, which is a DCT-based approach, the affine IFS is not a paradigm for an arbitrary waveform. A typical SNR versus compression ratio curve is shown in Figure 10.1.

In this section, we will look at a different way to construct the maps that code each range block. The goal of this method will be to make the maps more flexible to represent *any* range block. One consequence of this approach is that the reconstructed data will no longer have to have the self-similar characteristic that is a feature of fractals. This is not necessarily a detriment, as not all images are truly piecewise self-similar. As with any coding technique, compression is achieved by reducing the dimensionality of the data being coded. By finding a lower-dimensional representation of the data that maintains all of the characteristics of the original, the storage requirements are reduced. The problem is to find a set of basis vectors to best represent the signal in the sense of achieving the highest fidelity with good compression.

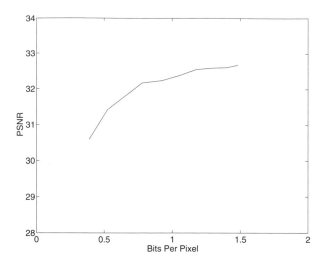

Figure 10.1: Typical SNR versus compression ratio for IFS coded image.

In the examples given in Chapters 8 and 7, each range block has been represented with a 2-dimensional approximation. For example, the kernel

$$K(\mathbf{x}) = \beta_2 \mathbf{x} + \beta_1 \mathbf{u} \tag{10.1}$$

consists of two basis vectors, $\mathbf{x}$ and $\mathbf{u}$. A more accurate approximation of the target range block may potentially be found by increasing the dimensionality of the kernel. Such an approach can reduce the compression. However, if care is taken in selecting the dimension for each range block, a more accurate model can be built. For example, some blocks may require only one dimension to accurately represent them, thereby achieving an even higher compression ratio without sacrificing signal quality. Thus, the goal is to represent each range block in the minimum number of dimensions to achieve a desired reconstructed signal quality.

The idea of increasing the complexity of the transformation may be thought of as using a linear combination of two separate domain blocks. Thus, the kernel can be written as

$$L_n = \beta_2^{(1)} P_n D_n F_n^{(1)} + \beta_2^{(2)} P_n D_n F_n^{(2)}, \tag{10.2}$$

where n is the range block number and the superscript is an index to the domain block, not an exponent. In fact, this idea may be carried to even larger numbers of domain blocks,

$$L_n = \sum_{i=1}^{\gamma} \beta_2^{(n_i)} P_n D_n F_n^{(i)}. \tag{10.3}$$

The problem with this approach is the search for the appropriate domain blocks rapidly becomes infeasible for as few as two domain blocks in the map using a simple search. However, by returning to the interpretation of the range blocks as being vectors in an $M_r = B_n \times B_n$ dimensional vector space, this problem becomes more manageable.

At this point it should be mentioned that there are other ways to increase the flexibility of the maps. For the 1-dimensional case a variety of nonlinearities have been used with success [35, 81, 80]. However, these approaches do not work as well for the 2-dimensional case and will not be discussed further [78].

The set of potential domain blocks is not necessarily restricted to a nonoverlapping tiling of the image as the range blocks are. Instead, the domain blocks may overlap one another. Therefore, there are many more potential domain blocks than range blocks. With this potential overlap in the domain blocks, there are

$$N_D = (L - 2L_R + 1)^2 \qquad (10.4)$$

domain blocks to choose from for each range block, where the image being coded is of size $L \times L$ and the range blocks are of size $L_R \times L_R$. For this value, we assume that the domain blocks are decimated by two in each dimension of the image.

10.1 Orthonormal Basis Approach

In this orthonormal basis approach, a set of orthonormal basis vectors will be created by the Gram-Schmidt procedure and the range blocks will be coded by projecting the blocks onto this basis [33, 78, 79]. The goal in determining the orthonormal basis will be to create a basis that allows each range block to be accurately represented with a minimum number of the basis vectors. By reducing the dimensionality of the representation of the range blocks, compression can be achieved. In addition, the amount of compression will vary automatically with the complexity of the range block. More complicated range blocks will require a higher-dimensional subspace to be represented. In this manner, this new model will be able to accommodate widely differing range blocks. First, we will discuss the generation of the orthonormal basis vectors and the encoding and decoding processes. Once the basic algorithm has been introduced, several approaches will be discussed for selecting a set of domain blocks to use.

Let the elements of a range block, whose upper left corner is at (x_{R_i}, y_{R_i}), be referenced as $r_{x_{R_i}, y_{R_i}}[x, y]$. Then each of the elements in the range block can be ordered sequentially to form a vector of length L_R^2 as outlined earlier,

$$\mathbf{r}_i = \left[r_{x_{R_i}, y_{R_i}}[0, 0], \ r_{x_{R_i}, y_{R_i}}[1, 0], \ldots, \ r_{x_{R_i}, y_{R_i}}[L_R - 1, L_R - 1] \right]^T, \qquad (10.5)$$

Each range block can then be interpreted as a vector in an L_R^2-dimensional vector space. A similar interpretation can be applied to the decimated domain blocks to form domain vectors $\hat{\mathbf{d}}_i$. Thus, for the typical 8×8 block size, we are working in a 64-dimensional vector space.

The simple affine map that transforms a single domain block into a range block with a scaling factor and a translation may also be placed in this setting and viewed as the linear combination of two vectors in this space,

$$\hat{\mathbf{r}}_i = b_i \cdot \mathbf{v_1} + s_i \cdot \hat{\mathbf{d}_i}, \qquad (10.6)$$

where $\mathbf{v_1}$ is a vector with all ones in it, and $\hat{\mathbf{d}}_\mathbf{i}$ is the decimated domain block in vector form. It is also possible to have other fixed vectors in addition to $\mathbf{v_1}$. For example, the map may be written as,

$$\hat{\mathbf{r}}_i = b_i \cdot \mathbf{v_1} + g_{i1} \cdot \mathbf{v_x} + g_{i2} \cdot \mathbf{v_y} + s_i \cdot \hat{\mathbf{d}}_\mathbf{i}, \tag{10.7}$$

where $\mathbf{v_x}$ and $\mathbf{v_y}$ are vectors which provide the "tilt" in each of the x and y axes, respectively,

$$\mathbf{v_x} = [\,0\,1\,2\,3\,4\,5\,6\,7\,0\,1\,2\,3\,4\,5\,6\,7\,\cdots 0\,1\,2\,3\,4\,5\,6\,7\,]^T, \tag{10.8}$$

$$\mathbf{v_y} = [\,0\,0\,0\,0\,0\,0\,0\,0\,1\,1\,1\,1\,1\,1\,1\,1\,\cdots 7\,7\,7\,7\,7\,7\,7\,7\,]^T. \tag{10.9}$$

The advantage of viewing the problem in this manner is that now we can take advantage of well-known vector space properties. In particular, we are interested in forming an orthonormal basis from the domain vectors. With these vectors, the range blocks can be encoded with a simple projection operation, and the map parameters will be the weights for this orthonormal basis. The set of basis vectors will consist of several vectors which are determined *a priori*, as well as those derived from the domain blocks. As long as the maps form contraction mappings we are assured of convergence and better control over the reconstructed image error.

Because the domain blocks are decimated early in the process, we are assured of a contraction in two of the three dimensions present in the block. To achieve a contraction in all three dimensions, domain vectors will be chosen which are *large* in an L^2 sense. Selection of *large* vectors will help in achieving contractive maps because large maps will have a better chance of having a large component in the direction of the range vector being coded. For example, consider in the 1-dimensional case, two domain vectors [0.1] and [1.3]. The larger vector [1.3] has a better chance of using a contractive map in forming an arbitrary range vector, say [1.0].

For a range block of size $L_R \times L_R$, let $M_r = L_R^2$ be the length of the range and decimated domain vectors. The collections of all such vectors will be denoted by $\{\mathbf{r}_i\}_{i=1}^{N_R}$ for the range vectors and $\{\hat{\mathbf{d}}_i\}_{i=1}^{N_D}$ for the decimated domain vectors. The set of three fixed basis vectors can be orthonormalized to form the first three of the required M_r orthonormal basis vectors. The remaining basis vectors will be chosen to span the $N_S = (M_r - 3)$-dimensional subspace S^0 orthogonal to the space spanned by these three vectors.

A projection operator for the subspace spanned by a vector $\mathbf{v}_i$ can be written:

$$P_{\mathbf{v}_i} = \mathbf{v}_i (\mathbf{v}_i^T \mathbf{v}_i)^{-1} \mathbf{v}_i^T. \tag{10.10}$$

Therefore, the operator to project a vector into the desired subspace S^0 is given by

$$P_{S^0} = I - (P_{\mathbf{v}_1} + P_{\mathbf{v}_2} + P_{\mathbf{v}_3}), \tag{10.11}$$

where I is the identity matrix, and $\mathbf{v}_1$, $\mathbf{v}_2$, and $\mathbf{v}_3$ are the *a priori* basis vectors. Using this projection operator, the projected versions of the range vectors can be computed as:

$$\mathbf{s}_j^0 = P_{S^0} \mathbf{r}_j. \tag{10.12}$$

Figure 10.2 illustrates this process for a simple 1-dimensional *a priori* vector subspace in a 2-dimensional example.

The three fixed vectors $\mathbf{v}_1$, $\mathbf{v}_2$, $\mathbf{v}_3$ and the N_S domain vectors form a set of M_r vectors which will span the space of the range vectors. If the selected N_S domain vectors are denoted as $\{\mathbf{b}_i\}_{i=1}^{N_S}$, then this set of vectors can be written as a matrix B, where

$$B = \left[\,\mathbf{v}_1,\ \mathbf{v}_2,\ \mathbf{v}_3,\ \mathbf{b}_1,\ \ldots,\ \mathbf{b}_{N_S}\,\right]. \tag{10.13}$$

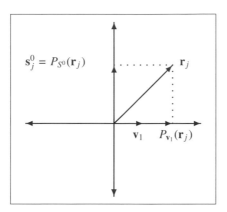

Figure 10.2: An example of the projection of a range vector onto a subspace orthogonal to an *a priori* basis vector subspace. This example uses only one *a priori* vector $\mathbf{v}_1$ for a simplified 2-dimensional example.

To achieve rapid computation of the weights with which these vectors represent each range vector, these basis vectors are orthonormalized with the Gram-Schmidt procedure, resulting in a Q matrix containing the orthonormal basis vectors for the space.

Once the set of basis vectors is determined, the coding process is straightforward. Thus, to represent a given range vector $\mathbf{r}_i$ with a set of weights $\mathbf{w}_i$, the following must hold,

$$\mathbf{r}_i = Q\mathbf{w}_i. \tag{10.14}$$

Therefore, premultiplying by Q^T yields

$$\mathbf{w}_i = Q^T\mathbf{r}_i, \tag{10.15}$$

which is used to determine the weights. These weights can be quickly calculated for each $\mathbf{r}_i$, and these two equations define the basic encoding and decoding process.

The remaining task to implement the coder is an algorithm to select the N_S domain vectors to use. The desired characteristics are that the vectors should:

- provide contractive mappings,

- be as orthogonal as possible, and

- form a set of basis vectors, which allows each range vector to be represented best with as few vectors as possible.

Several methods were developed based on these three criteria and are discussed below.

Covariance Method

Rather than search through the relatively large set of domain vectors, the range vectors are analyzed to determine the optimal basis vector directions. Then the domain vectors are searched to find the largest vector in each of these directions. This approach has two advantages: first, we are primarily interested in representing the range vectors; thus, the directions for the basis vectors should be based on the range vectors themselves. Second, there are far fewer range vectors than domain vectors, and so the computational task of determining the basis vectors is reduced.

The range vectors are searched iteratively to find the "best" direction for the next basis vector. After each direction is selected, the remaining range vectors are projected into the subspace orthogonal to this vector. The algorithm begins by projecting all of the range vectors into the subspace S^0 perpendicular to the subspace spanned by the *a priori* vectors. At the k-th iteration, the i-th projected range vector is noted by $\mathbf{s}_i^k$ which resides in a corresponding subspace S^k. The optimal basis vector direction is determined by taking the $\mathbf{s}_i^k$ vector with the largest correlation to all of the other $\mathbf{s}_i^k$ vectors. Thus, the vector $\mathbf{s}_l^k$ which maximizes the equation,

$$C_i = \sum_{j=1, j \neq i}^{N_R} |\langle \mathbf{s}_i^k, \mathbf{s}_j^k \rangle| \tag{10.16}$$

is selected, where $|\langle \mathbf{s}_j^k, \mathbf{s}_i^k \rangle|$ is the absolute value of the inner product of $\mathbf{s}_j^k$ and $\mathbf{s}_i^k$.

Once each basis vector direction is determined, the remaining $\mathbf{s}_i^k$ vectors are projected onto the subspace orthogonal to $\mathbf{s}_l^k$, using a projection operator of the form:

$$P_{S_k} = I - \mathbf{s}_l^k (\mathbf{s}_l^{k^T} \mathbf{s}_l^k)^{-1} \mathbf{s}_l^{k^T}. \tag{10.17}$$

The chosen basis vector direction is saved as $\mathbf{t}_k$ and the process is repeated until the necessary N_S vectors are obtained. Essentially the Gram-Schmidt procedure is performed on the range vectors. However, the vector used in each step is the vector with the largest correlation with the other remaining vectors. In this manner, we find the set of N_S selected vectors, or *direction vectors*, $\{\mathbf{t}_i\}_{i=1}^{N_S}$ which best represents the subspace S^0. In addition, these direction vectors are orthogonal. Thus, the last two criteria in the list above are satisfied. Figure 10.3 illustrates this process for a simple 2-dimensional, 3-vector case. In this example, $\mathbf{s}_3^k$ is chosen as the vector with the largest correlation. Therefore, the remaining vectors are projected into the subspace S^{k+1}. Effectively, the component in the remaining vectors which can be represented by $\mathbf{s}_3^k$ is removed.

However, domain vectors are needed for the contractive mappings, and a search must be performed through the domain vectors to find the best set of domain vectors for these direction vectors. Because the order in which the $\mathbf{t}_i$ vectors are selected is important, with the most significant vector coming first, the same order is used in finding the domain vectors. The domain vector with the largest component in the direction of the direction vector is used, where the projection of a domain vector $\hat{\mathbf{d}}_j$ onto a direction vector $\mathbf{t}_i$ may be written as:

$$P_{\mathbf{t}_i}(\hat{\mathbf{d}}_j) = \frac{|\hat{\mathbf{d}}_j \cdot \mathbf{t}_i|}{||\mathbf{t}_i||}, \tag{10.18}$$

where $|| \cdot ||$ indicates the L^2 norm. Because it is possible that one domain vector has the largest component on more than one direction vector, each domain vector is only allowed to be used once. Figure 10.4 shows how the domain vector with the largest component in the direction of the direction vector $\mathbf{t}_i$ is selected.

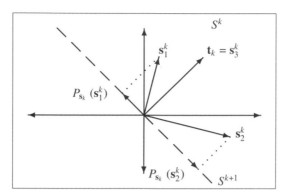

Figure 10.3: An example of choosing the vector with the largest correlation to the other vectors and projecting the remaining vectors into the orthogonal subspace. After the selection of $\mathbf{t}_k$, the remaining vectors are projected into the space S^{k+1}.

In summary, this algorithm proceeds as follows:

1. Loop N_S times to find the direction vectors, $\mathbf{t}_i$

   (a) Find vector with largest correlation using Equation (10.16).

   (b) Save vector as $\mathbf{t}_i$.

   (c) Project remaining into subspace using Equation (10.17).

2. Loop N_S times for each $\mathbf{t}_i$ to find best domain vector:

   (a) Find domain vector with largest component in direction of $\mathbf{t}_i$ using Equation (10.18).

   (b) Save vector and eliminate from list of domain vectors.

The final result is the set of domain vectors to use for encoding the image. The complete encoding algorithm is provided in Section 10.1.

K-Means-Based Approach

The previous approach is optimal in an incremental sense. That is, the optimal *additional* basis vector is selected at each iteration. This does not provide the best set of two, or n, vectors to represent the subspace S^0. For example, if each of the range vectors is going to be coded with a single arbitrary vector from a collection of 64 vectors, then a K-means approach would be optimal, with $k = 64$. Essentially, K-means is a clustering algorithm that guarantees convergence to a local minimum in grouping a set of vectors into k clusters. A complete

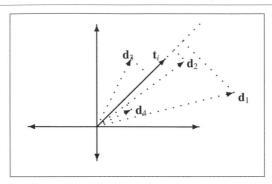

Figure 10.4: Selection of the domain vector with the largest projection on the direction vector t_i. In this example, d_1 is the selected domain vector.

discussion of the K-means algorithm can be found in [77]. Better overall results might be obtained by looking at the range vectors in groups. Thus, we do not try and find the best single vector to represent the entire set of range vectors. Instead, the range vectors are first clustered into a fixed number of groups, then each of these groups is represented by a single range vector.

The K-means algorithm groups the vectors into clusters. Thus the vectors r_j and $-r_j$ are very different as perceived by this algorithm. However, because we are interested in finding a set of basis vectors, it would be preferable to treat these two vectors as equivalent. A simple modification to the data prior to running the K-means algorithm eliminates this problem. All of the vectors are forced to reside in the same half-space. The selected half-space is the one that included the positive axes, and any vector which is outside of this half-space is set to its negative. This can easily be checked by examining the sign of the sum of the elements in each vector,

$$h_j = \sum_{i=1}^{M_r} r_j[i]. \tag{10.19}$$

If $h_j < 0$, then the negative of that vector is used.

Figure 10.5 illustrates this modification for a simple 2-dimensional example. In this example, vectors c and d on the left side of the dashed line are changed to their negative to form the corresponding vectors c and d in the right half-plane.

In addition, the magnitude of each of the range vectors is not significant, therefore the vectors are normalized prior to running the K-means routine. Once the k means are determined, the domain vectors are searched to find the set that matches these direction vectors.

Search Method

While the covariance method is optimal in the sense that each additional basis vector best represents the remaining subspace, there is a cost associated with this optimality. Performing the Gram-Schmidt process on the range vectors and the ensuing search through the domain vectors can be computationally expensive for large images. Similarly, the K-means algorithm begins to be computationally taxing with larger numbers of vectors. A sub-optimal approach was developed which is based on finding a smaller set of domain vectors from which the basis vectors are selected.

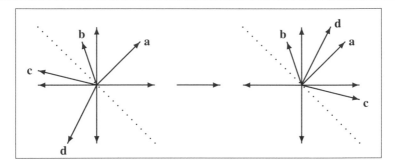

Figure 10.5: An example of moving vectors to the same half-space.

The set of domain vectors is first reduced by discarding those vectors whose projection in the subspace S^0 has a magnitude less than a threshold T_e. This forms a set of projected "large" vectors $\{\mathbf{e}_i\}$. Once the domain vectors are reduced to this subset, the set of N_S vectors is chosen to span the subspace S^0. The first N_S vectors are selected as,

$$\mathbf{b}_i = \mathbf{e}_i \tag{10.20}$$

for $i = 1, \ldots, N_S$. Then, a *quality factor*, defined by

$$q = \sum_{i=1}^{N_S} \sum_{j=i+1}^{N_S} |\mathbf{b}_i \cdot \mathbf{b}_j|, \tag{10.21}$$

is determined for this set of basis vectors. Essentially, the desired set of vectors should be as orthogonal as possible. This is accomplished by finding the set of vectors that minimize q. Once an initial set is collected, each of the remaining $\mathbf{e}_i$'s are substituted with the closest $\mathbf{b}_i$, and a new q is calculated. If the quality factor is reduced, then this new basis vector is kept, otherwise the next $\mathbf{e}_i$ is examined, until all of the $\mathbf{e}_i$ have been checked. The final set of $\mathbf{e}_i$'s is a set of nearly orthonormal basis vectors for the subspace S^0. These can be combined with the *a priori* vectors to form the B matrix as in Equation (10.13).

Searchless Method

Because any search over the domain vectors will tend to be time consuming for large images, a searchless approach was implemented which simply takes the required number of domain vectors from the domain blocks evenly spaced across the image. While this method does not insure that the set of vectors will span the space, and consequently the performance can be expected to be less than the other methods, it is extremely fast for small blocks. In coding an image with 8×8 blocks, after the Gram-Schmidt procedure is complete, there are 64 multiplies to code each pixel.

Encoding

The encoding method is based on Equation (10.15), which provides the weights that specify the vector in the rotated coordinate system. In order to achieve a compression, the weights must be quantized. Ideally we would like to discard most of the weights. The encoding algorithm consists of the following steps.

---

1. Determine the domain vectors to use by one of the above methods.

2. Form the B matrix and use the Gram-Schmidt procedure to get Q.

3. Determine the map parameters $\mathbf{w}_i$ with Equation (10.15).

4. Save those weights that exceed a threshold T_w.

The final encoded image consists of the indices for the N_S domain vectors and the quantized weights for each map which exceeded the threshold. By adjusting the threshold, the accuracy of the reconstructed image can be controlled. In addition, the quantization approach can be different for the *a priori* vectors, which will tend to have a different distribution as compared to the weights for the $\mathbf{b}_i$ vectors.

Decoding

In order to reconstruct the $\mathbf{r}_i$, which will provide the reconstructed image, Equation (10.14) is used. This requires the Q matrix, which can be obtained by performing a Gram-Schmidt procedure on the B matrix.

The reconstruction process is much simpler than the encoding procedure, and begins with any initial image V_0. It iteratively performs the following steps.

---

1. Gather the basis vectors $\hat{\mathbf{d}}_i$ from the image.

2. Form B as given in Equation (10.13).

3. Perform the Gram-Schmidt procedure on B to get Q.

4. Compute each $\mathbf{r}_i = Q\mathbf{w}_i$ and save in the image.

5. Go to step 1.

The image will converge in a few iterations.

10.2   Quantization

In order to compute the compression ratios for this approach, the parameters need to be quantized for each map. After the quantization of the map parameters a comparison of coding results is given.

Each of the maps for all of the methods consists of weights which must be quantized for storage. A uniform quantization method was not always efficient due to the distribution of these parameters, and therefore, a Lloyd-Max quantizer was implemented. The Lloyd-Max quantizer is basically a 1-dimensional K-means algorithm [56], which was discussed in Section 10.1.

10.3 Construction of Coders

The Orthonormal Basis Coder was implemented with 8×8 pixel range blocks. The coded image consists of the index for the domain blocks that constitute the basis vectors, followed by the indices and weights for the vectors that are used in each of the maps. Because the weights for the first three *a priori* basis vectors are always saved, the indices do not need to be saved for these weights.

Table 10.1: Image coding results for the Lenna image.

Search Method	BPP	PSNR
Orthonormal basis: Covariance	0.44	30.5
Orthonormal basis: K-means	0.47	30.2
Orthonormal basis: Search method	0.49	29.6
Orthonormal basis: Searchless method	0.50	27.0

10.4 Comparison of Results

Each of the approaches was tested with the standard Lenna image and the results are given in Table 10.1. In Figures 10.6 and 10.7 the orthonormal basis approach results are given prior to and after quantizing, respectively. This illustrates the noise artifact introduced with this method, as well as the extremely high quality of the image. To see how this method works, the basis vectors are shown in Figures 10.8 and 10.9. In the first figure, the blocks are shown prior to the Gram-Schmidt process, and in the second figure the orthonormal vectors and their noiselike appearance is illustrated.

It is interesting that the image reconstructed using all of the weights and without any quantization is not perfect (see Exercise 15). As it turns out, examination of the weights in the mappings shows that many are not contractions. While the existence of some expansive maps does not necessarily cause a problem as discussed in Section 8.3, the maps here apparently have an excessive number. It is possible to look at the actual contraction factors for the maps through the relationship between the B and Q matrices. The matrix R relates the orthonormal basis to the original domain vectors: $B = QR$. Therefore, $Q = BR^{-1}$, and Equation (10.14) can be written

$$\mathbf{r}_i = BR^{-1}\mathbf{w}_i = B(R^{-1}\mathbf{w}_i). \tag{10.22}$$

Thus, the contraction factors are the elements of the vector $R^{-1}\mathbf{w}_i$.

Figure 10.6: Lenna image coded with orthonormal basis method prior to quantizing.

Figure 10.7: Lenna image coded with orthonormal basis method at 0.44 bpp, PSNR = 30.5 dB.

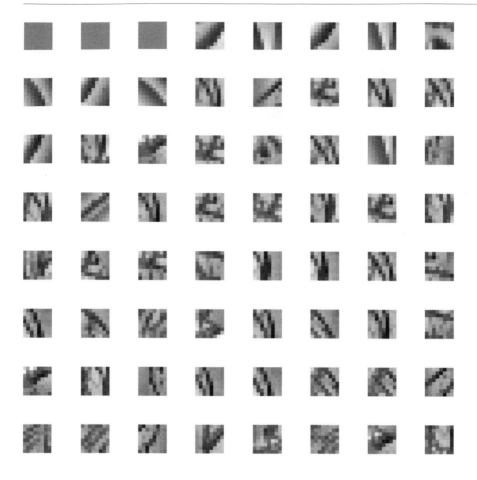

Figure 10.8: Basis vector blocks for covariance method: before Gram-Schmidt.

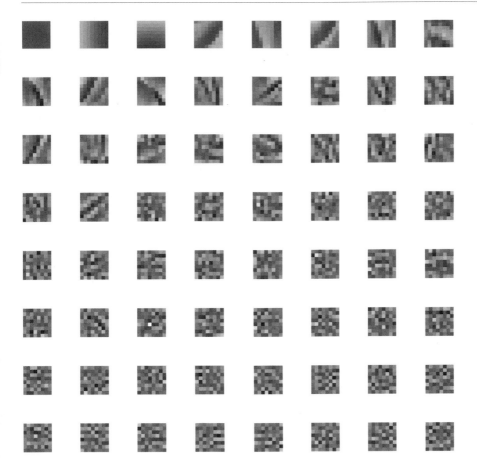

Figure 10.9: Basis vector blocks for covariance method: orthonormalized vectors.

The error in the reconstructed images is concentrated around the edges in the image and is a "salt and pepper" type of high-frequency noise, as can be seen in Figure 10.6. In an attempt to attenuate this noise, the weights which were not contraction mappings were scaled down to be contraction mappings during the reconstruction process. Then, just prior to the last iteration of the reconstruction, the weights were restored to their full magnitude. Unfortunately, this reduced the final PSNR even further.

To facilitate comparisons with other coding methods, several different images were coded using the orthonormal approach with the covariance method. Table 10.2 provides these results.

Table 10.2: Coding results for several images.

Image	Orthonormal Basis w/ Covariance	
	PSNR	bpp
Peppers	30.8	0.551
Boat	28.4	0.622
Mandrill	19.7	0.672
Moffett	28.0	0.972

10.5   Conclusion

This approach is still in its infancy. As mentioned above, there is no guarantee of a contraction in the maps. In addition, the computational cost has not been the focus of these efforts. Much work still needs to be done in the area of selecting the domain blocks from which the basis is formed. The four methods discussed here are just a starting point. See Section C.2 also.

Acknowledgment

This work was supported in part under the Joint Services Electronics Program, Grant No. DAAH-0493G0027.

Chapter 11

A Convergence Model

B. Bielefeld and Y. Fisher

Irrespective of the underlying model, most fractal compression methods, including most of those discussed in this book, use the following approach: A contractive map $\tau : F \to F$ is sought on a complete metric space F of all possible images. The Contractive Mapping Fixed-Point Theorem then implies that $|\tau| = \lim_{n\to\infty} \tau^{\circ n}$ exists and is unique. The transformation τ then serves as an encoding of the image $|\tau|$. The difficulty lies in finding τ and, when the goal is image compression, specifying it compactly.

In this chapter, we apply a new notation to this model with L^p spaces and their associated metrics. We then note that it is not necessary to use the Contractive Mapping Fixed-Point Theorem, and give several examples. The main results of this chapter must be considered preliminary, however, because they rely on several simplifying assumptions. In spite of this weakness, we believe the approach to be promising from both an applied and theoretical point of view.

11.1 The τ Operator

In particular, we consider an image as an element of $f \in L^p(I^2)$ where $I^2 = [0, 1] \times [0, 1]$ is the unit square. In this model $f(x)$ represents the grey level at the point $x \in I^2$. Thus, it only makes sense to have f take values in $I = [0, 1]$, since then we can take 0 to mean black and 1 to be white, with grey levels in between. However, in practice we will not restrict f with the understanding that f will be truncated when needed. The goal is to find functions $m : I^2 \to I^2$, $s : I^2 \to \mathbb{R}$ and $o : I^2 \to \mathbb{R}$ so that a given function f will be a fixed point for the operator $\tau_{m,s,o} : L^p(I^2) \to L^p(I^2)$ given by

$$\tau_{m,s,o}(f) = s(x)f(m(x)) + o(x).$$

In that case, m, s and o compose an encoding of f, and when these functions can be specified compactly then we can then encode f compactly. When τ has[1] a fixed point that attracts all of $L^p(I^2)$, then a simple way to find f is to iterate τ from any initial point until the fixed point is approximated.

It is possible to insist that τ be contractive, in which case the Contractive Mapping Fixed-Point Theorem implies that it will attract all of L^p, but this is not necessary. For example, we would be perfectly happy with pointwise or almost everywhere (a.e.) convergence of $\tau^{on}(f)$ without using any metric at all. In fact, it is arguable that a.e. convergence is visually more meaningful than convergence in L^p.

Finally, it is worth noting that the case of non-overlapping iterated function systems[2] with condensation can be handled as a special case[3] when we restrict the space of images to the characteristic functions of compact subsets of I^2 and restrict $o(x)$, $s(x) \in \{0, 1\}$ with compact support. See Example 1 below.

We begin with several examples. We then discuss a criterion for L^p convergence of the RIFS model of Section 2.7. The special case $m(x) = 2x$ is discussed, followed by a generalization to the standard fractal compression methods (e.g., those discussed in Chapter 1).

Examples.

The following examples illustrate the encoding idea. For $x = (u, v) \in I^2$, let

$$m(x) = m(u, v) = (2u(\bmod 1), 2v(\bmod 1)).$$

This is a four-to-one piecewise affine map, which we often denote $m(x) = 2x$. Let $s(x) = \alpha$ on the upper left subsquare, β on the upper right, γ on the lower left, and δ on the lower right.

Figure 11.1: The first three iterates of τ from Example 1. Here, black represent the value 1 and white the value 0.

Example 1: Let $f(x) = 1$, which is the characteristic function of I^2, and take $\alpha = \beta = \gamma = 1$ with $\delta = 0$ and $o(x) = 0$. Figure 11.1 shows the support of f, $\tau(f)$, $\tau^{o2}(f) = \tau(\tau(f))$ and $\tau^{o3}(f)$. Figure 11.2 shows an approximation of the support of $\lim_{n \to \infty} \tau^{on}(f)$, a Sierpinski triangle. Since $f(x)$ and $s(x)$ have compact support with $f(x)$, $s(x) \in \{0, 1\}$ and $o(x) = 0$, the image $\tau(f)$ is also a characteristic function and so are the iterates $\tau^{on}(f)$. In fact, in this case, The limit function, or attractor, is the characteristic function of a Sierpinski triangle, since τ is a representation of an iterated function system with this fixed point.

---

[1] We will not write the subscript when the context is clear.

[2] An IFS is nonoverlapping if there exists a non-empty bounded open set A such that the disjoint union $\cup w_i(A) \subset A$.

[3] Some typical IFS's define fractals with zero measure, which have a 0 characteristic function in L^p; so an L^p space is not an apt model.

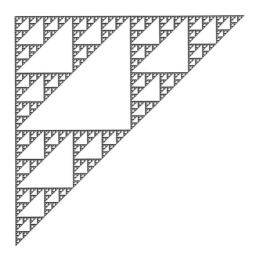

Figure 11.2: . A Sierpinski triangle generated as the fixed point of τ from Example 1.

Example 2: For $f \in L^p(I^2)$, we show that τ satisfies

$$||\tau(f) - \tau(g)||_p = \lambda_p ||f - g||_p,$$

where

$$\lambda_p = \left(\frac{|\alpha|^p + |\beta|^p + |\gamma|^p + |\delta|^p}{4} \right)^{1/p},$$

so that $\lambda_p < 1$ is the condition for τ to have a fixed point that attracts all of $L^p(I^2)$ (by the Contractive Mapping Fixed-Point Theorem). Compute,

$$||\tau(f) - \tau(g)||_p^p = \int_{I^2} |s(x)(f - g)(m(x))|^p = \lambda_p^p ||f - g||_p^p,$$

and the result follows. Figure 11.3 shows the fixed point for τ when $\alpha = -1.9$, $\beta = 0.8$, $\gamma = 0.8$, and $\delta = -0.8$ with $o(x) = 0.1$.

Remarks: When each of α, β, γ, and δ is less than 1, then $\lambda_p < 1$ for all p, and when they are all greater than 1, then τ diverges for all p. But otherwise, it is possible for $\tau^{\circ n}(f)$ to converge for some values of p and not for others. There are two distiguished values: a value p_0 such that τ converges in L^p for $p < p_0$, and τ diverges in L^p for $p > p_0$; and a value p_1 for which λ_p is a minimum. It is possible that these values are relevant measures of image content. For example, larger values of α, β, γ, and δ give smaller values of p_0; it may be that this corresponds to an image that contains a lot of variation.

It turns out that τ converges almost everywhere when

$$|\alpha\beta\gamma\delta| < 1.$$

This is proved later, but the following argument suggests why this is the case. Under iteration of τ, the iterates of $m(x)$ starting at almost all points x visit each quadrant an approximately equal

Figure 11.3: The fixed point of τ from Example 2.

number of times. At each quadrant, $f(m(x))$ is multiplied by α, β, γ, or δ, so the value of the factor of f after n steps is $|\alpha\beta\gamma\delta|^{n/4}|\alpha|^{\epsilon_1(n)}|\beta|^{\epsilon_2(n)}|\gamma|^{\epsilon_3(n)}|\delta|^{\epsilon_4(n)}$ where $\epsilon_i(n)/n \to 0$ as $n \to \infty$. Thus, when $|\alpha\beta\gamma\delta| < 1$, τ converges pointwise almost everywhere and when $|\alpha\beta\gamma\delta| > 1$ it diverges (equality is not resolved by this argument). In the example above, we have $\lambda_p > 1$ for all $p \geq 1$, so that τ diverges for all $p \geq 1$. However, τ still converges almost everywhere.

This means that L^p convergence, while interesting, may not be particularly apt in applications – in spite of the fact that L^2 is commonly used. It is possible for τ do diverge in L^p and still converge almost everywhere. In this case, the L^p model has *nothing* to say about the image compression problem, even though the encoding is well defined.

Finally, we note that even simple choices for m, s and o lead to very complicated behavior, as shown in Figures 11.4 and 11.5.

11.2 L^p **Convergence of the RIFS Model**

In this section we discuss the L^p convergence of the RIFS model of Section 2.7. Let $T(f) = s(x)f(m(x))$ denote the linear part of τ. We are interested in computing

$$||\tau(f) - \tau(g)||_p = ||T(f - g)||_p.$$

In this section we will make the following assumptions:

(a) $I^2 = \cup_{i=1}^{N} R_j$.

(b) $R_i \cap R_j$ has Lebesgue measure 0 for $i \neq j$.

(c) $m : R_j \to D_j$ is a homeomorphism.

(d) For every j there is a set of indices K_j such that $D_j = \cup_{k \in K_j} R_k$.

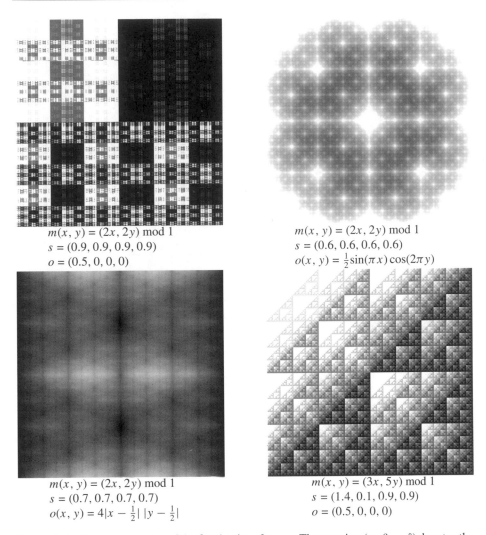

$m(x, y) = (2x, 2y) \bmod 1$
$s = (0.9, 0.9, 0.9, 0.9)$
$o = (0.5, 0, 0, 0)$

$m(x, y) = (2x, 2y) \bmod 1$
$s = (0.6, 0.6, 0.6, 0.6)$
$o(x, y) = \frac{1}{2}\sin(\pi x)\cos(2\pi y)$

$m(x, y) = (2x, 2y) \bmod 1$
$s = (0.7, 0.7, 0.7, 0.7)$
$o(x, y) = 4|x - \frac{1}{2}| \, |y - \frac{1}{2}|$

$m(x, y) = (3x, 5y) \bmod 1$
$s = (1.4, 0.1, 0.9, 0.9)$
$o = (0.5, 0, 0, 0)$

Figure 11.4: Various examples of the fixed point of $\tau_{m,s,o}$. The notation $(\alpha, \beta, \gamma, \delta)$ denotes the constant values of $s(x)$ or $o(x)$ on the four quadrants of I^2.

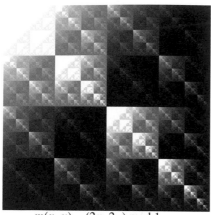 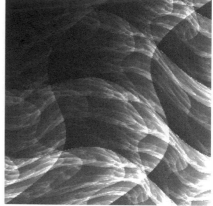

$m(x, y) = (2x, 2y) \bmod 1$
$s = (0, 9, 0.4, 0.4, 0.9)$
$o = (0.5, 0, 0, 0)$

$m(x, y) = (2x + y + \sin(2\pi y),$
$x - 3y + \cos(2\pi x)) \bmod 1$
$s(x, y) = x^2 + y^2$
$o = (0.2, 0.2, 0.2, 0.2)$

Figure 11.5: More examples of the fixed point of $\tau_{m,s,o}$. The notation $(\alpha, \beta, \gamma, \delta)$ denotes the constant values of $s(x)$ or $o(x)$ on the four quadrants of I^2.

(e) $s(x)$ is constant and equal to s_i on each R_i.

(f) $m(x)$ has constant jacobian equal to $j_i > 1$ on each R_i.

With these assumptions, we claim that this model is equivalent to the RIFS model of Section 2.7.

Let $f \in L^p(I^2)$ be some initial image. Let $\mathbf{r}(f)$ and $\mathbf{d}(f)$ be the N vectors given by

$$\mathbf{r}_i(f) = \int_{R_i} |f|^p$$

$$\mathbf{d}_i(f) = \int_{D_i} |f|^p.$$

We can then use condition (d) above to compute $\mathbf{d}(f)$ as a matrix times $\mathbf{r}(f)$. That is, since

$$\mathbf{d}_i(f) = \sum_{k \in K_i} \int_{R_k} |f|^p,$$

we can use the $N \times N$ *transition matrix* $A = (a_{ij})$ given by

$$a_{ij} = \begin{cases} 1 & \text{if } j \in K_i \\ 0 & \text{otherwise} \end{cases}$$

to write

$$\mathbf{d}(f) = A\mathbf{r}(f).$$

We also define the $N \times N$ diagonal matrix $C = (c_{ii})$ given by $c_{ii} = \frac{|s_i|^p}{j_i}$. We use $\mathbf{1}$ to denote the N vector containing all 1s.

Then we can compute

$$r_i(T(f)) = \int_{R_i} |s(x)f(m(x))|^p \tag{11.1}$$

$$= \int_{R_i} |s_i|^p |f(m(x))|^p \tag{11.2}$$

$$= \frac{|s_i|^p}{j_i} \int_{D_i} |f(x)|^p \tag{11.3}$$

Thus, we have that $\mathbf{r}(T(f)) = C\mathbf{d}(f)$, and so $\mathbf{r}(T(f)) = CA\mathbf{r}(f)$. Iterating this equation then gives

$$\mathbf{r}(T^{\circ n}(f)) = (CA)^n \mathbf{r}(f).$$

The p norm of f is given by

$$\|f\|_p^p = \int_{I^2 = \cup_i R_i} |f|^p = \sum_{i=1}^N \int_{R_i} |f|^p \tag{11.4}$$

$$= \mathbf{1} \cdot \mathbf{r}(f) \tag{11.5}$$

and so

$$\|T^{\circ n}(f)\|_p^p = \mathbf{1} \cdot \mathbf{r}(T^{\circ n}(f)) = \mathbf{1} \cdot (CA)^n \mathbf{r}(f).$$

We are interested in bounding $\|T^{\circ n}(f)\|_p$, so that means we want to know the growth rate of $(CA)^n \mathbf{r}(f)$. In the following section, we will state some results that will allow us to bound this growth rate in some special cases. The general case remains to be derived.

The Perron-Frobenius Theorem

Definition 11.1 *A matrix (or vector) $A = (a_{ij})$ is said to be* positive, *denoted $A > 0$, if $a_{ij} > 0$ for each i, j. A square matrix A is called* primitive *if A^n is positive for some n.*

Theorem 11.1 (Finite-Dimensional Perron-Frobenious Theorem) *Let Q be a primitive matrix. Then Q possesses an eigenvalue λ called the* leading eigenvalue *such that*

(a) $\lambda > |\mu|$ for all eigenvalues $\mu \neq \lambda$.

(b) The eigenspace for λ is 1-dimensional.

(c) The eigenvector $\mathbf{q}$ for λ, satisfies $\mathbf{q} > 0$.

Proof: See [32]. ∎

Lemma 11.1 *Let Q be a primitive matrix. Then for any $\mathbf{x}$ there exists $\mathbf{h} > 0$ such that*

$$\lim_{n \to \infty} \lambda^{-n} Q^n \mathbf{x} = (\mathbf{h} \cdot \mathbf{x})\mathbf{q},$$

where λ and $\mathbf{q}$ are the leading eigenvalue and eigenvector.

Proof: Let Y be the eigenspace for the all the eigenvalues other than λ, and $\mathbf{h}_1$ be a non zero vector perpendicular to Y. We can write

$$\mathbf{x} = a\mathbf{q} + \mathbf{y},$$

where $y \in Y$. Taking the dot product with $\mathbf{h}_1$ we get

$$(\mathbf{h}_1 \cdot \mathbf{x}) = a(\mathbf{h}_1 \cdot \mathbf{q}) + 0.$$

We can thus solve for a to get

$$a = \left(\frac{\mathbf{h}_1 \cdot \mathbf{x}}{\mathbf{h}_1 \cdot \mathbf{q}} \right).$$

We now let

$$\mathbf{h} = \left(\frac{\mathbf{h}_1}{\mathbf{h}_1 \cdot \mathbf{q}} \right).$$

Then we have

$$\mathbf{x} = (\mathbf{h} \cdot \mathbf{x})\mathbf{q} + \mathbf{y}.$$

Multiplying by $\lambda^{-n} Q^n$ we get

$$\lambda^{-n} Q^n \mathbf{x} = (\mathbf{h} \cdot \mathbf{x})\lambda^{-n} Q^n \mathbf{q} + \lambda^{-n} Q^n \mathbf{y} = (\mathbf{h} \cdot \mathbf{x})\mathbf{q} + \lambda^{-n} Q^n \mathbf{y}.$$

Since y has no component in the eigenspace spanned by $\mathbf{q}$ we have $\lambda^{-n} Q^n \mathbf{y} \to 0$, giving the desired result.

It still remains to show that $\mathbf{h} > 0$. Let $\mathbf{e}_i$ be the vector with a 1 in the i-th component and 0 everywhere else. Since $\lambda^{-n} Q^n \mathbf{e}_i \geq 0$ for all n we must have the limit $(\mathbf{h} \cdot \mathbf{e}_i)\mathbf{q} \geq 0$. Since $\mathbf{q} > 0$, we must have that $\mathbf{h} \cdot \mathbf{e}_i \geq 0$ showing $\mathbf{h} \geq 0$. We now prove by contradiction that $\mathbf{h} \cdot \mathbf{e}_i \neq 0$. If $\mathbf{h} \cdot \mathbf{e}_i = 0$ then $\mathbf{e}_i$ is in Y. Since Q is primitive we must have for large n that $Q^n \mathbf{e}_i > 0$. Since Y is invariant under Q we have that $Q^n \mathbf{e}_i$ is in Y and thus $Q^n \mathbf{e}_i \cdot \mathbf{h} = 0$. This is a contradiction since the dot product of a strictly positive vector with a positive non-zero vector is strictly positive. ∎

We are now in a position to give a theorem about the growth rate of the operator T in the case where CA is a primitive matrix.

Theorem 11.2 *If CA is a primitive matrix with leading eigenvalue λ_p, then there exist $c_1 > 0$ and $c_2 > 0$ which are independent of f such that*

$$c_1 \lambda^n \|f\|_p \leq \|T^n f\|_p \leq c_2 \lambda^n \|f\|_p,$$

where $\lambda = \lambda_p^{1/p}$.

Proof: Recall that

$$\|T^n f\|_p^p = \mathbf{1} \cdot (CA)^n \mathbf{r}(f) \quad \text{and} \quad \|f\|_p^p = \mathbf{1} \cdot \mathbf{r}(f).$$

Let Q be the transpose of the matrix CA. We then have that

$$\mathbf{1} \cdot (CA)^n \mathbf{r}(f) = Q^n \mathbf{1} \cdot \mathbf{r}(f).$$

It thus suffices to prove that

$$c_1^p \mathbf{1} \cdot \mathbf{r}(f) \leq \lambda_p^{-n} Q^n \mathbf{1} \cdot \mathbf{r}(f) \leq c_2^p \mathbf{1} \cdot \mathbf{r}(f).$$

The matrix Q is also primitive and has the same leading eigenvalue λ_p as CA. Applying the previous corollary we have

$$\lambda_p^{-n} Q^n \mathbf{1} = (\mathbf{h} \cdot \mathbf{1})\mathbf{q} + \mathbf{y}_n,$$

where $\mathbf{q} > 0$, $\mathbf{h} > 0$, and $\mathbf{y}_n \to 0$ as $n \to \infty$. Since $\mathbf{1} > 0$ we have that every vector in the sequence $\lambda_p^{-n} Q^n \mathbf{1} > 0$. In addition the limit $(\mathbf{h} \cdot \mathbf{1})\mathbf{q} > 0$. Thus, every component of every vector is bounded away from 0 and infinity. So there exist $c_1^p > 0$ and $c_2^p > 0$, so that every component of the sequence is between them. Sybolically we can write

$$c_1^p \mathbf{1} \leq \lambda_p^{-n} Q^n \mathbf{1} \leq c_2^p \mathbf{1}.$$

Since $\mathbf{r}(f) \geq 0$, it makes sense to take dot products of the above with $\mathbf{r}(f)$, giving

$$c_1^p \mathbf{1} \cdot \mathbf{r}(f) \leq \lambda_p^{-n} Q^n \mathbf{1} \cdot \mathbf{r}(f) \leq c_2^p \mathbf{1} \cdot \mathbf{r}(f),$$

finishing the theorem. ∎

When Q is not primitive, the situation is more complicated, as the following examples show:

Example 3: The matrix

$$A = \begin{bmatrix} 1 & 1 \\ 0 & 1 \end{bmatrix}$$

is not primitive with eigenvalue 1. In this case, the above theorem fails, since, in general, the growth rate $A\mathbf{x}$ is bounded by $n\mathbf{x}$, not $1^n = 1$.

Example 4: It is easy to see that the matrix

$$\begin{bmatrix} 2 & 0 \\ 0 & 3 \end{bmatrix}$$

has two growth rates.

The details of the general case can be found in [32].

11.3   Almost Everywhere Convergence

We now focus on deriving conditions for almost everywhere convergence of τ. Again, we must restrict to a special case. However, this should not be necessary, so that an interesting and complete theory can be derived.

We begin with pointwise convergence. We define

$$S_n(x) = \prod_{i=0}^{n-1} s(m^{\circ i}(x)),$$

with $S_0(x) = 1$.

Lemma 11.2 *Suppose $|s(x)| < r < 1$ and suppose that f and o are bounded. Then $\tau_{m,s,o}^{on}(f)$ converges pointwise to some bounded function in $L^p(I^2)$.*

Proof: We compute,

$$
\begin{aligned}
\tau_{m,s,o}(f) &= s(x)f(m(x)) + o(x) \\
\tau_{m,s,o}^{o2}(f) &= s(x)\big(s(m(x))f(m^{o2}(x)) + o(m(x))\big) + o(x) \\
&= s(x)s(m(x))f(m^{o2}(x)) + s(x)o(m(x)) + o(x) \\
&\vdots \qquad \vdots \\
\tau_{m,s,o}^{on}(f) &= S_n(x)\,f(m^{on}(x)) + \sum_{i=0}^{n-1} S_i(x)\,o(m^{oi}(x))
\end{aligned}
$$

But $S_n(x) < r^n$, so $\tau_{m,s,o}^{on}(f)$ converges.   ∎

In fact, having $|s(x)| < r < 1$ is sufficient for L^p convergence as well; but this condition does not imply contractivity of τ, just eventual contractivity. The pointwise convergence is just as good (if not better) as a theoretical model, and considerably simpler. The hypothesis that f is bounded is necessary, however, as the following example shows.

Example 5: Suppose f is not bounded. Let $x_n \in I^2$ be a sequence such that $r^n f(x_n)$ diverges to ∞. Then if we let $m(x_n) = x_{n+1}$ and $m(x) = x_1$ for all other x, we see that the leading term in $\tau_{m,s,o}^{on}(f)$ diverges.

The following section provides some background from ergodic theory.

The Ergodic Theorem

Definition 11.2 *We say that a μ measurable map $T : X \to X$ is measure preserving, and that μ is invariant under T, if for every $A \subset X$ we have $\mu(A) = \mu(T^{-1}(A))$. A set A is called T-invariant if $T^{-1}(A) = A$. T is ergodic with respect to μ if every T-invariant set has μ measure 0 or 1.*

Theorem 11.3 (Birkhoff) *If T is measure preserving for μ, and if $f \in L^1(\mu)$, then the limit*

$$
\tilde{f}(x) = \lim_{n \to \infty} \frac{1}{n} \sum_{j=0}^{n-1} f(T^{oj}(x))
$$

exists for μ almost all x, and $\tilde{f}(x)$ is in $L^1(\mu)$.

Proof: See [61] Theorem II.1.1.   ∎

Theorem 11.4 *If T is measure preserving and ergodic with respect to μ, then $\tilde{f}(x) = \int f\,d\mu$ for μ almost all x.*

Proof: See [61] Proposition II.2.1.   ∎

These theorems will give results about μ almost all points, but from an engineering point of view we are interested in almost all points with respect to Lebesgue measure. We will thus only be interested in measures μ which have the same σ–algebra and sets of measure zero as Lebesgue measure in which case we will say μ is *regular* with respect to Lebesgue measure. Such measures are characterized by Radon-Nikodym Theorem (see [61] 0.4.1) as follows.

$$\mu(A) = \int_A f(x)dx,$$

where $f(x) \geq 0$ and the set $f^{-1}(0)$ has Lebesgue measure 0.

So we will be interested in μ-measure preserving, ergodic maps which are regular with respect to Lebesgue measure. A simple example given in the next section is when μ is Lebesgue measure.

The Case $m(x) = 2x$ **mod** 1

In this section we consider the special case $m(x) = 2x$ mod 1. It should be noted that this corresponds to the IFS consisting of the four transformations

$$w_1 \begin{bmatrix} x \\ y \end{bmatrix} = \begin{bmatrix} \frac{1}{2} & 0 \\ 0 & \frac{1}{2} \end{bmatrix} \begin{bmatrix} x \\ y \end{bmatrix} + \begin{bmatrix} 0 \\ 0 \end{bmatrix}$$

$$w_2 \begin{bmatrix} x \\ y \end{bmatrix} = \begin{bmatrix} \frac{1}{2} & 0 \\ 0 & \frac{1}{2} \end{bmatrix} \begin{bmatrix} x \\ y \end{bmatrix} + \begin{bmatrix} 0 \\ \frac{1}{2} \end{bmatrix}$$

$$w_3 \begin{bmatrix} x \\ y \end{bmatrix} = \begin{bmatrix} \frac{1}{2} & 0 \\ 0 & \frac{1}{2} \end{bmatrix} \begin{bmatrix} x \\ y \end{bmatrix} + \begin{bmatrix} \frac{1}{2} \\ 0 \end{bmatrix}$$

$$w_4 \begin{bmatrix} x \\ y \end{bmatrix} = \begin{bmatrix} \frac{1}{2} & 0 \\ 0 & \frac{1}{2} \end{bmatrix} \begin{bmatrix} x \\ y \end{bmatrix} + \begin{bmatrix} \frac{1}{2} \\ \frac{1}{2} \end{bmatrix}.$$

In particular, the results of this section can be easily generalized to any non-overlapping IFS.

First, note that $m(x)$ preserves Lebesgue measure: if $A \subset I^2$, then $m^{-1}(A)$ consists of four smaller copies of A, each with one forth the measure of A. Next, note that $m(x)$ is ergodic. We can therefore apply the theorems above to show that:

Theorem 11.5 *Assume $m(x) = 2x$, $\log|s(x)| \in L^1(I^2)$ and $o(x)$ is bounded. Let*

$$\log \gamma = \int_{I^2} \log|s(x)|d\mu,$$

where μ is the Lebesgue measure. The iterates $\tau^{\circ n}(f_0)$ converge to a fixed point f, independent of f_0, almost everywhere if $\gamma < 1$, and diverge almost everywhere (with the exception of the case $f_0(x) = o(x) = 0$) if $\gamma > 1$.

Proof: By Theorems 11.3 and 11.4, we know that $\lim \frac{1}{n}\log|S_n(x)|$ converges a.e. to the constant $\log(\gamma)$. If $\gamma < 1$ then for almost every x there exist $C > 0$ and $\lambda < 1$ (C and λ depend on x) so that $|S_n(x)| < C\lambda^n$. Similarly, if $\gamma > 1$, there exist $C > 0$ and $\lambda > 1$ so that $|S_n(x)| > C\lambda^n$.

The iterates

$$\tau^{\circ n}(f_0)(x) = S_n(x) f_0(m^{\circ n}(x)) + \sum_{i=0}^{n-1} S_i(x) o(m^{\circ i}(x)).$$

will thus converge for almost every x if $\gamma < 1$, and diverge for almost every x if $\gamma > 1$ (with the exception of the case $f_0(x) = o(x) = 0$).

What happens in the case $\int_{I^2} \log s(x) d\mu = 0$? In this case the growth rate for S_n could (without further proof) be n or $1/n$. The first case diverges, the second converges. ∎

The General Case

Now assume that $m(x)$ preserves some measure μ that is regular with respect to Lebesgue measure and that $m(x)$ is ergodic with respect to μ. We get the same theorem as we did in the previous section with no significant change in the proof.

Theorem 11.6 *Assume* $\log |s(x)| \in L^1(I^2, \mu)$ *and* $o(x)$ *is bounded. Let*

$$\log \gamma = \int_{I^2} \log |s(x)| d\mu.$$

The iterates $\tau^{\circ n}(f_0)$ *converge a.e. to a fixed point* f, *independent of* f_0, *if* $\gamma < 1$, *and diverge a.e. (with the exception of the case* $f_0(x) = o(x) = 0$) *if* $\gamma > 1$.

The question is, given a map m how do we find a μ satisfying all the conditions above? Results from dynamical systems theory imply that such maps are common. They are called expanding maps.

Definition 11.3 *A partition* $\mathcal{P}$ *of a set* X *is a set of subsets of* X *such that*

$$X = \bigcup_{P \in \mathcal{P}} P,$$

subject to the condition that for $A \neq B$ *both in* $\mathcal{P}$ *we have* $A \cap B$ *has measure 0.*

For example, the set of ranges described earlier form a partition of I^2.

Definition 11.4 *We say a map* $m : I^2 \to I^2$ *is* expanding *if there exist a sequence of partitions of* I^2, $\mathcal{P}_0, \mathcal{P}_1, \ldots$, *such that*
1) For every $n \geq 0$ *and* $P \in \mathcal{P}_{n+1}$, $m(P)$ *is a union (mod sets of measure 0) of elements of* $\mathcal{P}_n$, *and* m *restricted to* P *is one to one.*
2) The map m *is eventually expanding. That is, there exist* $K > 0$ *and* $\lambda > 1$ *such that for every* $n \geq 0$, *for every* $P \in \mathcal{P}_n$, *and for every* x *and* y *in* P,

$$|m^n(x) - m^n(y)| \geq K\lambda^n |x - y|.$$

3) The map m *is primitive. That is there exists a* $k > 0$ *such that for any two elements* P *and* Q *of* $\mathcal{P}_0$ *we have* $m^{-k}(P) \cap Q$ *is not empty.*
4) The function $\log J$ *is Hölder continuous. That is, the jacobian* J *exists on every element of* $\mathcal{P}_0$, *and there exist* $C > 0$ *and* $0 < \gamma < 1$ *such that for any* x *and* y *in any element of* $\mathcal{P}_n$ *we have*

$$\left| \frac{J(y)}{J(x)} - 1 \right| \leq C |m(x) - m(y)|^\gamma.$$

Let us take a closer look at each of these conditions. A typical way of satisfying condition 1 would be to have $m(P)$ be an element of $\mathcal{P}_n$ rather than a union of many elements. Thus, take $\mathcal{P}_0$ to be a partition of I^2 into small sets. The elements of $\mathcal{P}_{n+1}$ will then be the components of the inverse image of the elements of $\mathcal{P}_n$. For example in the RIFS model described earlier we let $\mathcal{P}_0$ be the partition into ranges. Since every range R_i maps to a union of other ranges $R_{i_1}, \ldots, R_{i_k}$ in a 1-to-1 way, the range R_i will be made up of a union of inverse images $m^{-1}(R_{i_j})$. Those inverse images will form $\mathcal{P}_1$.

Condition 2 is the standard condition used throughout the book that the map is eventually expanding (i.e., the inverses are eventually contracting). The easiest way to satisfy this is to just have the map be expanding after 1 step.

Condition 3 says that the map mixes things up well. In the RIFS model this condition is exactly the same as the condition that CA is primitive. The value for k is the smallest value for which $(CA)^k > 0$.

Condition 4 is just a technical condition needed for the proof of the following theorem. If J is bounded away from 0 and differentiable on each element of each partition, the condition will be satisfied.

Theorem 11.7 *If m is an expanding map, there exists a unique measure μ which is regular with respect to Lebesgue measure, and m preserves μ and is ergodic with respect to μ.*

Proof: See [61] theorem III.1.3.   ∎

The next question is how do we find μ and compute

$$\int_{I^2} \log |s(x)| d\mu?$$

First, partition I^2 into very small sets P_i. The pixels used to make the image would be an obvious choice for P_i. Using Birkhoff's ergodic theorem we can estimate $\mu(P_i)$ as

$$\frac{1}{N} \sum_{n=0}^{N-1} \chi_{P_i}(m^n(x)),$$

where χ_{P_i} is the characteristic function for P_i and x is a random starting point. In other words iterate m from a random starting point, and count how many iterates land in each pixel. Next, divide each count by the total number of iterates (which should be very large) to get the measure of each pixel. Note that the counts can be used to create a computer picture of the measure. The integral can be estimated by

$$\int_{I^2} \log |s(x)| d\mu \approx \sum_i \log |s(x_i)| \mu(P_i),$$

where x_i is a point in pixel P_i.

11.4   Decoding by Matrix Inversion

When we deal with an $M \times N$ pixel image $f(i, j)$, $0 \le i < M$, $0 \le j < N$ we write the finite-dimensional version of τ as

$$\tau(f(i, j)) = s(i, j) f(m(i, j)) + o(i, j).$$

When decoding, we take an initial pixel image f^0 and compute

$$f^n = \tau(f^{n-1}) \tag{11.6}$$

until we are sufficiently close to the fixed point.

Denote by $(i \div N)$ the integral number of times N divides i and let $(i \bmod N)$ denote the remainder. Suppose we wish to decode an image $f(i, j)$ of resolution $M \times N$. We can write f as an MN column vector

$$f_i = f(i \div N, i \bmod N).$$

Encode both s and m as an $MN \times MN$ matrix by letting

$$S_{i,j} = \begin{cases} s(m(i \div N, i \bmod N)) & \text{if } ((N, 1) \cdot m(i \div N, i \bmod N)) = j \\ 0 & \text{otherwise} \end{cases},$$

where $\cdot$ is the standard dot product, $(a, b) \cdot (c, d) = ac + bd$. Set the MN column vector

$$O_i = o(i \div N, i \bmod M).$$

Equation (11.6) can then be written as the matrix equation

$$f^n = S f^{n-1} + O.$$

Note that by assuming that $m(i, j)$ exists, we are assuming that each pixel of $\tau(f)$ depends on just one pixel value of f. In that case, S contains at most one non-zero number per row. This is a fine assumption, though in practice it is also useful to allow a pixel of $\tau(f)$ to depend on several pixel values.[4]

Then

$$f^n = S^n f^0 + \sum_{j=1}^{n} S^{j-1} O.$$

When the first term is 0 in the limit, and when $I - S$ is invertible, the fixed point is

$$f^\infty = \sum_{j=0}^{\infty} S^j O = (I - S)^{-1} O,$$

where I is the identity matrix.

When each pixel value $f^n(i, j)$ depends on only one other pixel value $f^{n-1}(m(i, j))$, this matrix is very sparse and can be readily inverted.

A non-optimized version of this method was implemented with expected results and decoding times comparable to the iterative decoding method.

---

[4]In this case, the notation should be modified to

$$\tau(f)(x) = s(x) \sum_i f(m_i(x)) + o(x),$$

where the sum can be used to average over the different pixels refered to by several functions $m_i(x)$.

Chapter 12

Least-Squares Block Coding by Fractal Functions

F. Dudbridge

The image coding scheme described in this chapter represents an attempt to achieve image compression using, as closely as possible, the "classical" theory of strictly self-similar fractals. Image compression is achieved by establishing a partition of an image into adjoining rectangular blocks, and coding each block individually by an iterated function system together with a type of fractal function. The method differs from other fractal compression techniques in that an image is regarded as an assembly of essentially independent blocks, so that any block will have the same code regardless of the rest of the image. In this respect the method is necessarily inferior to others when encoding an entire image, but the work is of interest since it provides a nearly optimal code for a single block, which may be obtained in linear time. Moreover, this work has the theoretical interest of being the only direct solution to an inverse problem in fractal geometry yet known.

We will set up a theory of fractal functions, and will describe how a fractal code may be associated with a grey-scale image block. It will then be possible, using a least-squares technique, to obtain fractal codes for any given block. We will show how this may be achieved, and we will give an efficient decoding algorithm for constructing the fractal approximation block. Certain implementation issues will also be addressed.

12.1 Fractal Functions

We consider functions defined on particular sub-attractors of iterated function systems. In the graphical case, these sub-attractors will correspond to pixels. Let K be a complete normed vector space and let $w_1, \ldots, w_N$ be an IFS of order N with attractor $A \subset K$. Let m be a nonnegative integer, termed the *resolution*. We consider the set

$$P_m = \{A_{k_1 \ldots k_m}; k_1, \ldots, k_m = 1, \ldots, N\},$$

where $A_{k_1 \ldots k_m}$ denotes $w_{k_1} \circ \cdots \circ w_{k_m}(A)$. A *grey-scale function* is any function $f : P_m \to \mathbb{R}$. The space of all grey-scale functions at resolution m is denoted by $\mathcal{F}_m$. We define fractal functions in a similar way to IFS, by introducing a contraction mapping on $\mathcal{F}_m$. We use mappings of the form

$$Mf(A_{k_1 \cdots k_m}) = \int_{A_{k_1 \cdots k_m}} (c_{k_1} \cdot x + t_{k_1}) dx + s_{k_1} \sum_{i=1}^{N} f(A_{k_2 \cdots k_m i})$$

The parameters of the mapping are the s_k and t_k, which are real constants, and the c_k which are elements of K. This may be interpreted graphically as meaning that the new value of a pixel depends upon the current values of those pixels mapped into it by the IFS, and on the position of the pixel. Note that M is affine in f.

To obtain a least-squares approximation with respect to the parameters c_k, s_k and t_k, we define the Euclidean metric on $\mathcal{F}_m$ and then derive the contractivity condition in that metric.

Definition 12.1 *Let* $w_1, \ldots, w_N, A, m, \mathcal{F}_m$ *be as above. The* Euclidean metric at resolution *m is defined by*

$$d_m(f, g) = \left\{ \sum_{k_1 \cdots k_m = 1}^{N} \left[(f - g)(A_{k_1 \ldots k_m}) \right]^2 \right\}^{\frac{1}{2}}$$

for two grey-scale functions $f, g \in \mathcal{F}_m$.

Theorem 12.1 *Let M be a mapping on $\mathcal{F}_m$, of the above form, defined for an IFS $w_1, \ldots, w_N$. If* $\left| \sum_{k=1}^{N} s_k \right| < 1$, *then M is eventually contractive in the Euclidean metric at any resolution.*

Proof: First notice that

$$M^{\circ m} f(A_{k_1 \cdots k_m}) = s_{k_1} \sum_{i_1=1}^{N} M^{\circ (m-1)} f(A_{k_2 \cdots k_m i_1}) + \text{a term independent of } f$$

$$= s_{k_1} \cdots s_{k_m} \sum_{i_1 \cdots i_m = 1}^{N} f(A_{i_1 \cdots i_m}) + \cdots$$

Now

$$M^{\circ (m+1)} f(A_{k_1 \cdots k_m}) = s_{k_1} \cdots s_{k_m} \sum_{i_1 \cdots i_m = 1}^{N} Mf(A_{i_1 \cdots i_m}) + \cdots$$

$$= s_{k_1} \cdots s_{k_m} \sum_{i_1 \cdots i_m = 1}^{N} s_{i_1} \sum_{i_{m+1}=1}^{N} f(A_{i_2 \cdots i_{m+1}}) + \cdots$$

and similarly we can write

$$M^{\circ(m+n)} f(A_{k_1 \ldots k_m}) = s_{k_1} \cdots s_{k_m} \sum_{i_1 \cdots i_{m+n}=1}^{N} s_{i_1} \cdots s_{i_n} f(A_{i_{n+1} \cdots i_{m+n}}) + \cdots$$

In the Euclidean metric at resolution m, consider

$$d_m(M^{\circ(m+n)} f, M^{\circ(m+n)} g) = \left\{ \sum_{k_1 \cdots k_m=1} \left[(M^{\circ(m+n)} f - M^{\circ(m+n)} g)(A_{k_1 \cdots k_m}) \right]^2 \right\}^{\frac{1}{2}}$$

$$= \left\{ \sum_{k_1 \cdots k_m=1}^{N} \left[s_{k_1} \cdots s_{k_m} \sum_{i_1 \cdots i_{m+n}=1}^{N} s_{i_1} \cdots s_{i_n}(f - g)(A_{i_{n+1} \cdots i_{m+n}}) \right]^2 \right\}^{\frac{1}{2}}$$

$$= \left\{ \sum_{k_1 \cdots k_m=1}^{N} \left(s_{k_1} \cdots s_{k_m} \right)^2 \left[\sum_{i_1 \cdots i_n=1}^{N} s_{i_1} \cdots s_{i_n} \right]^2 \left[\sum_{p \in P_m} (f - g)(p) \right]^2 \right\}^{\frac{1}{2}}.$$

Now

$$\left[\sum_{p \in P_m} (f - g)(p) \right]^2 \le N^m d_m(f, g)^2.$$

from the inequality $\left[\sum_{i=1}^{n} x_i \right]^2 \le n \sum_{i=1}^{n} x_i^2$. Also,

$$\sum_{i_1 \cdots i_n=1}^{N} s_{i_1} \cdots s_{i_n} = \left[\sum_{k=1}^{N} s_k \right]^n.$$

Thus

$$d_m(M^{\circ(m+n)} f, M^{\circ(m+n)} g) \le N^{\frac{m}{2}} \left[\sum_{k_1 \cdots k_m=1}^{N} \left(s_{k_1} \cdots s_{k_m} \right)^2 \right]^{\frac{1}{2}} \left| \sum_{k=1}^{N} s_k \right|^n d_m(f, g).$$

Hence, if $\left| \sum_{k=1}^{N} s_k \right| < 1$, one can find a sufficiently large n so that

$$N^{\frac{m}{2}} \left[\sum_{k_1 \cdots k_m=1}^{N} \left(s_{k_1} \cdots s_{k_m} \right)^2 \right]^{\frac{1}{2}} \left| \sum_{k=1}^{N} s_k \right|^n < 1,$$

implying that $M^{\circ(m+n)}$ is contractive on $\mathcal{F}_m$. Hence, M is eventually contractive. ∎

A grey-scale mapping that is eventually contractive in all Euclidean metrics at all resolutions is termed *attractive*. It is found in practice that the condition on M above is sufficiently generous that the codes we obtain for image blocks are almost always attractive. By the Contractive Mapping Fixed-Point Theorem, there is a unique *invariant function* for each M, and the recursive definition of this function implies a fractal character similar to that for the attractor of an IFS. In fact, under a more restrictive contractivity condition, the parameters defining a fractal function may be used to construct an IFS in a higher-dimensional space, whose attractor is a fractal "graph" [21]. The Collage Theorem may now apply as before:

$$d_m(f, g) \le (1 - s)^{-1} d_m(g, Mg),$$

where g is a given grey-scale function, and f is the fixed point of mapping M. The theorem may now be applied to encode an image block.

12.2   Least-Squares Approximation

The fractal code for an image block consists of an iterated function system $w_1, \ldots, w_N$ of affine transformations, and the grey-scale mapping M defined above. This section describes how a fractal code for a given grey-scale function is obtained, before applying the method to the specific example of an image block.

If a function $g \in \mathcal{F}_m$ is given on the set P_m of attractor A, we assume that an IFS W is known whose attractor is A. The problem is then to find the grey-scale mapping M whose invariant function f best approximates g. A least-squares approach is used at resolution m, so that M is sought to minimize $d_m(f, g)$.

From the Collage Theorem, a fractal approximant to g may be found by minimizing $d_m(g, Mg)$ with respect to the parameters of M. This does not guarantee the best possible fractal match, since only part of the upper bound for $d_m(f, g)$ is minimized, and one assumes that the contractivity factor for M does not significantly affect the bound. Furthermore, if M is only eventually contractive, it is also assumed that M will also minimize $d_m(g, M^{\circ n} g)$ for suitably large n, which is not necessarily true. However, in spite of these assumptions, the method does give good approximations.

In image compression applications, the attractor A will usually be a rectangle, such as that given by the IFS

$$w_1 \begin{bmatrix} x \\ y \end{bmatrix} = \begin{bmatrix} \frac{1}{2} & 0 \\ 0 & \frac{1}{2} \end{bmatrix} \begin{bmatrix} x \\ y \end{bmatrix} + \begin{bmatrix} -\frac{x_0}{2} \\ \frac{y_0}{2} \end{bmatrix}$$

$$w_2 \begin{bmatrix} x \\ y \end{bmatrix} = \begin{bmatrix} \frac{1}{2} & 0 \\ 0 & \frac{1}{2} \end{bmatrix} \begin{bmatrix} x \\ y \end{bmatrix} + \begin{bmatrix} \frac{x_0}{2} \\ \frac{y_0}{2} \end{bmatrix}$$

$$w_3 \begin{bmatrix} x \\ y \end{bmatrix} = \begin{bmatrix} \frac{1}{2} & 0 \\ 0 & \frac{1}{2} \end{bmatrix} \begin{bmatrix} x \\ y \end{bmatrix} + \begin{bmatrix} -\frac{x_0}{2} \\ -\frac{y_0}{2} \end{bmatrix}$$

$$w_4 \begin{bmatrix} x \\ y \end{bmatrix} = \begin{bmatrix} \frac{1}{2} & 0 \\ 0 & \frac{1}{2} \end{bmatrix} \begin{bmatrix} x \\ y \end{bmatrix} + \begin{bmatrix} \frac{x_0}{2} \\ -\frac{y_0}{2} \end{bmatrix}$$

and illustrated in Figure 12.1.

In this case we have $K = \mathbb{R}^2$, and the parameter c_k in the grey-scale mapping is a Cartesian pair (a_k, b_k). We can now minimize $d_m(g, Mg)$ by minimizing $d_m(Mg, g)^2$, using standard calculus:

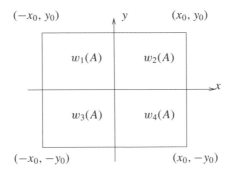

Figure 12.1: Tiling of a rectangle by reduced rectangles.

$$d_m(Mg, g)^2 = \sum_{k_1 \cdots k_m = 1}^{N} \left[(Mg - g)(A_{k_1 \cdots k_m}) \right]^2$$

$$= \sum_{k_1 \cdots k_m = 1}^{N} \left[\int_{A_{k_1 \cdots k_m}} a_{k_1} x + b_{k_1} y + t_{k_1} dx dy \right.$$

$$\left. + s_{k_1} \sum_{i=1}^{N} g(A_{k_2 \cdots k_m i}) - g(A_{k_1 \cdots k_m}) \right]^2.$$

We can economize on notation by letting $v(p) = \sum_{i=1}^{N} g(A_{k_2 \cdots k_m i})$ and $k = k_1$ when $p = A_{k_1 \cdots k_m}$.
Then

$$d_m(Mg, g)^2 = \sum_{p \in P_m} \left[\int_p (a_k x + b_k y + t_k) \, dx dy + s_k v(p) - g(p) \right]^2$$

$$= \sum_{k=1}^{N} \sum_{p \in w_k(P_{m-1})} \left[\int_p (a_k x + b_k y + t_k) \, dx dy + s_k v(p) - g(p) \right]^2$$

This may be minimized by minimizing

$$\sum_{p \in w_k(P_{m-1})} \left[\int_p (a_k x + b_k y + t_k) \, dx dy + s_k v(p) - g(p) \right]^2$$

for each k, because these sums are independent for a nonoverlapping attractor. We are therefore
minimizing the sum separately for each tile of the attractor. This yields the system

$$\begin{bmatrix} \sum_p (\int_p x)^2 & \sum_p \int_p x \int_p y & \sum_p \int_p x \int_p 1 & \sum_p v(p) \int_p x \\ \sum_p \int_p x \int_p y & \sum_p (\int_p y)^2 & \sum_p \int_p y \int_p 1 & \sum_p v(p) \int_p y \\ \sum_p \int_p x \int_p 1 & \sum_p \int_p y \int_p 1 & \sum_p (\int_p 1)^2 & \sum_p v(p) \int_p 1 \\ \sum_p v(p) \int_p x & \sum_p v(p) \int_p y & \sum_p v(p) \int_p 1 & \sum_p (v(p))^2 \end{bmatrix} \begin{bmatrix} a_k \\ b_k \\ t_k \\ s_k \end{bmatrix} = \begin{bmatrix} \sum_p g(p) \int_p x \\ \sum_p g(p) \int_p y \\ \sum_p g(p) \int_p 1 \\ \sum_p g(p) v(p) \end{bmatrix}, \tag{12.1}$$

which can be solved for each k to obtain the approximating grey-scale operator. Note that all the sums are for $p \in w_k(P_{m-1})$.

Now consider a square image block of side R pixels, where R is a power of 2. The coordinates of the pixels in the block are normalized so that the block is represented by a square centered on the origin, with a corner at $(\frac{R}{2}, \frac{R}{2})$. Then A is a square; each $A_k = w_k(A)$ matches a quadrant of the block, and each $A_{k_1 \dots k_m}$ matches an individual pixel, where $m = \log_2(R)$; see Figure 12.2. Thus, if M is minimized at resolution $\log_2(R)$, we are performing least-squares optimization at the sampling resolution.

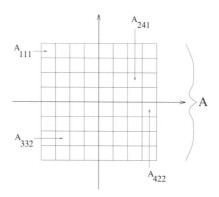

Figure 12.2: Tiling of an 8×8 block at resolution 3.

In this case P_m is the set of pixels in the image block, and the summations over $w_k(P_{m-1})$ are over the pixels in a quadrant. Each $g(p)$ is the value of the corresponding pixel in the given block, and $v(p)$ is the sum of the pixel values in $w_k^{-1}(p)$.

To simplify the solution of the Equation (12.1), we can move the origin to the center of each quadrant. We will then have $\sum_p \int_p x \int_p 1 = 0$ and $\sum_p \int_p y \int_p 1 = 0$, also $\sum_p \int_p x \int_p y = 0$, making the inversion of the matrix easier. We then obtain the best grey-scale mapping having the form

$$Mg(p) = \int_p a_k(x - x_k) + b_k(y - y_k) + t_k dx dy + s_k v(p),$$

where (x_k, y_k) is the center of the quadrant k, so on renormalizing we just subtract the obtained value of $a_k x_k + b_k y_k$ from the obtained value of t_k to get the correct value of t_k.

The method can now be illustrated on an image block taken from the image of the Mandrill which appears in Appendix E. The block is on the edge of the mouth on the left-hand side, about halfway down the image:

Figure 12.3: The decoded Mandrill image.

99	96	121	168	151	119	96	65
110	141	167	156	115	60	48	47
153	153	109	58	64	54	50	84
113	99	54	28	30	40	57	79
79	63	47	19	44	72	52	76
69	45	70	80	90	124	113	138
85	124	129	149	134	140	121	124
124	154	159	139	106	103	82	94

A rendering of this block is given in Figure 12.4.

This is an 8×8 block, so the appropriate IFS is given by Equation (12.1) with $(x_0, y_0) = (4, 4)$. To obtain least-squares per pixel, we approximate at resolution 3. Let us find a_k, b_k, t_k and s_k for $k = 1$. The sums in the matrix equation are then for $p = A_{1k_1k_2}$, k_1 and k_2 taking values between 1 and 4.

$$v(A_{k11}) = 99 + 96 + 110 + 141 = 446$$

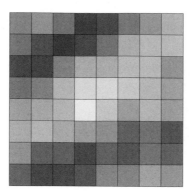

Figure 12.4: Rendering of the image block.

$$
\begin{aligned}
v(A_{k12}) &= 121 + 168 + 167 + 156 = 612 \\
v(A_{k13}) &= 153 + 153 + 113 + 99 = 518 \\
v(A_{k14}) &= 109 + 58 + 54 + 28 = 249
\end{aligned}
$$

for each k, for example; we use all these quantities to obtain

$$
\sum_{p\in w_1(P_2)} (v(p))^2 = 446^2 + 612^2 + 518^2 + 249^2 + \cdots = 2,619,798.
$$

The computation of the other terms in this summation is left as an exercise to the reader. Shifting the origin to $(-2, 2)$, we have

$$
\int_{A_{111}} x\,dx\,dy = \int_{-2}^{-1} \int_{1}^{2} x\,dx\,dy = -3/2.
$$

The other integrals can be worked out similarly, so that Equation (12.1) for $k = 1$ is

$$
\begin{bmatrix}
20 & 0 & 0 & -675 \\
0 & 20 & 0 & -290 \\
0 & 0 & 16 & 6,132 \\
-675 & -290 & 6,132 & 2,619,798
\end{bmatrix}
\begin{bmatrix}
a_1 \\
b_1 \\
t_1 \\
s_1
\end{bmatrix}
=
\begin{bmatrix}
-116.5 \\
335.5 \\
1,825 \\
645,457
\end{bmatrix}
$$

with the solution

$$
a_1 = -13.20,\ b_1 = 13.61,\ s_1 = -0.22,\ t_1 = 197.81.
$$

This is for the origin at $(-2, 2)$, so the correct solution for $k = 1$ comes from subtracting $-2a_1 + 2b_1$ from t_1, giving $t_1 = 144.20$.

The matrix in the left-hand side above will be the same for all the other values of k, because of the symmetry of A. We therefore need only calculate the right-hand side for the other k in the same manner as for $k = 1$. This gives

$$\begin{bmatrix} 20 & 0 & 0 & -675 \\ 0 & 20 & 0 & -290 \\ 0 & 0 & 16 & 6,132 \\ -675 & -290 & 6,132 & 2,619,798 \end{bmatrix} \begin{bmatrix} a_2 & a_3 & a_4 \\ b_2 & b_3 & b_4 \\ t_2 & t_3 & t_4 \\ s_2 & s_3 & s_4 \end{bmatrix} = \begin{bmatrix} -138.5 & 54.5 & 51.5 \\ 346.5 & -663.5 & -238.5 \\ 1,159 & 1,535 & 1,613 \\ 470,582 & 613,269 & 583,884 \end{bmatrix}$$

The complete solution, after normalization of the origin, is then

$$\begin{array}{llll} a_1 = -13.20 & b_1 = 13.61 & s_1 = -0.22 & t_1 = 144.20 \\ a_2 = -3.21 & b_2 = 18.92 & s_2 = 0.11 & t_2 = -1.23 \\ a_3 = 5.12 & b_3 = -32.15 & s_3 = 0.07 & t_3 = 14.72 \\ a_4 = -2.43 & b_4 = -14.08 & s_4 = -0.15 & t_4 = 134.40 \end{array}$$

Note that the sum of the s_k is -0.19, so that the grey-scale operator is attractive.

To encode blocks which are not squares whose side is a power of 2, the same approach can be used, but the sets p do not now correspond exactly to pixels. It would still be preferable to approximate at the closest resolution to $\log_4$ (total number of pixels), since this ensures that, on average, p is the same size as a pixel. For each p, one must find the pixels intersecting p and estimate $g(p)$ using a weighted sum of pixel values. The principle is illustrated in Figure 12.5. More calculation is required than for the simple case, and the result does not have quite the same accuracy.

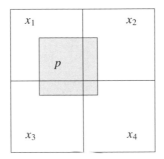

Figure 12.5: A case where p does not exactly match a pixel.

Another approach would be to regard each pixel as an approximate p, and compute the sums for the linear equations in the same way as for the square block. Thus, instead of finding all the sets p and computing weighted sums, one can just perform summations over pixels again. This approach is not as theoretically rigorous as the former method, but its computational requirement is simpler.

12.3 Construction of Fractal Approximation

Having obtained the fractal code for an image block, consisting of an IFS with an associated grey-scale operator, a method is required for constructing the approximation to the image block, that is, a decoding algorithm. A simple method exists for finding the fractal pixel values that is non-iterative and thus exact and fast.

Consider again the square block of side R, a power of 2. Then, at resolution $m = \log_2(R)$, each set $p \in P_m$ corresponds to a pixel in the block. If f is the invariant function for the grey-scale operator M, one must evaluate $f(p)$ for each p, and round to the nearest integer.

Let $f_{k_1 \cdots k_n} = \sum_{k_{n+1} \cdots k_m = 1}^{N} f(A_{k_1 \cdots k_m})$ when $n < m$. Thus f_{k_1} is the sum of all the pixel values in quadrant k_1. We have

$$
\begin{aligned}
f_{k_1} &= \sum_{k_2 \cdots k_m = 1}^{N} f(A_{k_1 \cdots k_m}) \\
&= \sum_{k_2 \cdots k_m = 1}^{N} (Mf)(A_{k_1 \cdots k_m}) \\
&= \sum_{k_2 \cdots k_m = 1}^{N} \int_{A_{k_1 \cdots k_m}} a_{k_1} x + b_{k_1} y + t_{k_1} \, dx dy + s_{k_1} \sum_{i=1}^{N} f(A_{k_2 \cdots k_m i}) \\
&= a_{k_1} \int_{A_{k_1}} x \, dx dy + b_{k_1} \int_{A_{k_1}} y \, dx dy + t_{k_1} \int_{A_{k_1}} 1 \, dx dy + s_{k_1} \sum_{i=1}^{N} f_i.
\end{aligned}
$$

Hence, summing for $k_1 = 1, \ldots, N$ we have

$$
\sum_{k=1}^{N} f_k = \frac{\sum_{k=1}^{N} a_k \int_{A_k} x + b_k \int_{A_k} y + t_k \int_{A_k} 1}{1 - \sum_{k=1}^{N} s_k}.
$$

This quantity, the total sum of pixel values in the decoded block, may be found directly from the fractal code. Moreover, we can then use it in the above to obtain f_{k_1} for each k_1. Now we can find $f_{k_1 k_2}$ in a similar way, because $f_{k_1} = \sum_{k_2 = 1}^{N} f_{k_1 k_2}$, and so, in a recursive fashion, we obtain all $f(p)$.

We may now apply this method to the fractal code obtained in the last section. We have

$$
\sum_{k=1}^{N} a_k \int_{A_k} x = -13.20 \times -32 - 3.21 \times 32 + 5.12 \times -32 - 2.43 \times 32 = 78.22
$$

and similarly

$$
\sum_{k=1}^{N} b_k \int_{A_k} y = 2520.11
$$

$$
\sum_{k=1}^{N} t_k \int_{A_k} 1 = 4673.45
$$

so that

$$
\sum_{k=1}^{N} f_k = \frac{78.22 + 2520.11 + 4673.45}{1 + (-0.19)} = 6132
$$

to the nearest integer, which matches the original block. Also,

$$f_1 = \int_0^4 \int_{-4}^0 (-13.20x + 13.61y + 144.20)\,dx\,dy - 0.22 \times 6132 = 1825,$$

and

$$f_2 = 1159$$
$$f_3 = 1535$$
$$f_4 = 1613$$

to nearest integers, which also match the original block. In fact, one may show that the fractal code to any given block will be exact at resolutions 0 and 1. The code is only exact at resolution 2 when this is the sampling resolution of the block. Now let us find the four pixel values in the top left of the block, which are f_{11k} for $k = 1, \ldots, 4$. We first need

$$f_{11} = \int_2^4 \int_{-4}^{-2} (-13.20x + 13.61y + 144.20)\,dx\,dy - 0.22 \times 1825 = 500,$$

and

$$f_{12} = 540$$
$$f_{13} = 454$$
$$f_{14} = 332$$

and now we can find

$$f_{111} = \int_3^4 \int_{-4}^{-3} (-13.20x + 13.61y + 144.20)\,dx\,dy - 0.22 \times 500 = 129,$$

and

$$f_{112} = 107$$
$$f_{113} = 125$$
$$f_{114} = 139$$

These compare with the original values of 99, 96, 110, and 141. The reader may wish to extend the calculations to the other pixels; to nearest integers, the fractal approximation block is

129	107	122	130	118	120	102	88
125	139	148	139	95	78	63	58
155	143	116	85	54	50	54	59
90	66	68	62	61	63	52	45
48	56	52	50	66	58	74	87
77	74	71	75	87	103	116	166
95	100	110	120	130	129	117	103
144	153	153	156	109	100	108	110

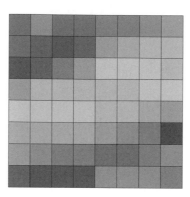

Figure 12.6: Decoded approximation to example block.

which has a PSNR of 22.7 dB of from the original block. The approximation block is rendered in Figure 12.6.

For more general rectangles, the same method is used to find the $f_{k_1 \cdots k_m}$, but these do not directly give pixel values. The value of a pixel will instead be a weighted sum over those sets p which intersect the pixel, as in Figure 12.5. The weights are determined by the sizes of the intersections. Again, more calculation is required, and the rendering is not exact, although still accurate.

Finally, consider the practical problem of storing the coefficients of the grey-scale operator. These are all unbounded real numbers; even the attractiveness condition $\left| \sum_{k=1}^{N} s_k \right| < 1$ does not impose any bound on the individual s_k. It is therefore difficult to assign a fixed number of bits to each coefficient, particularly as the decoding algorithm may be unstable with respect to M. However, a practical solution to this problem is possible. The fractal code for an image block consists of sixteen real numbers, and a resolution 2 block has sixteen pixels. By decoding the fractal function to resolution 2, we have an alternative representation for the code which proves to be more stable with respect to truncation error. Since the coding algorithm is exact at resolution 2, we may recover the original code at the decode stage by performing least-squares on the reduced resolution block.

In Figure 12.3 we give the decoded fractal approximation to the Mandrill's face. The original image is at 512×512 pixel resolution, with 8 bpp. The image is coded in 8×8 blocks; in each case the fractal code has been decoded to resolution 2 and then quantized to eight bits. The resultant bit rate is two bits per pixel, with a PSNR of 22.5 dB. The bit rates achieved by this method are usually inferior to those of other methods, at a given SNR, but may improve with further research. The method retains the advantage of linear complexity, which may prove useful in real-time applications.

12.4   Conclusion

We have described the theory and methods for encoding an image block using a standard iterated function system together with an affine grey-scale operator which defines a fractal approximation to the block. The methods are most exact for square blocks whose side length is a power of 2, but are easily extended to apply to general rectangular blocks. Encoding is achieved by solving a linear equation for each of four transformations in the IFS, and the coefficients of these equations are computed in linear time with the number of pixels in the block. The decoding algorithm is recursive and accurate, and is non-iterative, unlike other fractal construction algorithms. In order to make the storage of fractal codes more practical, it is preferable at the encoding stage to decode to resolution 2, and at the decoding stage to encode the resolution 2 block to recover the fractal code. This work is expected to be most applicable to real-time situations such as video compression and virtual reality, because of the computational efficiency of both encoding and decoding. One may also regard this work as completing the study of standard iterated function systems applied to image compression.

Acknowledgments

This work was supported in part by a SERC-NATO postdoctoral fellowship. The author wishes to thank Henry Abarbanel of the Institute for Nonlinear Science for the use of research facilities, and also thanks G. Dudbridge and A. P. Gaskell for their help in preparing the text. Yuval Fisher is also thanked for identifying an inconsistency in the original theory.

Chapter 13

Inference Algorithms for WFA and Image Compression

K. Culik II and J. Kari

In this chapter we will survey the encoding (inference) and decoding algorithms for weighted finite automata developed in [41] and [42], and their application to image-data compression.

In Section 13.1 we give a formal definition of finite resolution and a multiresolution image. Then we introduce our tool for image definition, namely, weighted finite automata (WFA). In general, WFA compute real functions of n variables – more precisely, functions $[0, 1]^n \to \mathbb{R}$. Their mathematical properties were studied in [39].

In Section 13.2 we discuss the "theoretical" inference algorithm from [41] that for a given multiresolution image infers a WFA that (approximately) regenerates the image. The inferred WFA has the minimal number of states but a relatively large number of edges. It is useful for theoretical applications [40], but not for efficient image compression. In [38] we have shown that WFA can easily implement Debauchies' wavelets or the whole wavelet transforms. We tried to explore this for image compression: First, we expressed the wavelet transform as a WFA and then simplified it using a modified version of our inference algorithm.

This approach produced better results than wavelets alone, but only marginally so.

In Section 13.3 we describe an efficient decoding algorithm for WFA, i.e., an algorithm that given a WFA and $k \geq 1$ generates the $2^k \times 2^k$ resolution image. Essentially, the same algorithm can also be used to generate the zooming image into any "binary" subsquare of the unit square. It is an important feature of WFA image representation that it is not only multiresolutional, i.e., it allows the regeneration of the image at any finite resolution, and it also allows efficient zooming. Section 13.4 is the most important one. Here we discuss the recursive inference algorithm for WFA introduced recently in [42]. Using this algorithm alone (without wavelets) we have obtained results comparable or better than wavelets alone. However, the best results are obtained by computing first the wavelets coefficients, representing them as an image (Mallat

Figure 13.1: The addresses of quadrants.

form) and then applying our new inference algorithm to this image and approximating it by a WFA. The decoding is the reverse of this process, using first the decoding algorithm for WFA and then the decoding algorithm for wavelets.

13.1 Images and Weighted Finite Automata

By a finite-resolution image we mean a digitized grey-scale picture that consists of $2^m \times 2^m$ pixels (typically, $7 \leq m \leq 11$), each of which takes a real value (in practice, digitized to a value between 0 and $2^k - 1$, typically, $k = 8$). By a multiresolution image, we mean a collection of compatible $2^n \times 2^n$ resolution images for $n = 0, 1, \ldots$. We will assign to each pixel at $2^n \times 2^n$ resolution a word of the length n over the alphabet $\Sigma = \{0, 1, 2, 3\}$. Each letter of Σ refers to one quadrant of the unit square as shown in Figure 13.1. We assign ε as the address of the root of the quadtree representing an image. Each letter of Σ is the address of a child of the root. Every word in $\Sigma^\star$ of length k, say w, is then an address of a unique node of the quadtree at depth k. The children of this node have addresses $w0$, $w1$, $w2$, and $w3$. Therefore, in our formalism a multiresolution image is a real function on $\Sigma^\star$. The compatibility of the different resolutions is formalized by requiring that $f : \Sigma^\star \to \mathbb{R}$ is an average preserving function. A function $f : \Sigma^\star \to \mathbb{R}$ is average preserving (ap) if

$$f(w) = \frac{1}{4} [f(w0) + f(w1) + f(w2) + f(w3)] \tag{13.1}$$

for each $w \in \Sigma^\star$.

An ap-function f is represented by an infinite labeled quadtree. The root is labeled by $f(\varepsilon)$, its children from left to right by $f(0)$, $f(1)$, $f(2)$, $f(3)$, etc. Intuitively, $f(w)$ is the average greyness of the subsquare w for a given grey-tone image.

We consider the set of functions $f : \Sigma^\star \to \mathbb{R}$ as a vector space. The operations of sum and multiplication with a real number are defined in a natural way :

$$\begin{aligned}(f_1 + f_2)(w) &= f_1(w) + f_2(w), \text{ for any } f_1, f_2 : \Sigma^\star \to \mathbb{R} \text{ and } w \in \Sigma^\star, \\ (cf)(w) &= cf(w), \text{ for any function } f : \Sigma^\star \to \mathbb{R} \text{ and } c \in \mathbb{R}.\end{aligned}$$

The set of ap-functions forms a linear subspace, because any linear combination of ap-functions is average preserving. The sum of two ap-functions represents the image obtained by summing the grey values of the two images, and the multiplication with a real number corresponds to the change of the contrast.

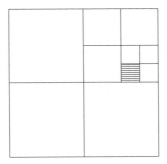

Figure 13.2: The subsquare specified by the string 320.

11	13	31	33
10	12	30	32
01	03	21	23
00	02	20	22

Figure 13.3: The addresses of subsquares of resolution 4×4.

We can view the addresses of the nodes of the quadtree as the addresses of the corresponding subsquares of the unit square $[0, 1]^2$. For example, the whole square $[0, 1]^2$ has the address ε and the marked square in Figure 13.2 has the address 320. All subsquares with their addresses for resolution 4×4 are shown in Figure 13.3.

By an infinite-resolution image we mean a local-greyness function $g : [0, 1]^2 \to \mathbb{R}$. For every integrable local-greyness function g we can find the corresponding multiresolution function $f : \Sigma^\star \to \mathbb{R}$ by computing $f(w)$ as the integral over the square with the address w, divided by the size of the square $1/4^{|w|}$, for each $w \in \Sigma^\star$. Conversely, for a point $p \in [0, 1]^2$, $g(p)$ is the limit of the subsquare values containing p, if such a limit exists. Thus, not every multiresolution image can be converted into an infinite-resolution image.

The distance between two functions f and g (the error of approximating f by g) is usually measured by

$$\|f - g\|_p = \left[\int_0^1 \int_0^1 |f(x, y) - g(x, y)|^p dx dy \right]^{\frac{1}{p}} \tag{13.2}$$

In practice, one desires a metric that parallels the human perception, i.e., such that the image differences that seem larger to the human eye are numerically large and those perceived as insignificant are numerically small. The usual choice is Equation (13.2) with $p = 2$, which

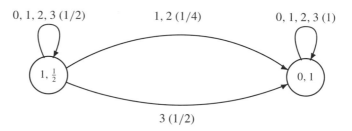

$$0, 1, 2, 3\ (1/2) \qquad\qquad 1, 2\ (1/4) \qquad\qquad 0, 1, 2, 3\ (1)$$

$$3\ (1/2)$$

Figure 13.4: A diagram for WFA A defining the linear greyness function f_A.

we will adapt. Therefore, we will consider the average square error also in the case of finite-resolution images.

A weighted finite automaton (WFA) A is specified by

1. Q a finite set of *states*,
2. Σ a finite alphabet (here we use the alphabet $\Sigma = \{0, 1, 2, 3\}$),
3. $W_a : Q \times Q \to \mathbb{R}$, for each $a \in \Sigma$, the weights at edges labeled by a,
4. $I : Q \to (-\infty, \infty)$, the *initial distribution*,
5. $F : Q \to (-\infty, \infty)$, the *final distribution*.

We say that $(p, a, q) \in Q \times \Sigma \times Q$ is an edge (transition) of A if $W_a(p, q) \neq 0$. This edge has label a and weight $W_a(p, q)$.

For $|Q| = n$ we will usually view W_a as an $n \times n$ matrix of reals and I, F as real vectors of size n.

A WFA defines a multiresolution image f_A by

$$f_A(a_1 a_2 \cdots a_k) = I\, W_{a_1} W_{a_2} \cdots W_{a_k} F$$

for each $k \geq 0$ and $a_1 a_2 \cdots a_k \in \Sigma^*$.

Example 1:   A WFA can be specified as a diagram with n nodes $\{1, \ldots, n\}$. There is an edge from node i to node j with label $a \in \Sigma$ and weight $r \neq 0$ iff $(W_a)_{ij} = r$. The initial and final distribution values are shown inside the nodes as illustrated in Figure 13.4 where $I = (1, 0)$, $F = (\frac{1}{2}, 1)$,

$$W_0 = \begin{pmatrix} \frac{1}{2} & 0 \\ 0 & 1 \end{pmatrix},\quad W_1 = \begin{pmatrix} \frac{1}{2} & \frac{1}{4} \\ 0 & 1 \end{pmatrix},$$

$$W_2 = \begin{pmatrix} \frac{1}{2} & \frac{1}{4} \\ 0 & 1 \end{pmatrix} \quad \text{and} \quad W_3 = \begin{pmatrix} \frac{1}{2} & \frac{1}{2} \\ 0 & 1 \end{pmatrix}.$$

The multiresolution image can be read from a diagram as follows: The weight of the path in the diagram is obtained by multiplying the weights of all transitions on the path, the initial distribution value of the first node, and the final distribution value of the last node on the path. Then, $f_A(w)$ is the sum of the weights of all paths whose labels form the word w. For example,

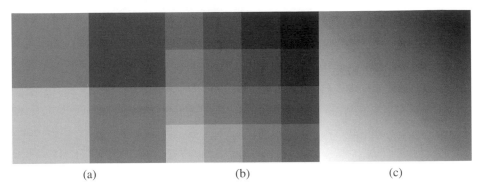

Figure 13.5: The image f_A at resolution (a) 2×2, (b) 4×4, and (c) 256×256.

$f_A(03) = I W_0 W_3 F = \frac{3}{8}$, or alternately $f_A(03) = $ sum of the weights of three paths labeled by $w = \frac{1}{8} + \frac{1}{4} + 0 = \frac{3}{8}$.

The image f_A for resolutions 2×2, 4×4, and 256×256 is shown in Figure 13.5.

If

$$(W_0 + W_1 + W_2 + W_3)F = 4F, \tag{13.3}$$

then $f_A(w0) + f_A(w1) + f_A(w2) + f_A(w3) = 4f_A(w)$ for all $w \in \Sigma^\star$. In other words, if Equation (13.3) holds then the multiresolution image f is *average preserving*. In this case we will also call WFA A *average preserving* (ap-WFA). Note that Equation (13.3) states that 4 is an eigenvalue of $W_0 + W_1 + W_2 + W_3$ and F is the corresponding eigenvector. All WFA considered here will be average preserving.

In the special case $F = (1, 1, \ldots, 1)$ considered in [39], Equation (13.3) is reduced to the requirement that for each state the sum of the weights of all the outgoing edges is 4.

The matrices W_a, $a \in \Sigma$, and the final distribution F define a multiresolution ψ_i for every state $i \in Q$ by

$$\psi_i(a_1 a_2 \cdots a_k) = (W_{a_1} W_{a_2} \cdots W_{a_k} F)_i .$$

Equivalently, for every $i \in Q$, $a \in \Sigma$, and $w \in \Sigma^\star$ we have

$$\psi_i(aw) = \sum_{j=1}^{n} (W_a)_{ij} \psi_j(w) . \tag{13.4}$$

We call ψ_i the *image of state i*. It is average preserving if the WFA A is. The final distribution value $F_i = \psi_i(\varepsilon)$, the average greyness of image ψ_i.

The transition matrices W_a, $a \in \Sigma$, specify how the four quadrants of each image ψ_i are expressed as linear combinations of $\psi_1, \psi_2, \ldots, \psi_n$, specifically, the image in quadrant a of ψ_i is expressed as $(W_a)_{i1}\psi_1 + (W_a)_{i2}\psi_2 + \cdots + (W_a)_{in}\psi_n$. The initial distribution I specifies how the multiresolution image f, computed by WFA A, is expressed as a linear combination of the images $\psi_1, \psi_2, \ldots, \psi_n$; clearly,

$$f_A = \sum_{j=1}^{n} I_j \psi_j .$$

Figure 13.6: The diminishing triangles.

For an arbitrary multiresolution image f over Σ and word $u \in \Sigma^\star$, f_u denotes the multiresolution image

$$f_u(w) = f(uw), \text{ for every } w \in \Sigma^\star.$$

f_u is the image defined by f inside the subsquare with address u.

Example 2: We will show intuitively how we can infer a WFA that generates a given image. The method follows the inference algorithm described in Section 13.2. Consider the multiresolution image f shown in Figure 13.6. We explain the construction of WFA A over $\Sigma = \{0, 1, 2, 3\}$ defining f. It is shown in Figure 13.7. (A is drawn as a labeled and weighted directed graph. The nodes of the graph represent the states. The edges represent non-zero elements of the transition matrices: If $(W_a)_{ij} = r \neq 0$ there is an edge in the graph from node i to node j with label a and weight r. Weights are shown in parentheses. To simplify the graph, multiple edges with the same weight but several labels are drawn as a single edge.)

As we know, each state i of the automaton defines an image ψ_i which in Figure 13.7 is shown for each state inside the box representing that state. In A, or in any WFA produced by the inference algorithm of [41], the images ψ_i of all states belong to $\{f_u \mid u \in \Sigma^\star\}$, that is, they are sub-images of f inside some subsquares of the unit square. We denote q_w the state with image f_w. Clockwise, starting from top left, WFA A has the states $q_\varepsilon, q_1, q_{10}, q_{100}, q_{00}$, and q_0.

We start by creating the state q_ε that represents f_ε. Now we have to "process" q_ε by expressing all quadrants of f_ε as linear combinations of existing states or as new states. We have $f_2 = f_\varepsilon$ and $f_3 = 0$ and two new states q_0 and q_1 representing f_0 and f_1, respectively. Thus, we draw the loop at q_ε with label 2 and weight 1, and edges from q_ε to q_0 and q_1 with weights 1 and labels 0 and 1, respectively.

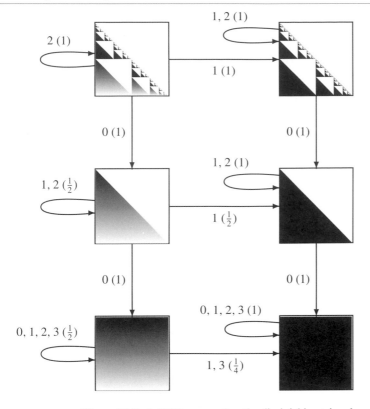

Figure 13.7: A WFA generating the diminishing triangles.

Next, we process q_1; the image f_1 is again self-similar, namely, $f_{12} = f_{11} = f_1$. Hence, we draw the loop at q_1 with labels 1 and 2 and weight 1. Quadrant 3 of f_1 is empty and quadrant 0 contains a new image, a black triangle f_{10}. Next we process q_0 which yields the first nontrivial linear combination, namely, $f_{01} = \frac{1}{2} f_0 + \frac{1}{2} f_{10}$. Hence, there are two edges, each with label 1 and weight $\frac{1}{2}$ starting at q_0: one loops back to q_0 and the other to q_{10}.

We repeat this process for each new created state and each $a \in \Sigma$. If it terminates, that is, if all created states have been processed so that each quadrant of each state is expressed as a linear combination of the states (implemented by the edges), then we have constructed a WFA that perfectly represents the given image. The initial distribution at state q_0 is 1 and at all other states is 0. The final distribution at each state is the average intensity of the image of that state. Note that WFA A is average preserving.

13.2 The Inference Algorithm for WFA

First, we consider briefly the deterministic WFA. We say that a weighted automaton A is deterministic if the underlying finite automaton obtained by omitting the weights is deterministic, formally if for each pair of state q and $a \in \Sigma$ there is at most one state p such that $W_a(q, p) \neq 0$.

Besides the existence of a short description of an image, it is also important that this description can be easily found from the given image (existence of an efficient encoding program) and that it is easy to regenerate the image from the description (existence of an efficient decoding program). Both of these requirements are excellently satisfied by the deterministic ap-WFA. An encoding algorithm restricted to the deterministic WFA is not difficult to construct; see [41]. The decoding (for resolution corresponding to the k-th level of the quadtree) can be done by multiplying the weights on every path from the root to a node at k-level (a pixel).

However, experiments have shown that deterministic ap-WFA are not powerful enough for encoding practical images (e.g., photographs) and there is a good theoretical reason for it. Indeed, in contrast to (non-probabilistic) generators for black-and-white images discussed in [37], the nondeterministic ap-WFA are much more powerful than the deterministic ones. The image f_A from Figure 13.5 is generated by the nondeterministic ap-WFA from Figure 13.4 but not by a deterministic ap-WFA. Actually, the deterministic ap-WFA can generate only countable unions of fractals or constant-level greyness functions, not smoothly growing greyness functions.

In [41] we gave an WFA inference algorithm which takes advantage of the power of nondeterminism. In [41] we described both a "theoretical" version of this algorithm which takes as an input an ideal infinite multiresolution image and a "practical" version that takes as an input a finite-resolution image. Here, we will describe only the theoretical multiresolution version since for the practical finite-resolution problem we now have a better recursive inference algorithm from [42] described in Section 13.4. The input of the (theoretical) inference algorithm is an ideal multiresolution image, i.e., an average-preserving function $f : \Sigma^* \to \mathbb{R}$ where $\Sigma = \{0, 1, 2, 3\}$.

We assume that the executor of the algorithm can find out if a given ap-function can be expressed as a linear combination of elements from a given finite collection of ap-functions. We also assume that we can compute the coefficients of such a linear combination as we have done in Example 13.1.

We recall the following notation introduced before Example 13.1. For an ap-function f and word $w \in \Sigma^*$, f_w denotes the ap-function defined by

$$f_w(u) = f(wu) \text{ for all } u \in \Sigma^*.$$

The function f_w represents thus the image that the ap-function f defines on the subsquare w. It is obtained from f by zooming in the subsquare.

Inference Algorithm:
For an image given by ap-function $f : \Sigma^* \to \mathbb{R}$, we construct an ap-WFA A such that $f_A = f$ provided such an ap-WFA exists. During the construction we use
N, the index of the last state created,
i, the index of the first unprocessed state,
$\gamma : Q \to \Sigma^*$, a mapping of states to subsquares.

  1. Set $N = 0, i = 0, F(q_0) = f(\varepsilon), \gamma(q_0) = \varepsilon$.

2. Process q_i, that is, for $w = \gamma(q_i)$ and each $a \in \{0, 1, 2, 3\}$ do

   (a) if there are $c_0, \ldots, c_N$ such that $f_{wa} = c_0\psi_0 + \cdots + c_N\psi_N$ where $\psi_j = f_{\gamma(q_j)}$ for $j = 0, \ldots, N$ then set $W_a(q_i, q_j) = c_j$ for $j = 0, \ldots, N$,

   (b) otherwise set $\gamma(q_{N+1}) = wa$, $F(q_{N+1}) = f(wa)$, where F is the final distribution of A, $W_a(q_i, q_{N+1}) = 1$, $N = N + 1$.

3. Set $i = i + 1$, if $i \leq N$, then go to 2.

4. Set $I(q_0) = 1$, $I(q_j) = 0$ for $j = 1, \ldots, N$, where I is the initial distribution of A.

It is easy to see that the algorithm terminates if and only if the set $\{f_w | w \in \Sigma^*\}$ generates a linear space of finite dimension. The number of states in the ap-WFA produced by the algorithm is the same as the dimension of the linear space. According to the next theorem the algorithm gives an ap-WFA with the minimal number of states defining the given image exactly, provided such an automaton exists.

Theorem 13.1 *Let* $f : \Sigma^* \to \mathbb{R}$ *be a function.*

   (i) *The function* f *can be defined using a WFA if and only if the set of functions defined in all the subsquares of* Σ^* *by* f *(that is, the set* $\{f_w | w \in \Sigma^*\}$*) generates a linear space of finite dimension* d.

   (ii) *If* f *is defined by a WFA and is average preserving, then it can be defined by an ap-WFA, and the ap-WFA produced by the inference algorithm has the smallest number of states (namely,* d *states) among all WFA defining* f.

Proof: (i) Assume first that f is defined by a WFA A. Let $Q = \{q_1, q_2, \ldots, q_s\}$ be the set of states of A. Consider the images (not necessarily ap) $\psi_1, \psi_2, \ldots, \psi_s$ of the states $q_1, \ldots, q_s$.

In the following, we show that for each $w \in \Sigma^*$ the function f_w defined by f in the square w can be expressed as a linear combination of the functions $\psi_1, \psi_2, \ldots, \psi_s$. We show this using induction on the length of w. Because $f_\varepsilon = f$ and

$$f = I(q_1)\psi_1 + I(q_2)\psi_2 + \cdots + I(q_s)\psi_s,$$

the claim is true for $w = \varepsilon$.

Assume then that the function defined by f in the subsquare w is

$$f_w = c_1\psi_1 + c_2\psi_2 + \cdots + c_s\psi_s,$$

where c_i are some real numbers. Let $a \in \Sigma$ be arbitrary. Obviously, $f_{wa} = (f_w)_a$, that is, the function defined by f in the subsquare wa is the same as the one defined by f_w in the subsquare a. Noting that the sub-image of ψ_i, for $i = 1, 2, \ldots, s$, in the subsquare a is

$$(\psi_i)_a = W_a(q_i, q_1)\psi_1 + W_a(q_i, q_2)\psi_2 + \ldots + W_a(q_i, q_s)\psi_s$$

we conclude that the function defined by f in the subsquare wa is

$$f_{wa} = (f_w)_a = \left(\sum_{i=1}^{s} c_i\psi_i\right)_a = \sum_{i=1}^{s} c_i(\psi_i)_a$$

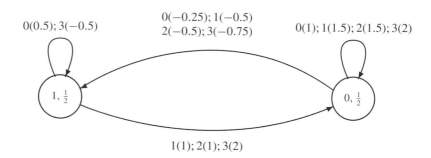

Figure 13.8: The ap-WFA for $z = \frac{x+y}{2}$ produced by the encoding algorithm.

$$= \sum_{i=1}^{s} c_i \cdot \sum_{j=1}^{s} W_a(q_i, q_j)\psi_j = \sum_{i,j=1}^{s} c_i W_a(q_i, q_j)\psi_j .$$

This is a linear combination of functions $\psi_1, \psi_2, \ldots, \psi_s$ as required.

We have shown that the functions defined by f in the subsquares $w \in \Sigma^*$ are all included in the linear space generated by the functions $\psi_1, \psi_2, \ldots, \psi_s$. This proves the first implication of (i). Note that if d denotes the dimension of the linear space generated by the set $\{f_w | w \in \Sigma^*\}$, then $d \leq s$.

Assume then conversely that the functions f_w defined by f in the subsquares $w \in \Sigma^*$ generate a linear space of finite dimension d. Then, using the inference algorithm, one can construct a WFA with d states defining the function f. Note that the inference algorithm can be used also for functions f which are not average preserving — in this case the WFA produced is not average preserving either. This completes the proof of (i).

(ii) Assume then that f is average preserving and that $\{f_w | w \in \Sigma^*\}$ generates a subspace of finite dimension d. In this case the inference algorithm produces an ap-WFA with d states which defines f. It was shown above that if s is the number of states in an arbitrary WFA defining f, then $s \geq d$. This proves (ii). ∎

Example 3:   Consider the linearly sloping ap-function f_A introduced in Example 1. Let us apply the inference algorithm to find a minimal ap-WFA generating f_A.

First, state q_0 is assigned to the square ε and we define $F(q_0) = \frac{1}{2}$. Consider then the four subsquares 0, 1, 2, 3. The image in the first subsquare 0 can be expressed as $\frac{1}{2} \cdot f_A$ (it is obtained from the original image f_A by decreasing the contrast by one-half) so that we define $W_0(q_0, q_0) = 0.5$.

The image in the subsquare 1 cannot be expressed as a linear combination of f_A so that we have to use a second state q_1. Define $W_1(q_0, q_1) = 1$ and $F(q_1) = \frac{1}{2}$ (the average greyness of the subsquare 1 is $\frac{1}{2}$). Let f_1 denote the image given by f_A into the square 1.

The image in the subsquare 2 is the same as in the subsquare 1, so that $W_2(q_0, q_1) = 1$. In the quadrant 3 we have image which can be expressed as $2 \cdot f_1 - \frac{1}{2} \cdot f_A$. We define $W_3(q_0, q_0) = -\frac{1}{2}$ and $W_3(q_0, q_1) = 2$. The outgoing transitions from state q_0 are now ready.

Consider then the images in the squares 10, 11, 12, and 13. They can be expressed as

$f_1 - \frac{1}{4} \cdot f_A, \frac{3}{2} \cdot f_1 - \frac{1}{2} \cdot f_A, \frac{3}{2} \cdot f_1 - \frac{1}{2} \cdot f_A$ and $2 \cdot f_1 - \frac{3}{4} \cdot f_A$, respectively. This gives us the ap-WFA of Figure 13.8.

The initial distribution I is defined by $I(q_0) = 1$ and $I(q_1) = 0$.

13.3   A Fast Decoding Algorithm for WFA

Assume we are given a WFA A by W_a, $a = 0, 1, 2, 3$, I and F, and we want to draw a finite-resolution approximation of the image it represents, that is, to compute $f_A(w)$ for all $w \in \Sigma^n$, for a given $n \geq 0$. The most straightforward approach is to find all paths of length n in the automaton A, to compute the weights of the paths and to add the weights to the greyness values of the corresponding pixels. However, in the worst case there can be $m \cdot (4m)^n$ paths of length n in A, where m is the number of states of A. So the time complexity of this simple algorithm can be very high if the number of states in the automaton is big.

The algorithm presented below is much faster. However, the space complexity is higher than in the algorithm above. There is a clear trade-off between time and space.

Using Equation (13.4) the algorithm computes values $\psi_p(w)$ for $p \in Q$ and $w \in \Sigma^*$, where ψ_p is the image of state p.

Decoding Algorithm:
Input: A WFA A specified by W_a, $a = 1, 2, 3$, I and F, and a non-negative integer n.
Output: The values $f_A(w)$ for all $w \in \Sigma^n$.

1. Set $\psi_p(\varepsilon) = F(p)$ for all $p \in Q$.

2. Do the following step 3 for $i = 1, 2, \ldots, n$:

3. For all $p \in Q$, $w \in \Sigma^{i-1}$ and $a \in \Sigma$ compute

$$\psi_p(aw) = \sum_{q \in Q} W_a(p, q) \cdot \psi_q(w).$$

4. For each $w \in \Sigma^n$ compute

$$f_A(w) = \sum_{q \in Q} I(q) \cdot \psi_q(w).$$

Let us analyze the time complexity of the decoding algorithm. Let m denote as before the number of states. The time complexities of steps 1, 3, and 4 of the algorithm are $O(m)$, $O(m^2 \cdot 4^i)$ and $O(m \cdot 4^n)$, respectively. Step 3 is repeated for $i = 1, 2, \ldots, n$ so that the time complexity of steps 2 and 3 together is $O(m^2(4 + 4^2 + \cdots + 4^n)) = O(m^2 \cdot 4^n)$. Altogether, the time needed by the algorithm is $O(m + m^2 \cdot 4^n + m \cdot 4^n) = O(m^2 \cdot 4^n)$. This means $O(m^2)$ operations for each pixel in the image in comparison with $O(m^{n+1})$ per pixel in the trivial algorithm.

On the other hand, space is needed for storing $O(m \cdot 4^n)$ numbers (the values $\psi_p(w)$), whereas in the trivial algorithm $O(4^n)$ numbers are enough.

In practical situations, the decoding algorithm can still be improved if many of the values $I(p)$ and weights $W_a(p, q)$ are zero. Let us call nodes q for which $I(q) \neq 0$ initial nodes. Compute for each node p the length of the shortest path from an initial node to p (along transitions with non-zero weights). Let *depth(p)* denote this length. Note that in the decoding algorithm the value $\psi_p(w)$ is needed only if $|w| \leq n - depth(p)$, otherwise it does not contribute anything to $f_A(u)$ for any $u \in \Sigma^n$. So in step 3 the values $\psi_p(aw)$ need to be computed only for nodes p satisfying $depth(p) \le n - i$.

The following theorem shows that zooming is easily done with WFA. We remind the notion from Section 13.1. For image (function) f and $w \in \Sigma^*$, f_w denotes the zoomed image in the subsquare w of the unit square.

Theorem 13.2 *Let A be a WFA given by matrices W_a, $a = 0, 1, 2, 3$, initial distribution I, and final distribution F. Let $w = a_1 \cdots a_r$, $I_w = I W_{a_1} \cdots W_{a_r}$, and A_w be the WFA given by matrices W_a, $a = 0, 1, 2, 3$, initial distribution I_w and final distribution F. Then $(f_A)_w = f_{A_w}$.*

Proof: The claim follows immediately from the definition of multiresolution image defined by WFA. Clearly, the quadtree f_{A_w} is the subtree of the quadtree f_A which has the node addressed w as the root. ∎

Therefore, in order to compute the zoomed image in subsquare with address w for WFA A we just compute the new initial distribution I_w and then use the decoding algorithm for A with I replaced by I_w.

If we want to zoom to an area other than a "binary" subsquare, then we can always approximate (cover) such an area by a union of "binary" subsquares. Note that zooming for WFA is much more efficient than for IFS [4] where the whole image has to be computed even if only a small part is to be shown.

13.4 A Recursive Inference Algorithm for WFA

In the following we describe a recursive inference algorithm that is intended for finite-resolution images. It produces a WFA with possibly a non-optimal number of states, but with sparse transition matrices. In practice, we are given an image, e.g., a grey-tone or color photograph, with finite resolution. In the terms of a quadtree we are given all the values at one level, say level k. By computing for each parent the average value of the labels of all its children, for all the nodes at the higher levels, we get the labels everywhere above the given resolution, and leave "don't cares" below it. By assigning a different state at level k and higher we trivially get (too large) ap-WFA that (perfectly) defines the given image. The practical problem therefore is not whether it is possible to encode a given image but rather whether we can get a good trade-off between the size of the automaton and the quality of the regenerated approximation of the given image. This trade-off, of course, depends on the "nonrandomness" or "regularity" of the image. It is a well-known fact in the descriptive (Kolmogorov) complexity of strings [15] that most strings are algorithmically random and cannot be encoded (compressed) by a shorter program. However, the "interesting strings" are not random and possibly can be compressed. The same holds for images.

Our algorithm needs to compute the distance d_k of two multiresolution functions at level k. This distance could be any additive function

$$d_k : \mathbb{R}^{\Sigma^k} \times \mathbb{R}^{\Sigma^k} \to \mathbb{R} ,$$

where additive means that there exists $d : \mathbb{R}^2 \to \mathbb{R}$ such that

$$d_k(f, g) = \sum_{w \in \Sigma^k} d(f(w), g(w))$$

for every $f, g \in \mathbb{R}^{\Sigma^*}$ and $k = 0, 1, \ldots$. Therefore the distance between two images at level k is the sum of the distances between their corresponding quadrants at level $k - 1$. In our implementation d_k is the square of the L^2-metric:

$$d_k(f, g) = \sum_{w \in \Sigma^k} [f(w) - g(w)]^2$$

Recursive Inference Algorithm:

The algorithm produces a small WFA A such that the k-th level of the input f and the function computed by A are close to each other. More precisely, we want the value of

$$d_k(f, f_A) + G \cdot \text{size}(A)$$

to be as small as possible, where size(A) denotes the storage space required to store the WFA A, and $G \in \mathbb{R}$ is a parameter. The parameter G controls the quality of the approximation. With large values of G a small automaton with poor approximation of f is produced. When G is made smaller, the approximation improves while the size of the automaton increases.

Table 13.1 contains an outline of the recursive inference algorithm. The global variable n indicates the number of states so far, and ψ_i denotes the multiresolution function of state i, $1 \le i \le n$. A call *make_wfa(i,k,max)* to the recursive function tries to approximate the multiresolution function ψ_i at level k as well as possible by adding new transitions and (possibly) new states to the automaton. The value of $cost = d_k(\psi_i, \psi_i') + G \cdot s$ is minimized, where ψ_i' is the obtained approximation of ψ_i and s is the increase in the size of the automaton caused by the new edges and states. If $cost > max$ the value ∞ is returned, otherwise $cost$ is returned.

Initially, before calling the recursive function for the first time, one sets $n \leftarrow 1$ and $\psi_1 \leftarrow f$, where f is the function that needs to be approximated at level k. Then one calls *make_wfa(1,k,∞)*. The call *make_wfa(i,k,max)* tries to approximate the functions $(\psi_i)_a$, $a \in \Sigma$, (the four quadrants of the image ψ_i when $\Sigma = \{0, 1, 2, 3\}$) in two different ways: by a linear combination of the functions of existing states (step 1 in Table 13.1), and by adding a new state and recursively calling *make_wfa* to approximate $(\psi_i)_a$ (steps 2 and 3). Whichever alternative yields a better result is chosen (steps 4 and 5).

The initial distribution of the WFA produced by the algorithm is $I_1 = 1$, $I_i = 0$ for all $i \ge 2$, and the final distribution is $F_i = \psi_i(\varepsilon)$ for all $i \ge 1$.

The compression achieved by the recursive algorithm can be further improved by introducing an *initial basis*. Before calling *make_wfa* for the first time, set $n \leftarrow N$ with fixed images $\psi_1, \psi_2, \ldots, \psi_N$. The functions in the basis do not need even to be defined by a WFA. The choice of the functions can, of course, depend on the type of images one wants to compress. Our initial basis resembles the codebook in vector quantization [71], which can be viewed as a very restricted version of our method.

Another modification of the algorithm that yields good results is to combine the WFA with a wavelet transformation (see [20]): Instead of applying the inference algorithm directly to the original image, one first makes a wavelet transformation on the image and writes the wavelet coefficients in the Mallat form [20]. The Mallat form can be understood as an image itself, on which our inference algorithm can be applied. In this case WFA can be understood as a fancy way of quantizing the wavelet coefficients. Decoding is, of course, done in the opposite order: first the decoding of the WFA, and then the inverse wavelet transformation. Because the wavelets we use are orthonormal, the L^2-error done on the Mallat form by the WFA is equal to the error caused to the original image.

Let us describe shortly how a WFA is stored efficiently to get a good final compression. There are two types of edges in the automaton: edges created in step 2 of the algorithm when

Table 13.1: Outline of the recursive inference algorithm for WFA. (See the text for details.)

---

make_wfa(i,k,max) :
If *max* < 0 then return(∞);
cost $\leftarrow$ 0;
If $k = 0$ *cost* $\leftarrow d_0(f, 0)$;
else do the steps 1–5 with $\psi = (\psi_i)_a$ for all $a \in \Sigma$:

    1. Find $r_1, r_2, \ldots r_n$ such that the value of

$$costl \leftarrow d_{k-1}(\psi, r_1\psi_1 + \cdots + r_n\psi_n) + G \cdot s$$

       is small, where s denotes the increase in the size of the automaton caused
       by adding edges from state i to states j with non-zero weights r_j and label
       a, and d_{k-1} denotes the distance between two multiresolution images at
       level $k - 1$.

    2. $n_0 \leftarrow n$, $n \leftarrow n + 1$, $\psi_n \leftarrow \psi$ and add an edge from state i to the new state
       n with label a and weight 1. Let s denote the increase in the size of the
       automaton caused by the new state and edge.

    3. $cost2 \leftarrow G \cdot s + make\_wfa(n,k-1,min\{max-cost,costl\}-G \cdot s)$;

    4. If $cost2 \le costl$ then $cost \leftarrow cost + cost2$;

    5. If $costl < cost2$ then $cost \leftarrow cost + costl$, remove all outgoing transitions from
       states $n_0+1, \ldots, n$ (added during the recursive call), as well as the transition
       from state i added on step 2. Set $n \leftarrow n_0$ and add the transitions from state
       i with label a to states $j = 1, 2, \ldots, n$ with weights r_j whenever $r_j \neq 0$.

If $cost \le max$ return($cost$)
else return(∞);

---

a new state is added, and edges created in step 1, which express the linear combinations. The
former ones form a tree, and they can be stored trivially using 4 bits per state, where each bit
indicates for one label which of the two alternatives was chosen in steps 4–5. Even fewer bits
are enough if a coding with variable-length codewords is used. For example, most states are
typically leaves of the tree, that is, the alternative with linear combinations was chosen for all
quadrants, and therefore it is wise to use a 1-bit codeword instead of 4 bits for leaves.

For the second type of edge, both the weight and the endpoints need to be stored. According
to our experiments the weights are normally distributed. Using this fact, an effective encoding
with variable length codewords has been done. Storing the endpoints of the edges is equivalent
to storing four sparse binary matrices. This has been done using run-length coding. If an initial
basis is used, some states in the initial basis tend to be used more frequently than others. In
particular, according to our experiments, if the completely black square is one of the images in
the basis, it is used more frequently than other states. In such cases we further improved the
encoding of the matrices.

Figure 13.9: 160 times compressed Lenna.

The recursive inference algorithm alone seems to be competitive with any other method for image compression, especially for high compression rates. Its combination with wavelets seems to be the best high-compression method available. It not only gives a very good ratio between compression rate and the quality of the regenerated images, but also it is relatively simple and time-efficient compared to the other fractal methods.

In Figure 13.9 are examples of 160 times compressed images of Lenna (resolution 512 × 512). The figure shows, clockwise starting at topleft, the original image, the regenerated images using WFA only, the combination of Debauchies' wavelets W6 with WFA, and the W6 wavelets only. The quality of the images as signal-to-noise ratios are 25.38 dB for WFA, 25.99 dB for the combination of WFA and W6, and 24.91 dB for W6 alone.

Acknowledgments

Reported research was supported by the National Science Foundation under Grant No. CCR-9202396.

Appendix A

Sample Code

Y. Fisher

This appendix contains the sample code that was used in to generate the results of Chapter 3. Before the actual code listing, we present a UNIX style manual page describing the use of the code, its peculiarities, and its shortcomings. The code itself consists of some 1500 lines of vanilla C. The C language has been both praised and derided: the good thing about C, goes the saying, is that it is so portable, but its main drawback is its lack of portability. This code has been compiled on several standard UNIX workstations, but no guarantee of any type is made about its compilability, freedom from bugs, or even its ability to compress images. After the code listings, we present a detailed line-by-line explanation of how the code works.

A.1 The Enc Manual Page

NAME

enc - use a quadtree-partition-based fractal method to encode a file of consecutive byte data representing an image.

SYNOPSIS

enc [-t *tolerance*] [-m *minimum recursion depth*] [-M *maximum recursion depth*]
[-S *stripchars*] [-w *width*] [-h *height*] [-d *domain pool step size*]
[-D *domain pool type (0, 1, 2)*] [-s *scaling bits*] [-o *offset bits*]
[-N *maximum scale*] [-HepfF] [*input file name*] [*outputfilename*]

DESCRIPTION

Enc encodes a file of sequential byte values which represent an image in row first order. Various flags can alter the time and fidelity of the resulting encoding, which can be

259

decoded with the program *dec*. All input flags are optional, including the input and output file names. Output files typically have an extension of .trn, to denote a *transform* file containing the coefficients of the partitioned iterated transform which encodes the image.

Ranges are selected in the image using a recursive quadtree partitioning of square sub-images of the input image. For each range, a domain sub-image is sought which is twice the range side length. If a domain is not found which can be mapped onto a range with rms error less than the tolerance set with the -t flag, then the range is partitioned into four quadrants, and the process is repeated. The maximum and minimum recursion depth are set using the -m and -M flags, and these refer to the maximal square sub-image. For example, a 256 × 256 pixel image with -m 4 and -M 6 will have ranges of maximum size 16 and minimum size 4 (note that maximum range size corresponds to the minimum recursion depth). A 300 × 400 image with the same flag values will have the same size ranges, since the largest square sub-image which fits in a 300 × 400 image is 256 × 256.

The -D, -e, and -d affect the selection of domains that are compared with ranges. The -D flag can take values of 0, 1, and 2, which select among different schemes (described below) for defining a pool of domains for comparison with each range. The number of domains is also affected by the -d flag, which takes a value (0–15) that determines the domain density. The -e flag changes the meaning of the -d value from a divisor to a multiplier (see below). In general, -D and -e are not of interest to the casual user. With no -D or -e flag, higher -d flag values give better encodings that take longer to compute.

The -f and -F flag determine how many domains are searched for each range. The -f flag searches 24 classes and the -F flag searches 3. Both cause the program to run longer and result in better encodings. The -p flag causes only positive scalings to be used, which means that only one class (as opposed to 2 normally) is searched. It causes the program to run faster and give somewhat poorer encodings.

The -s and -o flags are not of interest to the casual user.

OPTIONS

-t *tolerance.* A real value (typically in the range 2–15) which is a loose target for the final rms error of the encoded image. Lower values result in better-looking encoded images which are larger (i.e., have lower compression ratio).

-m *minimum recursion depth.* This is the minimum number of times that the largest square sub-image will be recursively quadtree partitioned. Using larger values will result in higher compression ratios when the file is easy to encode.

-M *maximum recursion depth.* This is the maximum number of times that the largest square sub-image will be quadtree partitioned recursively. Using larger values will help encode the fine detail in the image at the cost of smaller compression ratio. Using too large a value will result in waste of memory, since ranges of size 2 × 2 or less should not be encoded.

-S *stripchars.* The number of bytes skipped at the beginning of the input data file. This is useful with some image formats which store the image after a header that can be skipped. Care should be taken that the image data correspond to the actual grey-scale values (and not a reference through a color lookup table).

-w *width*. The width of the input image, assumed to be the height if -w is not used.

-h *height*. The height of the input image, assumed to be the width if -h is not used.

-d *domain pool step size*. The domain pool step size. See -D.

-D *domain pool type (0, 1, 2)*. This flag selects a method for determining the domain pool. The domain pool consists of sub-images of the image which are equally spaced vertically and horizontally. The step size is determined by the value to this flag, the value to the -d flag, and the -e flag. Below, d denotes the domain size, and s denotes the domain pool step size – the value in the -d flag.

    0. Use sub-images of the image of size $d \times d$ centered on a grid with vertical and horizontal spacing of d/s. Here, the largest domains get the largest grid size.

    1. Use sub-images of the image of size $d \times d$ centered on a grid with vertical and horizontal spacing of t. The value t depends on the maximum and minimum quadtree partitions. The grid size is essentially inverted with respect to domain size from the 0 case above. The largest domains get the smallest grid, and vice versa.

    2. Use sub-images of the image of size $d \times d$ centered on a grid with vertical and horizontal spacing of s. Here the grid size is fixed.

-e Change the division to a multiplication in the domain pool definition above.

-s *scaling bits*. The number of bits used to quantize the scale factor of the affine transformation of the pixel values between the domain and range. While the default of 5 bits is roughly optimal, high-fidelity encodings give better results with more bits (e.g., 6) and low-fidelity encodings give better results with fewer bits (e.g., 4). This value is not stored in the encoding, and so files encoded with this flag must be decoded using the -s flag in *dec*.

-o *offset bits*. The number of bits used to quantize the offset of the affine transformation of the pixel values between the domain and range. While the default of 7 bits is roughly optimal, high-fidelity encodings give better results with more bits (e.g., 8) and low-fidelity encodings give better results with fewer bits (e.g., 6). This value is not stored in the encoding, and so files encoded with this flag must be decoded using the -o flag in *dec*.

-N *maximum scale*. The maximum allowable scale factor that can be used. Using values greater than the default value of 1.0 may result in encodings which don't converge in very rare cases. However, encodings with up to -N 3.0 can be marginally better than without. This value is not stored in the encoding, and so files encoded with this flag must be decoded using the -N flag in *dec*.

-p Use positive scale values only. This also causes the program to search only one domain class out of the 72. Normally, two classes are searched. This results in faster encoding times with worse encodings.

-f Search 24 domain classes. This results in slower encoding times with better encodings. The -f flag can be combined with the -F flag. The -p flag shouldn't be used with the -f flag.

-F Search 3 domain classes. This results in slower encoding times with better encodings. The -F flag can be combined with the -f and -p flags.

-H Print out a help message showing a brief explanation of the input options and their default values.

FILES

The default input file name is *lenna.dat*. The default output file name is *lenna.trn*. The *dec* program decodes files encoded with *enc*.

BUGS

- There is no dummy proofing of input flags.

- The input file format should accept various image formats rather than raw byte data.

- The encoding of the domain position, the scaling, and the offset is not efficient.

- The output file name should be the input file name with the extension changed to be .trn.

- Using a high -d flag can cause the program to not work properly, and there is no warning of this. The value of the flag is divided into the domain size, so if the value is too large and the domains too small, the domain lattice may attempt to have zero spacing.

A.2   The Dec Manual Page

NAME

dec - decode a file encoded by *enc* and output a file of consecutive byte data representing an image.

SYNOPSIS

dec [-n *number of iterations*] [-f *scale factor*] [-i *file name*] [-N *maximum scale*]
[-s *scaling bits*] [-o *offset bits*] [-HPp] [*input file name*] [*outputfilename*]

DESCRIPTION

Dec decodes a file containing output data from *enc* and outputs a file of sequential byte values that represent an image in row first order. Input files typically have an extension of .trn, to denote a *transform* file containing the coefficients of the partitioned IFS that encodes the image. Output files typically have an .out extension.

Since the representation of images encoded using *enc* contains no resolution information, the image can be decoded at any resolution. Images decoded at resolutions greater than the encoded resolution will have artificial detail automatically created. The -f flag changes the output image size. The output image is typically postprocessed to eliminate some blocky artifacts that are side effects of the encoding method. This can be suppressed with the -p flag. Finally, if the input data file was encoded using the -N, -s, or -o flags of *enc*, the same flags must be passed to *dec*.

OPTIONS

-n *number of iterations.* An integer specifying the number of iterations to perform in approximating the fixed point, which is the output image. This typically does not need to be changed from the default of 10, unless a large maximum scale factor was used, (see -N) and the convergence is slow.

-i *file name.* A file containing image data (bytes in row sequential order) which is used as the initial image from which the partitioned iterated function is iterated. The image size should be the same as the output size of the input image.

-f *scale factor.* A real value which scales the output image. For example, -f 2.0 will yield an image that is 2 times the side length (and 4 times the area) of the original encoded image.

-s *scaling bits.* The number of bits used to quantize the scale factor of the affine transformation of the pixel values between the domain and range. This flag needs to be used only if the input data was encoded with a non-default -s flag in *enc*.

-o *offset bits.* The number of bits used to quantize the offset of the affine transformation of the pixel values between the domain and range. This flag needs to be used only if the input data was encoded with a non-default -o flag in *enc*.

-N *maximum scale.* The maximum allowable scale factor that was used during the encoding. Encodings made with this flag must be decoded with this flag also.

-p Suppress postprocessing of the output image.

-P Output data that can be plotted to show the range partition of the image. The data contain sequential coordinates of endpoints of line segments, with blank lines separating segments that are not contiguous. The data appear in the output file, and are suitable for plotting with *plot* or *gnuplot*.

-H Print out a help message showing a brief explanation of the input options and their default values.

FILES

The default input file name is *lenna.trn*. The default output file name is *lenna.out*. The *enc* program encodes files that are decoded with *dec*.

BUGS

- There is no check to see if the iterations have converged.
- There is no dummy proofing of input flags.
- The output file format should allow various image formats rather than raw byte data.
- As in *enc*, the encoding of the domain position, the scaling, and the offset is not efficient.
- The output file name should be the input file name with the extension changed to be .out.
- There are various methods of decoding images rapidly, of which this is one of the slowest.
- There should be a stripchar option for the -i option.

A.3 Enc.c

```
 1  /**************************************************************************/
 2  /* Encode a byte image using a fractal scheme with a quadtree partition  */
 3  /*                                                                        */
 4  /*          Copyright 1993,1994 Yuval Fisher. All rights reserved.        */
 5  /*                                                                        */
 6  /* Version 0.03 3/14/94                                                   */
 7  /**************************************************************************/
 8
 9  #include <stdio.h>
10  #include <math.h>
11
12  #define DEBUG 0
13  #define GREY_LEVELS 255
14
15  #define bound(a)    ((a) < 0.0 ? 0 : ((a)>255.0? 255 : a))
16  #define IMAGE_TYPE unsigned char /* may be different in some applications */
17
18  /* various function declarations to keep compiler warnings away. ANSI     */
19  /* prototypes can go here, for the hearty.                                */
20  void fatal();
21  char *malloc();
22  char *strcpy();
23
24  /* The following #define allocates an hsize x vsize  matrix of type TYPE */
25  #define matrix_allocate(matrix, hsize, vsize, TYPE) {\
26      TYPE *imptr; \
27      int _i; \
28      matrix = (TYPE **)malloc((vsize)*sizeof(TYPE *));\
29      imptr = (TYPE*)malloc((long)(hsize)*(long)(vsize)*sizeof(TYPE));\
30      if (imptr == NULL) \
31          fatal("\nNo memory in matrix allocate."); \
32      for (_i = 0; _i<vsize; ++_i, imptr += hsize) \
33          matrix[_i] = imptr; \
34  }
35
36  #define swap(a,b,TYPE)              {TYPE _temp; _temp=b; b=a; a= _temp;}
37
38  IMAGE_TYPE **image;          /* The input image data                     */
39  double     **domimage[4];    /* Decimated input image used for domains    */
40
41  double max_scale = 1.0;      /* Maximum allowable grey level scale factor */
42
43  int     s_bits = 5,          /* Number of bits used to store scale factor */
44          o_bits = 7,          /* Number of bits used to store offset       */
45          min_part = 4,        /* Min and max _part determine a range of    */
46          max_part = 6,        /* Range sizes from hsize>>min to hsize>>max */
47          dom_step = 1,        /* Density of domains relative to size       */
48          dom_step_type = 0,   /* Flag for dom_step a multiplier or divisor */
49          dom_type = 0,        /* Method of generating domain pool 0,1,2..  */
50          only_positive = 0,   /* A flag specifying use of positive scaling */
51          subclass_search = 0, /* A flag specifying classes searched        */
52          fullclass_search = 0,/* A flag specifying classes searched        */
53          *bits_needed,        /* Number of bits to encode domain position. */
54          zero_ialpha,         /* The const ialpha when alpha = 0           */
55          max_exponent;        /* The max power of 2 side of square image   */
56                               /* that fits in our input image.            */
57
58
59                               /* The class_transform gives the transforms  */
60                               /* between classification numbers for        */
61                               /* negative scaling values, when brightest   */
62                               /* becomes darkest, etc...                   */
63  int     class_transform[2][24] = {23,17,21,11,15,9,22,16,19,5,13,3,20,10,18,
64                                    4,7,1,14,8,12,2,6,0,
65                                    16,22,10,20,8,14,17,23,4,18,2,12,11,21,5,
66                                    19,0,6,9,15,3,13,1,7};
```

```
67
68                                  /* rot_transform gives the rotations for    */
69                                  /* domains with negative scalings.          */
70  int     rot_transform[2][8] = {7,4,5,6,1,2,3,0, 2,3,0,1,6,7,4,5};
71
72  struct domain_data {
73        int *no_h_domains,      /* The number of domains horizontally for   */
74            *no_v_domains,      /* each size.                               */
75            *domain_hsize,      /* The size of the domain.                  */
76            *domain_vsize,      /* The size of the domain.                  */
77            *domain_hstep,      /* The density of the domains.              */
78            *domain_vstep;      /* The density of the domains.              */
79  struct domain_pixels {        /* This is a three (sigh) index array that   */
80        int dom_x, dom_y;       /* dynamically allocated. The first index is */
81        double sum,sum2;        /* the domain size, the other are two its    */
82        int sym;                /* position. It contains the sum and sum^2   */
83  } ***pixel;                   /* of the pixel values in the domains, which */
84  } domain;                     /* are computed just once.                   */
85
86
87  struct classified_domain {                /* This is a list which contains  */
88        struct domain_pixels *the;      /* pointers to  the domain data    */
89        struct classified_domain *next; /* in the structure above. There   */
90  } **the_domain[3][24];                  /* are three classes with 24 sub-  */
91                                          /* classes. Using this array, only */
92                                          /* domains and ranges in the same  */
93                                          /* class are compared..            */
94                                          /* The first pointer points to the */
95                                          /* domain size the the second to   */
96                                          /* list of domains.                */
97
98  FILE *output;                  /* Output FILE containing compressed data    */
99
100 main(argc,argv)
101 int argc;
102 char **argv;
103 /* Usage: quadfrac [tol [inputfilename [outputfilename [hsize [vsize]]]]]  */
104 {
105     /* Defaults are set initially */
106     double       tol = 8.0;             /* Tolerance value for quadtree.  */
107     char         inputfilename[200];
108     char         outputfilename[200];
109     int          i,j,k,
110                  hsize = -1,            /* The horizontal and vertical    */
111                  vsize = -1;            /* size of the input image.       */
112     long         stripchar=0;          /* chars to ignore in input file. */
113     FILE         *input;
114
115     inputfilename[0] = 1;  /* We initially set the input to this and */
116     outputfilename[0] = 1; /* then check if the input/output names    */
117                            /* have been set below.                    */
118
119     /* scan through the input line and read in the arguments */
120     for (i=1; i<argc; ++i)
121       if (argv[i][0] != '-' )
122           if (inputfilename[0] == 1)
123               strcpy(inputfilename, argv[i]);
124           else if (outputfilename[0] == 1)
125               strcpy(outputfilename, argv[i]);
126           else;
127       else { /* we have a flag */
128           if (strlen(argv[i]) == 1) break;
129           switch(argv[i][1]) {
130              case 't': tol = atof(argv[++i]);
131                      break;
132              case 'S': stripchar = atoi(argv[++i]);
133                          break;
134              case 'x':
```

```
135                    case 'w': hsize = atoi(argv[++i]);
136                             break;
137             case 'y':
138             case 'h': vsize = atoi(argv[++i]);
139                             break;
140             case 'D': dom_type = atoi(argv[++i]);
141                             break;
142             case 'd': dom_step = atoi(argv[++i]);
143                             if (dom_step < 0 || dom_step > 15)
144                                  fatal("\n Bad domain step.");
145                             break;
146             case 's': s_bits = atoi(argv[++i]);
147                             break;
148             case 'o': o_bits = atoi(argv[++i]);
149                             break;
150             case 'm': min_part = atoi(argv[++i]);
151                             break;
152             case 'M': max_part = atoi(argv[++i]);
153                             break;
154             case 'e': dom_step_type= 1;
155                             break;
156             case 'p': only_positive = 1;
157                             break;
158             case 'f': subclass_search = 1;
159                             break;
160             case 'F': fullclass_search = 1;
161                             break;
162             case 'N': max_scale = atof(argv[++i]);
163                             break;
164             case '?':
165             case 'H':
166             default:
167     printf("\nUsage: enc -[options] [inputfile [outputfile]]");
168     printf("\nOptions are: (# = number), defaults show in ()");
169     printf("\n -t # tolerance criterion for fidelity. (%lf)", tol);
170          printf("\n -m # minimum quadtree partitions.  (%d)",min_part);
171     printf("\n -M # maximum quadtree partitions. (%d)",max_part);
172     printf("\n -S # number of input bytes to ignore. (%ld)",stripchar);
173     printf("\n -w # width (horizontal size) of input data. (256)");
174     printf("\n -h # height (vertical size) of input data. (256)");
175     printf("\n -d # domain step size. (%d)", dom_step);
176     printf("\n -D # method {0,1,2} for domain pool (%d)",dom_type);
177     printf("\n -s # number of scaling quantizing bits. (%d)",s_bits);
178     printf("\n -o # number of offset quantizing bits. (%d)",o_bits);
179     printf("\n -N # maximum scaling in encoding. (%lf)",max_scale);
180     printf("\n -e   domain step as multiplier not divisor. (off)");
181     printf("\n -p   use only positive scaling (for speed). (off)");
182     printf("\n -f   search 24 domain classes (for fidelity). (off)");
183     printf("\n -F   search 3 domain classes (for fidelity). (off)");
184     fatal("\n       -F and -f can be used together.");
185             }
186         }
187
188     if (inputfilename[0] == 1) strcpy(inputfilename, "lenna.dat");
189     if (outputfilename[0] == 1) strcpy(outputfilename, "lenna.trn");
190
191     if (hsize == -1)
192         if (vsize == -1) hsize = vsize = 256;
193         else hsize = vsize;
194     else
195         if (vsize == -1) vsize = hsize;
196
197     /* allocate memory for the input image. Allocating one chunck saves */
198     /* work and time later.                                             */
199     matrix_allocate(image, hsize, vsize, IMAGE_TYPE)
200     matrix_allocate(domimage[0], hsize/2, vsize/2, double)
201     matrix_allocate(domimage[1], hsize/2, vsize/2, double)
202     matrix_allocate(domimage[2], hsize/2, vsize/2, double)
```

```
203        matrix_allocate(domimage[3], hsize/2, vsize/2, double)
204
205        /* max_ & min_ part are variable, so this must be run time allocated */
206        bits_needed = (int *)malloc(sizeof(int)*(1+max_part-min_part));
207
208        if ((input = fopen(inputfilename, "r")) == NULL)
209            fatal("Can't open input file.");
210
211        /* skip the first  stripchar  chars */
212        fseek(input, stripchar, 0);
213        i = fread(image[0], sizeof(IMAGE_TYPE), hsize*vsize, input);
214        fclose(input);
215
216        if (i < hsize*vsize)
217            fatal("Not enough input data in the input file.");
218        else
219            printf("%dx%d=%d pixels read from %s.", hsize,vsize,i,inputfilename);
220
221        /* allcate memory for domain data and initialize it */
222        compute_sums(hsize,vsize);
223
224        if ((output = fopen(outputfilename, "w")) == NULL)
225            fatal("Can't open output file.");
226
227        /* output some data into the outputfile.                      */
228        pack(4,(long)min_part,output);
229        pack(4,(long)max_part,output);
230        pack(4,(long)dom_step,output);
231        pack(1,(long)dom_step_type,output);
232        pack(2,(long)dom_type,output);
233        pack(12,(long)hsize,output);
234        pack(12,(long)vsize,output);
235
236    /* This is the quantized value of zero scaling.. needed later */
237        zero_ialpha = 0.5 + (max_scale)/(2.0*max_scale)*(1<<s_bits);
238
239        /* The following routine takes a rectangular image and calls the */
240        /* quadtree routine to encode square sum-images in it.         */
241        /* the tolerance is a parameter since in some applications different */
242        /* regions of the image may need to be compressed to different tol's */
243        printf("\nEncoding Image.....");
244        fflush(stdout);
245        partition_image(0, 0, hsize,vsize, tol);
246        printf("Done.");
247        fflush(stdout);
248
249        /* stuff the last byte if needed */
250        pack(-1,(long)0,output);
251
252        fclose(output);
253        i = pack(-2,(long)0,output);
254        printf("\n Compression = %lf from %d bytes written in %s.\n",
255                (double)(hsize*vsize)/(double)i, i, outputfilename);
256
257        /* Free allocated memory*/
258        free(bits_needed);
259        free(domimage[0]);
260        free(domimage[1]);
261        free(domimage[2]);
262        free(domimage[3]);
263        free(domain.no_h_domains);
264        free(domain.no_v_domains);
265        free(domain.domain_hsize);
266        free(domain.domain_vsize);
267        free(domain.domain_hstep);
268        free(domain.domain_vstep);
269        for (i=0; i <= max_part-min_part; ++i)
270            free(domain.pixel[i]);
```

```
271      free(domain.pixel);
272      free(image[0]);
273      for (i=0; i <= max_part-min_part; ++i)
274      for (k=0; k<3; ++k)
275       for (j=0; j<24; ++j) list_free(the_domain[k][j][i]);
276      return(0);
277  }
278
279  /* *********************************************************** */
280  /* free memory allocated in the list structure the_domain      */
281  /* *********************************************************** */
282  list_free(node)
283  struct classified_domain *node;
284  {
285      if (node->next != NULL)
286        list_free(node->next);
287      free(node);
288  }
289
290  /* *********************************************************** */
291  /* return the average pixel value of a region of the image.    */
292  /* *********************************************************** */
293  void average(x,y,xsize,ysize, psum, psum2)
294  int x,y,xsize,ysize;
295  double *psum, *psum2;
296  {
297      register int i,j,k;
298      register double pixel;
299      *psum = *psum2 = 0.0;
300      k = ((x%2)<<1) + y%2;
301      x >>= 1; y >>= 1;
302      xsize >>= 1; ysize >>= 1;
303      for (i=x; i<x+xsize; ++i)
304      for (j=y; j<y+ysize; ++j) {
305      pixel = domimage[k][j][i];
306        *psum += pixel;
307        *psum2 += pixel*pixel;
308      }
309  }
310
311  /* *********************************************************** */
312  /* return the average pixel value of a region of the image. This */
313  /* routine differs from the previous in one slight way. It does  */
314  /* not average 2x2 sub-images to pixels. This is needed for clas- */
315  /* sifying ranges rather than domain where decimation is needed.  */
316  /* *********************************************************** */
317  void average1(x,y,xsize,ysize, psum, psum2)
318  int x,y,xsize,ysize;
319  double *psum, *psum2;
320  {
321      register int i,j;
322      register double pixel;
323      *psum = *psum2 = 0.0;
324
325      for (i=x; i<x+xsize; ++i)
326      for (j=y; j<y+ysize; ++j) {
327      pixel = (double)image[j][i];
328        *psum += pixel;
329        *psum2 += pixel*pixel;
330      }
331  }
332
333  /* *********************************************************** */
334  /* Take a region of the image at x,y and classify it.            */
335  /* The four quadrants of the region are ordered from brightest to */
336  /* least bright average value, then it is rotated into one of the */
337  /* three cannonical orientations possible with the brightest quad */
338  /* in the upper left corner.                                     */
```

```
339    /* The routine returns two indices that are class numbers: pfirst */
340    /* and psecond; the symmetry operation that bring the square into */
341    /* cannonical position; and the sum and sum^2 of the pixel values */
342    /* *********************************************************** */
343    classify(x, y, xsize, ysize, pfirst, psecond, psym, psum, psum2, type)
344    int x,y,xsize,ysize,    /* position, size of subimage to be classified */
345         *pfirst, *psecond, /* returned first and second class numbers    */
346         *psym;             /* returned symmetry operation that brings the */
347                            /* subimage to cannonical position.           */
348    double *psum, *psum2;   /* returned sum and sum^2 of pixel values      */
349    int type;               /* flag for decimating (for domains) or not    */
350    {
351
352         int order[4], i,j;
353         double a[4],a2[4];
354         void (*average_func)();
355
356         if (type == 2) average_func = average; else average_func = average1;
357
358         /* get the average values of each quadrant                      */
359
360
361         (*average_func)(x,y,               xsize/2,ysize/2,   &a[0], &a2[0]);
362         (*average_func)(x,y+ysize/2,       xsize/2,ysize/2,   &a[1], &a2[1]);
363         (*average_func)(x+xsize/2,y+ysize/2, xsize/2,ysize/2, &a[2], &a2[2]);
364         (*average_func)(x+xsize/2,y,       xsize/2,ysize/2,   &a[3], &a2[3]);
365
366         *psum = a[0] + a[1] + a[2] + a[3];
367         *psum2 = a2[0] + a2[1] + a2[2] + a2[3];
368
369         for (i=0; i<4; ++i) {
370             /* after the sorting below order[i] is the i-th brightest  */
371             /* quadrant.                                               */
372             order[i] = i;
373             /* convert a2[] to store the variance of each quadrant     */
374             a2[i] -= (double)(1<<(2*type))*a[i]*a[i]/(double)(xsize*ysize);
375         }
376
377         /* Now order the average value and also in order[],  which will */
378         /* then tell us the indices (in a[]) of the brightest to darkest */
379         for (i=2; i>=0; --i)
380         for (j=0; j<=i; ++j)
381             if (a[j]<a[j+1]) {
382                 swap(order[j], order[j+1],int)
383                 swap(a[j], a[j+1],double)
384         }
385
386         /* because of the way we ordered the a[] the rotation can be */
387         /* read right off of order[]. That will make the brightest   */
388         /* quadrant be in the upper left corner. But we must still    */
389         /* decide which cannonical class the image portion belogs     */
390         /* to and whether to do a flip or just a rotation. This is    */
391         /* the following table summarizes the horrid lines below      */
392         /* order       class         do a rotation                    */
393         /* 0,2,1,3       0            0                                */
394         /* 0,2,3,1       0            1                                */
395         /* 0,1,2,3       1            0                                */
396         /* 0,3,2,1       1            1                                */
397         /* 0,1,3,2       2            0                                */
398         /* 0,3,1,2       2            1                                */
399
400         *psym = order[0];
401         /* rotate the values */
402         for (i=0; i<4; ++i)
403             order[i] = (order[i] - (*psym) + 4)%4;
404
405         for (i=0; order[i] != 2; ++i);
406         *pfirst = i-1;
```

```
407        if (order[3] == 1 || (*pfirst == 2 && order[2] == 1)) *psym += 4;
408
409        /* Now further classify the sub-image by the variance of its    */
410        /* quadrants. This give 24 subclasses for each of the 3 classes */
411        for (i=0; i<4; ++i) order[i] = i;
412
413        for (i=2; i>=0; --i)
414        for (j=0; j<=i; ++j)
415            if (a2[j]<a2[j+1]) {
416                swap(order[j], order[j+1],int)
417                swap(a2[j], a2[j+1],double)
418        }
419
420        /* Now do the symmetry operation */
421        for (i=0; i<4; ++i)
422            order[i] = (order[i] - (*psym%4) + 4)%4;
423        if (*psym > 3)
424            for (i=0; i<4; ++i)
425                if (order[i]%2) order[i] = (2 + order[i])%4;
426
427        /* We want to return a class number from 0 to 23 depending on */
428        /* the ordering of the quadrants according to their variance  */
429        *psecond = 0;
430        for (i=2; i>=0; --i)
431        for (j=0; j<=i; ++j)
432            if (order[j] > order[j+1]) {
433                swap(order[j],order[j+1], int);
434                if (order[j] == 0 || order [j+1] == 0)
435                    *psecond += 6;
436                else if (order[j] == 1 || order [j+1] == 1)
437                    *psecond += 2;
438                else if (order[j] == 2 || order [j+1] == 2)
439                    *psecond += 1;
440        }
441 }
442
443 /* ************************************************************ */
444 /* Compute sum and sum^2 of pixel values in domains for use in  */
445 /* the rms computation later. Since a domain is compared with   */
446 /* many ranges, doing this just once saves a lot of computation */
447 /* This routine also fills a list structure with the domains    */
448 /* as they are classified and creates the memory for the domain */
449 /* data in a matrix.                                            */
450 /* ************************************************************ */
451 compute_sums(hsize,vsize)
452 int hsize,vsize;
453 {
454        int i,j,k,l,
455            domain_x,
456            domain_y,
457            first_class,
458            second_class,
459            domain_size,
460            domain_step_size,
461            size,
462            x_exponent,
463            y_exponent;
464
465        struct classified_domain *node;
466
467        printf("\nComputing domain sums... ");
468        fflush(stdout);
469
470        /* pre-decimate the image into domimage to avoid having to    */
471        /* do repeated averaging of 2x2 pixel groups.                 */
472        /* There are 4 ways to decimate the image, depending on the   */
473        /* location of the domain, odd or even address.               */
474        for (i=0; i<2; ++i)
```

```
475        for (j=0; j<2; ++j)
476        for (k=i; k<hsize-i; k += 2)
477        for (l=j; l<vsize-j; l += 2)
478           domimage[((i<<1)+j][l>>1][k>>1] =
479                      ((double)image[l][k] + (double)image[l+1][k+1] +
480                      (double)image[l][k+1] + (double)image[l+1][k])*0.25;
481
482
483        /* Allocate memory for the sum and sum^2 of domain pixels    */
484        /* We first compute the size of the largest square that fits in */
485        /* the image.                                                */
486        x_exponent = (int)floor(log((double)hsize)/log(2.0));
487        y_exponent = (int)floor(log((double)vsize)/log(2.0));
488
489        /* exponent is min of x_ and y_ exponent */
490        max_exponent = (x_exponent > y_exponent ? y_exponent : x_exponent);
491
492        /* size is the size of the largest square that fits in the image */
493        /* It is used to compute the domain and range sizes.         */
494        size = 1<<max_exponent;
495
496        if (max_exponent < max_part)
497          fatal("Reduce maximum number of quadtree partitions.");
498        if (max_exponent-2 < max_part)
499          printf("\nWarning: so many quadtree partitions yield absurd ranges.");
500
501        i = max_part - min_part + 1;
502        domain.no_h_domains = (int *)malloc(sizeof(int)*i);
503        domain.no_v_domains = (int *)malloc(sizeof(int)*i);
504        domain.domain_hsize = (int *)malloc(sizeof(int)*i);
505        domain.domain_vsize = (int *)malloc(sizeof(int)*i);
506        domain.domain_hstep = (int *)malloc(sizeof(int)*i);
507        domain.domain_vstep = (int *)malloc(sizeof(int)*i);
508
509        domain.pixel= (struct domain_pixels ***)
510                  malloc(i*sizeof(struct domain_pixels **));
511        if (domain.pixel == NULL) fatal("No memory for domain pixel sums.");
512
513        for (i=0; i <= max_part-min_part; ++i) {
514           /* first compute how many domains there are horizontally */
515           domain.domain_hsize[i] = size >> (min_part+i-1);
516           if (dom_type == 2)
517                  domain.domain_hstep[i] = dom_step;
518           else if (dom_type == 1)
519                  if (dom_step_type == 1)
520                     domain.domain_hstep[i] = (size >> (max_part - i-1))*dom_step;
521                  else
522                     domain.domain_hstep[i] = (size >> (max_part - i-1))/dom_step;
523           else
524                  if (dom_step_type == 1)
525                     domain.domain_hstep[i] = domain.domain_hsize[i]*dom_step;
526                  else
527                     domain.domain_hstep[i] = domain.domain_hsize[i]/dom_step;
528
529           domain.no_h_domains[i] = 1+(hsize-domain.domain_hsize[i])/
530                                              domain.domain_hstep[i];
531           /* bits_needed[i][0] = ceil(log(domain.no_h_domains[i])/log(2.0));  */
532
533           /* now compute how many domains there are vertically. The sizes */
534           /* are the same for square domains, but not for rectangular ones */
535           domain.domain_vsize[i] = size >> (min_part+i-1);
536           if (dom_type == 2)
537                  domain.domain_vstep[i] = dom_step;
538           else if (dom_type == 1)
539                  if (dom_step_type == 1)
540                     domain.domain_vstep[i] = (size >> (max_part - i-1))*dom_step;
541                  else
542                     domain.domain_vstep[i] = (size >> (max_part - i-1))/dom_step;
```

```
543          else
544              if (dom_step_type == 1)
545                  domain.domain_vstep[i] = domain.domain_vsize[i]*dom_step;
546              else
547                  domain.domain_vstep[i] = domain.domain_vsize[i]/dom_step;
548
549          domain.no_v_domains[i] = 1+(vsize-domain.domain_vsize[i])/
550                                                  domain.domain_vstep[i];
551
552          /* Now compute the number of bits needed to store the domain data */
553          bits_needed[i] = ceil(log((double)domain.no_h_domains[i]*
554                              (double)domain.no_v_domains[i])/log(2.0));
555
556          matrix_allocate(domain.pixel[i], domain.no_h_domains[i],
557                      domain.no_v_domains[i], struct domain_pixels)
558
559      }
560
561      /* allocate and zero the list containing the classified domain data */
562      i = max_part - min_part + 1;
563      for (first_class = 0; first_class < 3; ++first_class)
564      for (second_class = 0; second_class < 24; ++second_class) {
565          the_domain[first_class][second_class] =
566                          (struct classified_domain **)
567                          malloc(i*sizeof(struct classified_domain *));
568          for (j=0; j<i; ++j)
569                      the_domain[first_class][second_class][j] = NULL;
570      }
571
572      /* Precompute sum and sum of squares for domains            */
573      /* This part can get faster for overlapping domains if repeated */
574      /* sums are avoided                                         */
575      for (i=0; i <= max_part-min_part; ++i) {
576          for (j=0,domain_x=0; j<domain.no_h_domains[i]; ++j,
577                  domain_x+=domain.domain_hstep[i])
578          for (k=0,domain_y=0; k<domain.no_v_domains[i]; ++k,
579                  domain_y+=domain.domain_vstep[i]) {
580  classify(domain_x, domain_y,
581                      domain.domain_hsize[i],
582                      domain.domain_vsize[i],
583      &first_class, &second_class,
584      &domain.pixel[i][k][j].sym,
585                          &domain.pixel[i][k][j].sum,
586                          &domain.pixel[i][k][j].sum2, 2);
587
588          /* When the domain data is referenced from the list, we need to */
589          /* know where the domain is.. so we have to store the position  */
590          domain.pixel[i][k][j].dom_x = j;
591          domain.pixel[i][k][j].dom_y = k;
592          node = (struct classified_domain *)
593                          malloc(sizeof(struct classified_domain));
594
595          /* put this domain in the classified list structure */
596          node->the = &domain.pixel[i][k][j];
597          node->next = the_domain[first_class][second_class][i];
598          the_domain[first_class][second_class][i] = node;
599  }
600      }
601
602      /* Now we make sure no domain class is actually empty.  */
603      for (i=0; i <= max_part-min_part; ++i)
604      for (first_class = 0; first_class < 3; ++first_class)
605      for (second_class = 0; second_class < 24; ++second_class)
606          if (the_domain[first_class][second_class][i] == NULL) {
607              node = (struct classified_domain *)
608                              malloc(sizeof(struct classified_domain));
609              node->the = &domain.pixel[i][0][0];
610              node->next = NULL;
```

```
611                the_domain[first_class][second_class][i] = node;
612         }
613
614      printf("Done.");
615      fflush(stdout);
616  }
617
618  /* ********************************************************** */
619  /* pack value using size bits and output into foutf */
620  /* ********************************************************** */
621  int pack(size, value, foutf)
622  int size; long int value;
623  FILE *foutf;
624  {
625       int i;
626       static int ptr = 1, /* how many bits are packed in sum so far */
627                  sum = 0, /* packed bits */
628                  num_of_packed_bytes = 0; /* total bytes written out */
629
630       /* size == -1 means we are at the end, so write out what is left */
631       if (size == -1 && ptr != 1) {
632           fputc(sum<<(8-ptr), foutf);
633           ++num_of_packed_bytes;
634           return(0);
635       }
636
637       /* size == -2 means we want to know how many bytes we have written */
638       if (size == -2)
639               return(num_of_packed_bytes);
640
641       for (i=0; i<size; ++i, ++ptr, value = value>>1, sum = sum<<1) {
642           if (value & 1) sum |= 1;
643
644            if (ptr == 8) {
645                fputc(sum,foutf);
646                ++num_of_packed_bytes;
647                sum=0;
648                ptr=0;
649            }
650        }
651  }
652
653  /* ********************************************************** */
654  /* Compare a range to a domain and return rms and the quantized */
655  /* scaling and offset values (pialhpa, pibeta).              */
656  /* ********************************************************** */
657  double compare(atx,aty, xsize, ysize, depth, rsum, rsum2, dom_x,dom_y,
658                                    sym_op, pialpha,pibeta)
659  int atx, aty, xsize, ysize, depth, dom_x, dom_y, sym_op, *pialpha, *pibeta;
660  double rsum, rsum2;
661  {
662      int i, j, i1, j1, k,
663          domain_x,
664          domain_y;        /* The domain position                 */
665
666      double pixel,
667             det,           /* determinant of solution */
668             dsum,          /* sum of domain values */
669             rdsum = 0,     /* sum of range*domain values   */
670             dsum2,         /* sum of domain^2 values   */
671             w2 = 0,        /* total number of values tested */
672             rms = 0,       /* root means square difference */
673             alpha,         /* the scale factor */
674             beta;          /* The offset */
675
676
677
678      /* xsize = hsize >> depth; */
```

```
679     /* ysize = vsize >> depth; */
680     w2 = xsize * ysize;
681
682     dsum = domain.pixel[depth-min_part][dom_y][dom_x].sum;
683     dsum2 = domain.pixel[depth-min_part][dom_y][dom_x].sum2;
684     domain_x = (dom_x * domain.domain_hstep[depth-min_part]);
685     domain_y = (dom_y * domain.domain_vstep[depth-min_part]);
686     k = ((domain_x%2)<<1) + domain_y%2;
687     domain_x >>= 1;
688     domain_y >>= 1;
689
690     /* The main statement in this routine is a switch statement which */
691     /* scans through the domain and range to compare them. The loop   */
692     /* center is the same so we #define it for easy modification      */
693 #define COMPUTE_LOOP              {                                    \
694         pixel = domimage[k][j1][i1];                                  \
695         rdsum += image[j][i]*pixel;                                   \
696         }
697
698     switch(sym_op) {
699         case 0: for (i=atx, i1 = domain_x; i<atx+xsize; ++i, ++i1)
700                 for (j=aty, j1 = domain_y; j<aty+ysize; ++j, ++j1)
701                     COMPUTE_LOOP
702                 break;
703         case 1: for (j=aty, i1 = domain_x; j<aty+ysize; ++j, ++i1)
704                 for (i=atx+xsize-1, j1 = domain_y; i>=atx; --i, ++j1)
705                     COMPUTE_LOOP
706                 break;
707         case 2: for (i=atx+xsize-1, i1 = domain_x; i>=atx; --i, ++i1)
708                 for (j=aty+ysize-1, j1 = domain_y; j>=aty; --j, ++j1)
709                     COMPUTE_LOOP
710                 break;
711         case 3: for (j=aty+ysize-1, i1 = domain_x; j>=aty; --j, ++i1)
712                 for (i=atx, j1 = domain_y; i<atx+xsize; ++i, ++j1)
713                     COMPUTE_LOOP
714                 break;
715         case 4: for (j=aty, i1 = domain_x; j<aty+ysize; ++j, ++i1)
716                 for (i=atx, j1 = domain_y; i<atx+xsize; ++i, ++j1)
717                     COMPUTE_LOOP
718                 break;
719         case 5: for (i=atx, i1 = domain_x; i<atx+xsize; ++i, ++i1)
720                 for (j=aty+ysize-1, j1 = domain_y; j>=aty; --j, ++j1)
721                     COMPUTE_LOOP
722                 break;
723         case 6: for (j=aty+ysize-1, i1 = domain_x; j>=aty; --j, ++i1)
724                 for (i=atx+xsize-1, j1 = domain_y; i>=atx; --i, ++j1)
725                     COMPUTE_LOOP
726                 break;
727         case 7: for (i=atx+xsize-1, i1 = domain_x; i>=atx; --i, ++i1)
728                 for (j=aty, j1 = domain_y; j<aty+ysize; ++j, ++j1)
729                     COMPUTE_LOOP
730                 break;
731     }
732
733     det = (xsize*ysize)*dsum2 - dsum*dsum;
734
735     if (det == 0.0)
736         alpha = 0.0;
737     else
738         alpha = (w2*rdsum - rsum*dsum)/det;
739
740     if (only_positive && alpha < 0.0) alpha = 0.0;
741     *pialpha = 0.5 + (alpha + max_scale)/(2.0*max_scale)*(1<<s_bits);
742     if (*pialpha < 0) *pialpha = 0;
743     if (*pialpha >= 1<<s_bits) *pialpha = (1<<s_bits)-1;
744
745     /* Now recompute alpha back */
746     alpha = (double)*pialpha/(double)(1<<s_bits)*(2.0*max_scale)-max_scale;
```

```
747
748         /* compute the offset */
749         beta = (rsum - alpha*dsum)/w2;
750
751         /* Convert beta to an integer */
752         /* we use the sign information of alpha to pack efficiently */
753         if (alpha > 0.0)   beta += alpha*GREY_LEVELS;
754         *pibeta = 0.5 + beta/
755                 ((1.0+fabs(alpha))*GREY_LEVELS)*((1<<o_bits)-1);
756         if (*pibeta< 0) *pibeta = 0;
757         if (*pibeta>= 1<<o_bits) *pibeta = (1<<o_bits)-1;
758
759         /* Recompute beta from the integer */
760         beta = (double)*pibeta/(double)((1<<o_bits)-1)*
761                 ((1.0+fabs(alpha))*GREY_LEVELS);
762         if (alpha > 0.0) beta  -= alpha*GREY_LEVELS;
763
764         /* Compute the rms based on the quantized alpha and beta! */
765         rms= sqrt((rsum2 + alpha*(alpha*dsum2 - 2.0*rdsum + 2.0*beta*dsum) +
766                     beta*(beta*w2 - 2.0*rsum))/w2);
767
768         return(rms);
769   }
770
771   /* ********************************************************** */
772   /* Recursively partition an image, computing the best transfoms */
773   /* ********************************************************** */
774   quadtree(atx,aty,xsize,ysize,tol,depth)
775   int atx, aty, xsize, ysize, depth;
776   double tol;  /* the tolerance fit  */
777   {
778       int i,
779           sym_op,                  /* A symmetry operation of the square */
780           ialpha,                  /* Intgerized scalling factor      */
781           ibeta,                   /* Intgerized offset               */
782           best_ialpha,             /* best ialpha found               */
783           best_ibeta,
784           best_sym_op,
785           best_domain_x,
786           best_domain_y,
787           first_class,
788           the_first_class,
789           first_class_start,       /* loop beginning and ending values   */
790           first_class_end,
791           second_class[2],
792           the_second_class,
793           second_class_start,      /* loop beginning and ending values   */
794           second_class_end,
795           range_sym_op[2],         /* the operations to bring square to  */
796           domain_sym_op;           /* cannonical position.               */
797
798       struct classified_domain *node;  /* var for domains we scan through */
799
800       double rms, best_rms,          /* rms value and min rms found so far */
801              sum=0, sum2=0;          /* sum and sum^2 of range pixels       */
802
803
804       /* keep breaking it down until we are small enough */
805       if (depth < min_part) {
806           quadtree(atx,aty, xsize/2, ysize/2, tol,depth+1);
807           quadtree(atx+xsize/2,aty, xsize/2, ysize/2,tol,depth+1);
808           quadtree(atx,aty+ysize/2, xsize/2, ysize/2,tol,depth+1);
809           quadtree(atx+xsize/2,aty+ysize/2, xsize/2, ysize/2,tol,depth+1);
810           return;
811       }
812
813       /* now search for the best domain-range match and write it out */
814       best_rms = 10000000000.0;      /* just a big number */
```

```
815
816     classify(atx, aty, xsize,ysize,
817               &the_first_class, &the_second_class,
818               &range_sym_op[0], &sum, &sum2, 1);
819
820
821     /* sort out how much to search based on -f and -F input flags */
822     if (fullclass_search) {
823           first_class_start = 0;
824           first_class_end = 3;
825     } else {
826           first_class_start = the_first_class;
827           first_class_end = the_first_class+1;
828     }
829
830     if (subclass_search) {
831           second_class_start = 0;
832           second_class_end = 24;
833     } else {
834           second_class_start = the_second_class;
835           second_class_end = the_second_class+1;
836     }
837
838     /* these for loops vary according to the optimization flags we set */
839     /* for subclass_search and fullclass_search==1, we search all the  */
840     /* domains (except not all rotations).                             */
841     for (first_class = first_class_start;
842           first_class < first_class_end; ++first_class)
843     for (second_class[0] = second_class_start;
844           second_class[0] < second_class_end; ++second_class[0]) {
845
846        /* We must check each domain twice. Once for positive scaling,  */
847        /* once for negative scaling. Each has its own class and sym_op */
848        if (!only_positive) {
849           second_class[1] =
850           class_transform[(first_class == 2 ? 1 : 0)][second_class[0]];
851           range_sym_op[1] =
852              rot_transform[(the_first_class == 2 ? 1 : 0)][range_sym_op[0]];
853        }
854
855        /* only_positive is 0 or 1, so we may or not scan             */
856        for (i=0; i<(2-only_positive); ++i)
857        for (node = the_domain[first_class][second_class[i]][depth-min_part];
858           node != NULL;
859           node = node->next) {
860           domain_sym_op = node->the->sym;
861           /* The following if statement figures out how to transform  */
862           /* the domain onto the range, given that we know how to get */
863           /* each into cannonical position.                          */
864           if (((domain_sym_op>3 ? 4: 0) + (range_sym_op[i]>3 ? 4: 0))%8 == 0)
865             sym_op = (4 + domain_sym_op%4 - range_sym_op[i]%4)%4;
866           else
867             sym_op = (4 + (domain_sym_op%4 + 3*(4-range_sym_op[i]%4))%4)%8;
868
869           rms = compare(atx,aty, xsize, ysize,  depth, sum,sum2,
870                                   node->the->dom_x,
871                                   node->the->dom_y,
872                                   sym_op, &ialpha,&ibeta);
873
874           if (rms < best_rms) {
875                   best_ialpha = ialpha;
876                   best_ibeta = ibeta;
877                   best_rms = rms;
878                   best_sym_op = sym_op;
879                   best_domain_x = node->the->dom_x;
880                   best_domain_y = node->the->dom_y;
881           }
882        }
```

```
883        }
884
885        if (best_rms > tol && depth < max_part) {
886            /* We didn't find a good enough fit so quadtree down */
887            pack(1,(long)1,output);  /* This bit means we quadtree'd down */
888            quadtree(atx,aty, xsize/2, ysize/2, tol,depth+1);
889            quadtree(atx+xsize/2,aty, xsize/2, ysize/2, tol,depth+1);
890            quadtree(atx,aty+ysize/2, xsize/2, ysize/2, tol,depth+1);
891            quadtree(atx+xsize/2,aty+ysize/2, xsize/2, ysize/2, tol,depth+1);
892        } else {
893            /* The fit was good enough or we just can't get smaller ranges */
894            /* So write out the data */
895            if (depth < max_part)            /* if we are not at the smallest range */
896                    pack(1,(long)0,output);/* then we must tell the decoder we    */
897                                           /* stopped quadtreeing                 */
898            pack(s_bits, (long)best_ialpha, output);
899            pack(o_bits, (long)best_ibeta, output);
900            /* When the scaling is zero, there is no need to store the domain */
901            if (best_ialpha != zero_ialpha) {
902               pack(3, (long)best_sym_op, output);
903               pack(bits_needed[depth-min_part], (long)(best_domain_y*
904                 domain.no_h_domains[depth-min_part]+best_domain_x), output);
905        }
906        }
907 }
908
909 /* ************************************************************ */
910 /* Recursively partition an image, finding the largest contained */
911 /* square and call the quadtree routine the encode that square.  */
912 /* This enables us to encode rectangular image easily.          */
913 /* ************************************************************ */
914 partition_image(atx, aty, hsize,vsize, tol)
915 int atx, aty, hsize,vsize;
916 double tol;
917 {
918    int x_exponent,     /* the largest power of 2 image size that fits */
919        y_exponent,     /* horizontally or vertically the rectangle.  */
920        exponent,       /* The actual size of image that's encoded.   */
921        size,
922        depth;
923
924    x_exponent = (int)floor(log((double)hsize)/log(2.0));
925    y_exponent = (int)floor(log((double)vsize)/log(2.0));
926
927    /* exponent is min of x_ and y_ exponent */
928    exponent = (x_exponent > y_exponent ? y_exponent : x_exponent);
929    size = 1<<exponent;
930    depth = max_exponent - exponent;
931    quadtree(atx,aty,size,size,tol,depth);
932    if (size != hsize)
933       partition_image(atx+size, aty, hsize-size,vsize, tol);
934
935    if (size != vsize)
936       partition_image(atx, aty+size, size,vsize-size, tol);
937 }
938
939 /* fatal is for when there is a fatal error... print a message and exit */
940 void fatal(s)
941 char *s;
942 {
943     printf("%s\n",s);
944     exit(-1);
945 }
```

A.4   Dec.c

```
1   /***************************************************************************/
2   /* Decode an image encoded with a quadtree partition based fractal scheme */
3   /*                                                                         */
4   /*          Copyright 1993,1994 Yuval Fisher. All rights reserved.         */
5   /*                                                                         */
6   /* Version 0.04 3/14/94                                                    */
7   /***************************************************************************/
8
9   /* The following belong in a encdec.h file, but nevermind...               */
10  /* ----------------------------------------------------------------------*/
11  #include <stdio.h>
12  #include <math.h>
13
14  #define DEBUG 0
15  #define GREY_LEVELS 255
16
17  #define IMAGE_TYPE unsigned char /* may be different in some applications */
18
19  /* The following #define allocates an hsize x vsize  matrix of type type */
20  #define matrix_allocate(matrix, hsize,vsize, TYPE) {\
21      TYPE *imptr; \
22      int _i; \
23      matrix = (TYPE **)malloc(vsize*sizeof(TYPE *));\
24      imptr = (TYPE*)malloc((long)hsize*(long)vsize*sizeof(TYPE));\
25      if (imptr == NULL) \
26          fatal("\nNo memory in matrix allocate."); \
27      for (_i = 0; _i<vsize; ++_i, imptr += hsize) \
28          matrix[_i] = imptr; \
29  }
30
31  #define bound(a)   ((a) < 0.0 ? 0 : ((a)>255.0? 255 : a))
32
33  /* various function declarations to keep compiler warnings away          */
34  void fatal();
35  char *malloc();
36  char *strcpy();
37
38  /* ---------------------------------------------------------------------- */
39
40  IMAGE_TYPE **image,*imptr,**image1;
41                              /* The input image data and a dummy        */
42
43  double max_scale = 1.0;     /* maximum allowable grey level scale factor */
44
45  int   s_bits = 5,           /* number of bits used to store scale factor */
46        o_bits = 7,           /* number of bits used to store offset       */
47        hsize = -1,           /* The horizontal size of the input image    */
48        vsize = -1,           /* The vertical size                         */
49        scaledhsize,          /* hsize*scalefactor                         */
50        scaledvsize,          /* vsize*scalefactor                         */
51        size,                 /* largest square image that fits in image   */
52        min_part = 3,         /* min and max _part determine a range of    */
53        max_part = 4,         /* Range sizes from hsize>>min to hsize>>max  */
54        dom_step = 4,         /* Density of domains relative to size        */
55        dom_step_type = 0,    /* Flag for dom_step a multiplier or divisor  */
56        dom_type = 0,         /* Method of generating domain pool 0,1,2..   */
57        post_process = 1,     /* Flag for postprocessing.                   */
58        num_iterations= 10,   /* Number of decoding iterations used.        */
59        *no_h_domains,        /* Number of horizontal domains.              */
60        *domain_hstep,        /* Domain density step size.                  */
61        *domain_vstep,        /* Domain density step size.                  */
62        *bits_needed,         /* Number of bits to encode domain position.  */
63        zero_ialpha,          /* the const ialpha when alpha = 0            */
64        output_partition=0,   /* A flag for outputing the partition         */
65        max_exponent;         /* The max power of 2 side of square image    */
66                              /* that fits in our input image.             */
```

```
67
68   struct transformation_node {
69          int rx,ry,            /* The range position and size in a trans.   */
70                 xsize, ysize,
71                 rrx,rry,
72                 dx,dy;         /* The domain position.                      */
73          int sym_op;           /* The symmetry operation used in the trans. */
74          int depth;            /* The depth in the quadtree partition.      */
75          double scale, offset; /* scalling and offset values.               */
76          struct transformation_node *next;       /* The next trans. in list */
77   } transformations, *trans;
78
79   FILE *input,*output,*other_input;
80
81
82   /* fatal is for when there is a fatal error... print a message and exit */
83   void fatal(s)
84   char *s;
85   {
86        printf("\n%s\n",s);
87        exit(-1);
88   }
89
90   /* ********************************************************* */
91   /* unpack value using size bits read from fin.               */
92   /* ********************************************************* */
93   long unpack(size, fin)
94   int size;
95   FILE *fin;
96   {
97        int i;
98        int value = 0;
99        static int ptr = 1; /* how many bits are packed in sum so far */
100       static int sum;
101
102
103       /* size == -2 means we initialize things */
104       if (size == -2) {
105            sum = fgetc(fin);
106            sum <<= 1;
107            return((long)0);
108       }
109
110       /* size == -1 means we want to peek at the next bit without */
111       /* advancing the pointer */
112       if (size == -1)
113            return((long)((sum&256)>>8));
114
115        for (i=0; i<size; ++i, ++ptr,  sum <<= 1) {
116            if (sum & 256) value |= 1<<i;
117
118             if (ptr == 8) {
119                 sum = getc(fin);
120                 ptr=0;
121            }
122         }
123       return((long)value);
124  }
125
126  main(argc,argv)
127  int argc;
128  char **argv;
129  {
130       /* Defaults are set initially */
131       double        scalefactor = 1.0;              /* Scale factor for output */
132       char          inputfilename[200];
133       char          outputfilename[200];
134       char          other_input_file[200];
```

```
135     int              i,j, x_exponent, y_exponent;
136     int              domain_size, no_domains;
137
138
139     inputfilename[0] = 1;    /* We initially set the input to this and */
140     outputfilename[0] = 1;   /* then check if the input/output names   */
141     other_input_file[0] = 1; /* have been set below.                   */
142
143     /* scan through the input line and read in the arguments */
144     for (i=1; i<argc; ++i)
145        if (argv[i][0] != '-' )
146              if (inputfilename[0] == 1)
147                 strcpy(inputfilename, argv[i]);
148              else if (outputfilename[0] == 1)
149                 strcpy(outputfilename, argv[i]);
150              else;
151        else { /* we have a flag */
152              if (strlen(argv[i]) == 1) break;
153              switch(argv[i][1]) {
154                 case 'i': strcpy(other_input_file,argv[++i]);
155                           break;
156                 case 'n': num_iterations = atoi(argv[++i]);
157                           break;
158                 case 'f': scalefactor = atof(argv[++i]);
159                           break;
160                 case 'P': output_partition = 1;
161                           break;
162                 case 'p': post_process = 0;
163                           break;
164                 case 's': s_bits = atoi(argv[++i]);
165                           break;
166                 case 'o': o_bits = atoi(argv[++i]);
167                           break;
168                 case 'N': max_scale = atof(argv[++i]);
169                           break;
170                 case '?':
171                 case 'H':
172                 default:
173     printf("\nUsage: dec -[options] [inputfile [outputfile]]");
174     printf("\nOptions are: (# = number), defaults show in ()");
175     printf("\n  -f # scale factor of output size. (%lf)", scalefactor);
176     printf("\n  -i file name. An initial image to iteration from.");
177     printf("\n  -n # no. of decoding iterations. (%d)", num_iterations);
178     printf("\n  -N # maximum allowed scaling. (%lf)",max_scale);
179     printf("\n  -s # number of scaling quantizing bits. (%d)",s_bits);
180     printf("\n  -o # number of offset quantizing bits. (%d)",o_bits);
181     printf("\n  -P   output the partition into the output file. (off)");
182     printf("\n  -p   supress artifact postprocessing. (off)");
183              fatal(" ");
184          }
185     }
186
187     if (inputfilename[0] == 1) strcpy(inputfilename, "lenna.trn");
188     if (outputfilename[0] == 1) strcpy(outputfilename, "lenna.out");
189
190     if ((input = fopen(inputfilename, "r")) == NULL)
191        fatal("Can't open input file.");
192
193     unpack(-2,input); /* initialize the unpacking routine */
194
195 /* read the header data from the input file. This should probably */
196     /* be put into one read which reads a structure with the info     */
197     min_part = (int)unpack(4,input);
198     max_part = (int)unpack(4,input);
199     dom_step = (int)unpack(4,input);
200     dom_step_type = (int)unpack(1,input);
201     dom_type = (int)unpack(2,input);
202     hsize = (int)unpack(12,input);
```

```
203       vsize = (int)unpack(12,input);
204
205       /* we now compute size */
206       x_exponent = (int)floor(log((double)hsize)/log(2.0));
207       y_exponent = (int)floor(log((double)vsize)/log(2.0));
208
209       /* exponent is min of x_ and y_ exponent */
210       max_exponent = (x_exponent > y_exponent ? y_exponent : x_exponent);
211       /* size is the size of the largest square that fits in the image */
212       /* It is used to compute the domain and range sizes.          */
213       size = 1<<max_exponent;
214
215       /* This is the quantized value of zero scaling */
216       zero_ialpha = 0.5 + (max_scale)/(2.0*max_scale)*(1<<s_bits);
217
218       /* allocate memory for the output image. Allocating one chunck saves  */
219       /* work and time later.                                        */
220   scaledhsize = (int)(scalefactor*hsize);
221   scaledvsize = (int)(scalefactor*vsize);
222       matrix_allocate(image, scaledhsize,scaledvsize, IMAGE_TYPE);
223       matrix_allocate(image1, scaledhsize, scaledvsize, IMAGE_TYPE);
224
225       if (other_input_file[0] != 1) {
226           other_input = fopen(other_input_file, "r");
227           i = fread(image[0], sizeof(IMAGE_TYPE),
228   scaledhsize*scaledvsize, other_input);
229           if (i < scaledhsize*scaledvsize)
230              fatal("Couldn't read input... not enough data.");
231           else
232              printf("\n%d pixels read from %s.\n", i,other_input_file);
233           fclose(other_input);
234       }
235
236       /* since max_ and min_ part are variable, these must be allocated */
237       i = max_part - min_part + 1;
238       bits_needed = (int *)malloc(sizeof(int)*i);
239       no_h_domains = (int *)malloc(sizeof(int)*i);
240       domain_hstep = (int *)malloc(sizeof(int)*i);
241       domain_vstep = (int *)malloc(sizeof(int)*i);
242
243       /* compute bits needed to read each domain type */
244       for (i=0; i <= max_part-min_part; ++i) {
245          /* first compute how many domains there are horizontally */
246          domain_size = size >> (min_part+i-1);
247          if (dom_type == 2)
248               domain_hstep[i] = dom_step;
249          else if (dom_type == 1)
250               if (dom_step_type ==1)
251                   domain_hstep[i] = (size >> (max_part - i-1))*dom_step;
252               else
253                   domain_hstep[i] = (size >> (max_part - i-1))/dom_step;
254          else
255               if (dom_step_type ==1)
256                   domain_hstep[i] = domain_size*dom_step;
257               else
258                   domain_hstep[i] = domain_size/dom_step;
259
260          no_h_domains[i] = 1+(hsize-domain_size)/domain_hstep[i];
261          /* bits_needed[i][0] = ceil(log(no_domains)/log(2));   */
262
263          /* now compute how many domains there are vertically */
264          if (dom_type == 2)
265               domain_vstep[i] = dom_step;
266          else if (dom_type == 1)
267               if (dom_step_type ==1)
268               domain_vstep[i] = (size >> (max_part - i-1))*dom_step;
269               else
270               domain_vstep[i] = (size >> (max_part - i-1))/dom_step;
```

```
271          else
272               if (dom_step_type ==1)
273                   domain_vstep[i] = domain_size*dom_step;
274               else
275                   domain_vstep[i] = domain_size/dom_step;
276
277          no_domains = 1+(vsize-domain_size)/domain_vstep[i];
278          bits_needed[i] = ceil(log((double)no_domains*(double)no_h_domains[i])/
279                          log(2.0));
280       }
281
282       if ((output = fopen(outputfilename, "w")) == NULL)
283               fatal("Can't open output file.");
284
285       /* Read in the transformation data */
286       trans = &transformations;
287       printf("\nReading transformations.....");
288       fflush(stdout);
289       partition_image(0, 0, hsize,vsize );
290       fclose(input);
291       printf("Done.");
292       fflush(stdout);
293
294   if (scalefactor != 1.0) {
295   printf("\nScaling image to %d x %d.", scaledhsize,scaledvsize);
296   scale_transformations(scalefactor);
297   }
298
299       /* when we output the partition, we just read the transformations */
300       /* in and write them to the outputfile                           */
301       if (output_partition) {
302           fprintf(output,"\n%d %d\n %d %d\n%d %d\n\n",
303                   0, 0, scaledhsize, 0, scaledhsize, scaledvsize);
304           printf("\nOutputed partition data in %s\n",outputfilename);
305           fclose(output);
306           return;
307       }
308
309       for (i=0; i<num_iterations; ++i)
310         apply_transformations();
311
312       if (post_process)
313         smooth_image();
314
315       i = fwrite(image[0], sizeof(IMAGE_TYPE), scaledhsize*scaledvsize, output);
316       if (i < scaledhsize*scaledvsize)
317           fatal("Couldn't write output... not enough disk space ?.");
318       else
319           printf("\n%d pixels written to output file.\n", i);
320
321       fclose(output);
322   }
323
324   /* ********************************************************** */
325   /* Read in the transformation data from *input.            */
326   /* This is a recursive routine whose recursion tree follows the */
327   /* recursion done by the encoding program.                 */
328   /* ********************************************************** */
329   read_transformations(atx,aty,xsize,ysize,depth)
330   int atx,aty,xsize,ysize,depth;
331   {
332       /* Having all these locals in a recursive procedure is hard on the */
333       /* stack.. but it is more readable.                                */
334       int i,j,
335           sym_op,              /* A symmetry operation of the square */
336           ialpha,              /* Intgerized scalling factor         */
337           ibeta;               /* Intgerized offset                  */
338
```

```
339      long domain_ref;
340
341      double alpha, beta;
342
343      /* keep breaking it down until we are small enough */
344      if (depth < min_part) {
345          read_transformations(atx,aty, xsize/2, ysize/2, depth+1);
346          read_transformations(atx+xsize/2,aty, xsize/2, ysize/2, depth+1);
347          read_transformations(atx,aty+ysize/2,xsize/2, ysize/2,  depth+1);
348          read_transformations(atx+xsize/2,aty+ysize/2,xsize/2,ysize/2,depth+1);
349          return;
350      }
351
352      if (depth < max_part && unpack(1,input)) {
353          /* A 1 means we subdivided.. so quadtree */
354          read_transformations(atx,aty, xsize/2, ysize/2, depth+1);
355          read_transformations(atx+xsize/2,aty, xsize/2, ysize/2, depth+1);
356          read_transformations(atx,aty+ysize/2, xsize/2, ysize/2, depth+1);
357          read_transformations(atx+xsize/2,aty+ysize/2,xsize/2,ysize/2,depth+1);
358      } else {
359          /* we have a transformation to read */
360          trans->next = (struct transformation_node *)
361                  malloc(sizeof(struct transformation_node ));
362          trans = trans->next;
363          ialpha = (int)unpack(s_bits,  input);
364          ibeta = (int)unpack(o_bits,  input);
365          alpha = (double)ialpha/(double)(1<<s_bits)*(2.0*max_scale)-max_scale;
366
367          beta = (double)ibeta/(double)((1<<o_bits)-1)*
368              ((1.0+fabs(alpha))*GREY_LEVELS);
369          if (alpha > 0.0) beta -= alpha*GREY_LEVELS;
370
371          trans->scale = alpha;
372          trans->offset = beta;
373          if (ialpha != zero_ialpha) {
374              trans-> sym_op = (int)unpack(3, input);
375              domain_ref = unpack(bits_needed[depth-min_part], input);
376              trans->dx = (double)(domain_ref % no_h_domains[depth-min_part])
377                                  * domain_hstep[depth-min_part];
378              trans->dy = (double)(domain_ref / no_h_domains[depth-min_part])
379                                  * domain_vstep[depth-min_part];
380          } else {
381              trans-> sym_op = 0;
382              trans-> dx  = 0;
383              trans-> dy = 0;
384          }
385          trans->rx = atx;
386          trans->ry = aty;
387          trans->depth = depth;
388
389          trans->rrx = atx + xsize;
390          trans->rry = aty + ysize;
391
392          if (output_partition)
393              fprintf(output,"\n%d %d\n %d %d\n%d %d\n\n",
394              atx,        vsize-aty-ysize,
395              atx,        vsize-aty,
396              atx+xsize, vsize-aty);
397
398      }
399  }
400
401  /* ********************************************************** */
402  /* Apply the transformations once to an initially black image.  */
403  /* ********************************************************** */
404  apply_transformations()
405  {
406      IMAGE_TYPE **tempimage;
```

```
407  int i,j,i1,j1,count=0;
408      double pixel;
409
410      trans = &transformations;
411      while (trans->next != NULL) {
412          trans = trans->next;
413  ++count;
414
415  /* Since the inner loop is the same in each case of the switch below */
416  /* we just define it once for easy modification.                     */
417  #define COMPUTE_LOOP                {                          \
418          pixel = (image[j1][i1]+image[j1][i1+1]+image[j1+1][i1]+    \
419                   image[j1+1][i1+1])/4.0;                          \
420          image1[j][i] = bound(0.5 + trans->scale*pixel+trans->offset);\
421          }
422
423      switch(trans->sym_op) {
424          case 0: for (i=trans->rx, i1 = trans->dx;
425                       i<trans->rrx; ++i, i1 += 2)
426                  for (j=trans->ry, j1 = trans->dy;
427                       j<trans->rry; ++j, j1 += 2)
428                      COMPUTE_LOOP
429                  break;
430          case 1: for (j=trans->ry, i1 = trans->dx;
431                       j<trans->rry; ++j, i1 += 2)
432                  for (i=trans->rrx-1,
433                       j1 = trans->dy; i>=(int)trans->rx; --i, j1 += 2)
434                      COMPUTE_LOOP
435                  break;
436          case 2: for (i=trans->rrx-1,
437                       i1 = trans->dx; i>=(int)trans->rx; --i, i1 += 2)
438                  for (j=trans->rry-1,
439                       j1 = trans->dy; j>=(int)trans->ry; --j, j1 += 2)
440                      COMPUTE_LOOP
441                  break;
442          case 3: for (j=trans->rry-1,
443                       i1 = trans->dx; j>=(int)trans->ry; --j, i1 += 2)
444                  for (i=trans->rx, j1 = trans->dy;
445                       i<trans->rrx; ++i, j1 += 2)
446                      COMPUTE_LOOP
447                  break;
448          case 4: for (j=trans->ry, i1 = trans->dx;
449                       j<trans->rry; ++j, i1 += 2)
450                  for (i=trans->rx, j1 = trans->dy;
451                       i<trans->rrx; ++i, j1 += 2)
452                      COMPUTE_LOOP
453                  break;
454          case 5: for (i=trans->rx, i1 = trans->dx;
455                       i<trans->rrx; ++i, i1 += 2)
456                  for (j=trans->rry-1,
457                       j1 = trans->dy; j>=(int)trans->ry; --j, j1 += 2)
458                      COMPUTE_LOOP
459                  break;
460          case 6: for (j=trans->rry-1,
461                       i1 = trans->dx; j>=(int)trans->ry; --j, i1 += 2)
462                  for (i=trans->rrx-1,
463                       j1 = trans->dy; i>=(int)trans->rx; --i, j1 += 2)
464                      COMPUTE_LOOP
465                  break;
466          case 7: for (i=trans->rrx-1,
467                       i1 = trans->dx; i>=(int)trans->rx; --i, i1 += 2)
468                  for (j=trans->ry, j1 = trans->dy;
469                       j<trans->rry; ++j, j1 += 2)
470                      COMPUTE_LOOP
471                  break;
472          }
473
474      }
```

```
475        tempimage = image;
476        image = image1;
477        image1 = tempimage;
478
479        printf("\n%d transformations applied.",count);
480    }
481
482    /*      This should really be done when they are read in.    */
483    /* *********************************************************** */
484    scale_transformations(scalefactor)
485    double scalefactor;
486    {
487        trans = &transformations;
488        while (trans->next != NULL) {
489            trans = trans->next;
490
491    trans->rrx *= scalefactor;
492    trans->rry *= scalefactor;
493    trans->rx *= scalefactor;
494    trans->ry *= scalefactor;
495    trans->dx *= scalefactor;
496    trans->dy *= scalefactor;
497        }
498    }
499
500    /* *********************************************************** */
501    /* Recursively partition an image, finding the largest contained */
502    /* square and call read_transformations .                      */
503    /* *********************************************************** */
504    partition_image(atx, aty, hsize,vsize )
505    int atx, aty, hsize,vsize;
506    {
507        int x_exponent,     /* the largest power of 2 image size that fits */
508            y_exponent,     /* horizontally or vertically the rectangle.   */
509            exponent,       /* The actual size of image that's encoded.    */
510            size,
511            depth;
512
513        x_exponent = (int)floor(log((double)hsize)/log(2.0));
514        y_exponent = (int)floor(log((double)vsize)/log(2.0));
515
516        /* exponent is min of x_ and y_ exponent */
517        exponent = (x_exponent > y_exponent ? y_exponent : x_exponent);
518        size = 1<<exponent;
519        depth = max_exponent - exponent;
520
521        read_transformations(atx,aty,size,size,depth);
522
523        if (size != hsize)
524            partition_image(atx+size, aty, hsize-size,vsize );
525
526        if (size != vsize)
527            partition_image(atx, aty+size, size,vsize-size );
528    }
529
530    /* *********************************************************** */
531    /* Scan the image and average the transformation boundaries.    */
532    /* *********************************************************** */
533    smooth_image()
534    {
535        IMAGE_TYPE pixel1, pixel2;
536    int i,j;
537    int w1,w2;
538
539    printf("\nPostprocessing Image.");
540        trans = &transformations;
541        while (trans->next != NULL) {
542            trans = trans->next;
```

```
543  if (trans->rx == 0 || trans->ry == 0)
544              continue;
545
546  if (trans->depth == max_part) {
547  w1 = 5;
548  w2 = 1;
549  } else {
550  w1 = 2;
551  w2 = 1;
552  }
553
554          for (i=trans->rx; i<trans->rrx; ++i) {
555                  pixel1 = image[(int)trans->ry][i];
556                  pixel2 = image[(int)trans->ry-1][i];
557                  image[(int)trans->ry][i] = (w1*pixel1 + w2*pixel2)/(w1+w2);
558                  image[(int)trans->ry-1][i] = (w2*pixel1 + w1*pixel2)/(w1+w2);
559          }
560
561          for (j=trans->ry; j<trans->rry; ++j) {
562                  pixel1 = image[j][(int)trans->rx];
563                  pixel2 = image[j][(int)trans->rx-1];
564                  image[j][(int)trans->rx] = (w1*pixel1 + w2*pixel2)/(w1+w2);
565                  image[j][(int)trans->rx-1] = (w2*pixel1 + w1*pixel2)/(w1+w2);
566          }
567      }
568  }
```

A.5   The Encoding Program

We will follow the code in execution order, as opposed to line-by-line. The line numbers discussed in the text appear in the margins to aid the reader who is just interested in particular sections of the code. The explanation here is detailed, but knowledge of C is assumed. Knowledge of the algorithm described in Chapter 3 is also assumed. Finally, the comments in the program complement the text below.

100–277   The main routine ranges between lines 100 and 277. It parses the command line, sets up the input/output, and calls the partition_image routine to do all the real work. Lines 120 through

120–186   186 parse the command line by looping through argc. Lines 188 and 189 set the default files
188–189
191–195   names, and the default image size is set in lines 191 to 195.

199-203   Lines 199–203 allocate space for the input image (in image) and for decimated copies of the

25–34, 199   image (in doimage[]). The macro matrix_allocate is defined on lines 25–34; it allocates a section of memory and assigns it to a handle so that a matrix can be doubly indexed. For example, after line 199, image can be indexed by image[row][column]. Matrix_allocate takes as arguments the handle, the number of columns, the number of rows, and the type of the elements of the matrix.

The doimage[] variable holds a copy of the image reduced by a factor or 2 in four ways. Each nonoverlapping group of 2×2 pixels is averaged into one value and stored into the variable. There are four ways of selecting the initial 2×2 group, depending on whether the row and column position are odd or even, and these four ways correspond to the indices of doimage. Later, when domains must be averaged to range size, there is no repetition of averaging, since the domains will be chosen from this preaveraged array.

206   Line 206 allocates memory for the bits_needed array which holds the number of bits needed to encode the domain position. Since the domain lattice may be different at different domain sizes, we precompute the number of bits required for each domain size and store this ahead of time.

Lines 208–219 open the input file, read in the file, and complain if some error has occurred. 208–219
The routine compute_sums on line 222 pre-computes various data for the domains, avoiding 222
recomputation later. It is discussed below. On lines 224-234, some data are packed into the 224-234
output file. The pack routine appears on lines 621–651, but we will not discuss it in detail. The 621–651
written data are essentially a header which allows the decoding program to know some of the
encoding parameters; the choice of parameters written, as opposed to those that are not passed
to the decoding program, is somewhat arbitrary.

The zero_ialpha variable is computed on line 237. It is the quantized value corresponding 237
to a zero scaling factor. It is needed for later comparison because zero scaling factors do not
require any domain data to be stored.

The routine partition_image is called on line 245, surrounded by some diagnostic printf 245
statements. It converts a rectangular image into a union of square regions and calls the encoding
routine on each. It is discussed below. Lines 258–275 free the allocated memory and conclude 258–275
the main routine.

We now discuss compute_sums, which ranges between lines 451 and 616. The variable 415–616
doimage[] is assigned values on lines 474-480. Next, memory is allocated for the domain 474-480
variable, which is defined on lines 72–84. This monster contains all the information about all 72–84
of the domains. Since domains come in several sizes, and since the domain lattice depends on
the domain size, and since the number of domain sizes depends on the number of range sizes
(which depends on the maximum and minimum partition sizes), everything in sight must be
dynamically allocated. In order to know the range sizes, we must know the size of the largest
square with size length a power of 2 that fits in the image. This is computed on lines 486–494. 486–494
Lines 496–499 complain if the maximum or minimum partition depths, given by max_part and 496–499
min_part are too small or large. The value of i on line 499 is the number of different domain
(or range) sizes, and arrays of this length are allocated on lines 502–507. The contents of these 502–507
arrays is described on lines 73–78. Domain.pixel is allocated on lines 509–511. This variable 73–78
is discussed later.   509-511

Lines 513-559 assign values to the various arrays just allocated. The dom_type is assigned 513-559
on line 140, and dom_step on line 142. Dom_type determines the type of domain pool, and 140, 142
dom_step along with dom_step_type, the domain pool lattice. There are variables for differing
vertical and horizontal lattices, but only the horizontal is used. If the program is modified to
handle rectangular image differently, the vertical lattice data would differ from the horizontal,
and this could be easily changed on lines 535–557.   535-557

The variable the_domain is defines on lines 87–90. It holds the classified domain data. 87–90
This means that after each domain is classified, it is added to a list in the right class. There
are two class numbers, the first and second class. The first class has 3 bins and the second 24.
The_domain is allocated and initialized on lines 563–570. Lines 575–600 classify the domains 563–570
and enter the data into a list structure in the_domain. Line 575 loops over the domain sizes; line 575–600
576 loops horizontally over the domains of a fixed size in the lattice (which is size dependent);
line 578 loops vertically.

Lines 580-586 call classify. This routine takes a domain (specified by its 580-586
position: domain_x,domain_y) and its size (specified by domain.domain_hsize[i]),
and it returns the classification of the domain (in first_class and second_class),
the symmetry operation needed to bring the domain into canonical position (in the structure
domain.pixel[i][j][k].sym), and the sum and sum squared of the pixel values in the domain. These are stored in the domain.pixel variable as well. Once this data is known, the

variable node is allocated and node->the is made to point to the domain data, while the node is entered into a linked list which is pointed to by the_domain[first_class][second_-class][i]. This way, when a range is found of a certain first and second class, all the domains with the same class can easily be compared to it by running through the linked list from this point. It is probably just as reasonable, however, to store the actual domain data in the linked list, rather than referencing it though the the field of the list elements.

343–441   We now consider the routine classify from lines 343–441. Along with the parameters discussed above, classify also takes one last parameter, type. This is used to distinguish between classifying domains (which must use the data in doimage) and ranges (which must use the data in image). This is done on line 356; the variable average_func points to the function which is used to compute the sum and sum squared of the pixels in the square at position xy.

356, 293–309   The functions average (lines 293–309) and average1 (lines 317–331) differ only by which
317–331   variable they take their data from.

361–364   The variables a and a2, computed on lines 361–364, hold the sum and sum squared of the pixels in the quadrants of the sub-image at position x,y with size xsize,ysize. These
369–375   are summed to be returned in psum and psum2. Lines 369–375 assign order, used later, and convert a2 to hold the variance rather than the sum-squared values. After ordering the a values
379–384   on lines 379–384, it is possible to read the symmetry operation on lines 400 and 407 right from
400, 406, 407 the order. The first class is set on line 406. Line 411 resets order, and lines 413–418 order
411, 413–418
430–440   a2 for computation of the second class on lines 430–440.

914–937   We can now continue with lines 914–937 containing partition_image, the next important routine called from main. This routine again computes the largest square with size a power
924–929   of 2 on lines 924–929. It then calls the quadtree routine on each square of that type in the (possibly rectangular) image. After each square is quadtree-ed, the remaining rectangles of
932–936   the image are handled by a recursive call to partition_image on lines 932–936.

774–907   The quadtree routine, lines 774–907, does the main work. Lines 805–810 call quadtree
805–810   recursively on each quadrant of the sub-image passed to quadtree if the current depth is below min_part. This means that the initial square passed to quadtree will be partitioned at least min_part many times, and each piece will be considered as a potential range. Line 816 classifies the potential range using classify. Note that the last parameter to classify is 1, signifying
816, 822–836   the range type. Lines 822–836 use the values set on lines 158 and 160 to determine how many
158, 160
481, 483, 856 classes are looped through on lines 481 and 483. Line 856 loops once or twice, depending on whether we are searching only positive scaling values or not. If the loop is repeated twice, lines
848–853   848–853 compute the class and symmetry operation needed when a domain is multiplied by a negative scaling value (and hence the ordering of its average quadrant values changes). Line
857, 869   857 runs through the linked list containing the classified domains. Line 869 calls compare, which takes a domain and range and returns the rms value and the quantized scaling and offset values in ialpha and ibeta.

874–880, 130   The best values found so far are stored in lines 874–880. If the best rms found is greater than tol, set on line 130, then the current range is partitioned into four pieces by calling quadtree
888–891   recursively on its quadrants on lines 888–891 (provided this will not make the depth greater
887   than max_part, the maximum number of allowed recursions). Line 887 outputs one bit set to 1
895–904   to signal the decoding program that another partition was made. Finally, on lines 895–904 the
901   data for the current range are written out. On line 901, we check if the scaling value is zero, in which case no information for the range position is written out.

To complete the description of the encoding program, we now discuss compare, from lines

657–769. To compute the regression we need the sum and sum squared of the pixels, as well 657–769
as the cross terms. The sum and sum squared are precomputed, and they are extracted on lines
682–683. The variable k, assigned on line 686, indexes the doimage[] array, depending on the 682–683, 686
parity of the domain position. The COMPUTE_LOOP macro on lines 693–696 is just shorthand 693–696
for the inner loop which takes place for each of the 8 possible orientations of mapping domain
onto range. These are handled in the switch statement on lines 698–731. From the cross term 698–731
drsum, and the precomputed sums and squared sums, we can compute the scaling alpha and
offset beta. These are quantized and clipped on lines 741–743 and 754–757. The rms value is 741–743
computed on line 765 from the unquantized values of alpha and beta, computed on lines 746 754–757, 765
and 760. 746, 760

A.6   The Decoding Program

The decoding program contains some overlap with the encoding program. We will not discuss
identical (e.g., lines 10–38) code fragments again. 10–38
    The main routine runs from line 126 to 297. It scans the command line, assigns the default 126–297
file names (lines 144–188), sets up the IO (lines 190–193), and reads, on lines 197–203, the 187–188
header information from the file to be decoded using the unpack routine. The unpack routine 190–193
is listed on lines 93–124; it is straightforward. Lines 206–213 compute the largest square image 197–203
with a side that is a power of 2. Line 216 computes the quantized value corresponding to zero 93–124
scaling. Lines 220–221 scale the image size by scaledsize, set on line 158. Lines 225–234 206–213, 216
read in an optional input image, depending on the value of other_input_file, set on line 154. 220–221
Lines 237–280 are similar to lines 501–559 in *enc.c*. 158, 225–234
    Partition_image, called on lines 289, is also almost identical to the routine with the 154, 237–280
same name in *enc.c*. It is listed in lines 504–528; the only change is that read_transfor- 289, 504–528
mations is used in place of quadtree. Lines 294–297 scale the transformations using 294–297
scale_transformations, and lines 301–307 output an initial piece of the quadtree parti- 301–307
tion when output_partition is set on line 160. Lines 309–310 apply the transformations to 160, 309–310
the initial image num_iterations times using apply_transformations. Num_iterations
is set on line 156. If postprocess is not set on line 162, then the image is postprocessed using 156, 162
smooth_image. Finally, the final image is written out with some diagnostics on lines 315–321. 315–321
We now turn our attention to each of the routines just mentioned.
    Read_transformations, lines 329–399 reads the input stream using unpack. It is anal- 329–399
ogous to quadtree, reading the transformation data when quadtree writes it, but otherwise
transversing the quadtree partition identically. It first partitions a virtual image recursively
when the partition depth is below min_part on lines 344–349. On line 352 it reads one bit 344–349, 352
from the input stream and decided if further partitioning is necessary on lines 354–357. If not, 354–357
it allocates space in a linked list containing the transformation data (line 360) and reads and sets
the transformation parameters (lines 363–390). The transformations variable is defined on 360, 363–390
lines 68–77 and set on line 286. It is the root of the linked list containing all the transformation 68–77
data. Lines 365 and 367 compute the scaling alpha and offset beta from their quantized values 286
ialpha and ibeta. Lines 392–397 output the quadtree partition when output_partition is 365, 367
set. 392–397
    Next we look at scale_transformations, listed on lines 484–498. This routine simply 484–498
multiplies the position of the range and domain by the scale factor. The range position ranges

from `trans->rx` to `trans->rrx` in the x-direction, and similarly for the y-direction. The size of the domain is twice the range size, so only its upper left corner must be multiplied. Doing this multiplication is sufficient to scale the image to any resolution.

404–480   The `apply_transformations` routine is listed on lines 404–480. It scans through the linked list starting at `transformations` and maps the pixels in the array `image` into the array

418–420   `image1`. This is done on lines 418–420, which define a macro that is identical for each of the 8 possible domain-range orientations handled in the `switch` statement of line 423. The pixels

418–419, 423   are decimated on lines 418–419, and they are scaled and offset on line 420. Each of the `case`
420   statements loops through the pixels of the range, applying the decimation and scaling and offset. So that we can apply the same routine again, the pointers to `image` and `image1` are swapped

475–477   on lines 475–477.

533–568   Finally, the `smooth_image` routine is listed on lines 533–568. This routine also scans

543   through the linked list starting at `transformations`. On line 543, transformations whose ranges have a boundary that is either at the left or top edge of the image are excluded from smoothing. This allows smoothing of the lower and right boundaries of every other range,

546–552   for a complete smoothing of all the boundaries. Lines 546–552 set the smoothing weights,

554–559   depending on the depth of the current range in the quadtree. Finally, lines 554–559 smooth

561–566   the horizontal edges (of both ranges adjacent to a boundary), while lines 561–566 smooth the vertical edges.

A.7   Possible Modifications

Aside from the list of "bugs" for each program in the manual pages, there are various optimizations which can be attempted on this code. Here is a list of possible modifications/optimizations/improvements:

Enc.c:

- If scaling values that are close to zero are forced to zero, then the bits needed to store the domain data would not need to be saved for a larger number of ranges.

- Rectangular image can be decomposed into subsquares which are not necessarily powers of 2 in size. It is probably a good idea to allow decomposition into squares which are power of 3, 4, or 5 as well.

- It is not necessary to use square domains and ranges at all, in fact. The code is largely set up to handle this already.

- When the image is such that the number of domains just barely requires an extra bit to index all the domains, it is probably better to just save the bit and ignore those domains.

- The bit allocation for the scaling should change with the -p option.

- When the domains overlap, it is possible to save summation of the same pixel values during the domain classification.

- A lookup table would save some multiplcations time (at least on some hardware architectures.

Dec.c:

- The scaling operations should be done when the transformations are read in.

- The code that is common to *enc.c* and *dec.c* should be combined in a separate file, and a header file should incorporate the common headers.

- There should be an option to check the difference between iterations and stop on a threshold condition.

Appendix B

Exercises

Y. Fisher

This appendix contains exercises that are meant to complement the main text in various places, and so the page on which an exercise is mentioned is listed at the end of each.

1. How many geometrically distinct attractors can be generated from the transformations in Figure 2.2 if rotations and flips are allowed also ? (Hint: See [69], page 246).

Pg. 3

2. Why is the orientation of the step in the Barnsley fern important ?

Pg. 3

3. When an IFS is composed of similitudes, that is, Lipschitz maps that satisfy $d(w_i(x), w_i(y)) = s_i d(x, y)$ for $s_i < 1$, and when the IFS is nonoverlapping, then the Hausdorff and Box dimension can be computed directly (see [36]).

In this case, the Hausdorff and box dimensions are the same; they are equal to the number D satisfying

$$\sum_i |s_i|^D = 1.$$

Figures 2.2 and 2.9 show such IFS's composed of similitudes. Compute the fractal dimension of their attractors.

Pg. 28

4. Why is $d(x, y) = x - y$ not a metric on the real line ? Show that in the plane, $d(x, y) = ||x-y||^2$ is not a metric. If L is the space of functions that are square integrable, why is $d(f, g) = \int (f-g)^2$ not a metric ?

Pg. 31

5. Show that the sequence $1, \frac{3}{2}, \frac{7}{5}, \ldots, \frac{p}{q}, \frac{p+2q}{p+q}, \ldots$ converges to $\sqrt{2}$. (Hint 1: Write $\begin{bmatrix} p_{n+1} \\ q_{n+1} \end{bmatrix} =$

$A \begin{bmatrix} p_n \\ q_n \end{bmatrix}$, where A is a matrix. Solve for the eigenvectors and eigenvalues of A to get expressions for p_n and q_n. Hint 2: Note that if $x = \frac{1}{2+x}$ then $x = \frac{1}{2+x} = \frac{1}{2+\frac{1}{2+x}}$, etc.)

Pg. 33

6. Is the map $\frac{p}{q} \mapsto \frac{q}{p+q}$ contractive in the Euclidean metric ? Does it converge to a fixed point ?

Pg. 34

7. Figure B.1 shows examples of IFS's with their fixed points, but without the orientation information. Determine the orientation of the mappings.

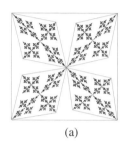 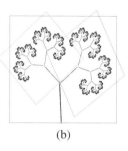 

(a)	(b)	(c)

Figure B.1: Three examples of IFS's with their attractors. The transformations of the IFS's are shown as the images of the square around the attractor.

Pg. 39

8. Show that the norm of

$$A = \begin{bmatrix} a & b \\ c & d \end{bmatrix}$$

is

$$||A|| = \frac{1}{2}\left(a^2 + b^2 + c^2 + d^2 + \sqrt{(a^2 + b^2 + c^2 + d^2)^2 - 4(ad - bc)^2}\right).$$

Pg. 46

9. Show that if

$$A = \begin{bmatrix} a & b \\ c & d \end{bmatrix},$$

then A can be decomposed into

$$A = A_s A_t A_u A_\theta,$$

its component scaling, stretching, skewing, and rotation as in Figure 2.14, with

$$s = \frac{ad - bc}{\sqrt{c^2 + d^2}}$$
$$t = \frac{c^2 + d^2}{ad - bc}$$

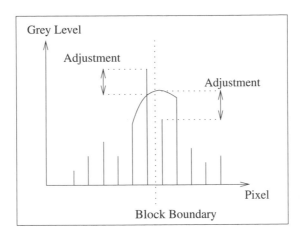

Figure B.2: In Ramstad and Lepsøy's postprocessing scheme, boundary pixels are forced to lie on a parabola that intersects their neighbor's value and that minimizes the distance to their original values.

$$u = \frac{ac + bd}{ad - bc}$$
$$\theta = \arccos\left(\frac{d}{\sqrt{c^2 + d^2}}\right).$$

Pg. 46

10. Find an IFS encoding the "bow tie" of Figure 2.15. What is the minimum number of transformations that is needed ?

Pg. 46

11. Construct a RIFS model of image encoding as in Section 2.7 but using domains instead of ranges. Note that the domains are not necessarily distinct, so that the space H may have less than n components.

Pg. 49

12. A postprocessing scheme due to Ramstad and Lepsøy appears in [52]. In this scheme the boundary pixels are forced to lie on a parabola that passes through the pixel values adjoining the boundary pixels and that minimizes the rms distance to the original boundary pixel values, as shown in Figure B.2.

Show that if four pixels have the values a_1, a_2, a_3, a_4 with the boundary of a range falling between values a_2 and a_3, then this scheme corresponds to substituting the value

$$\frac{1}{6}a_1 + \frac{1}{2}a_2 + \frac{1}{2}a_3 - \frac{1}{6}a_4$$

for a_2 and the value

$$-\frac{1}{6}a_1 + \frac{1}{2}a_2 + \frac{1}{2}a_3 + \frac{1}{6}a_4$$

for a_3.

Pg. 61

13. Show that the probability distribution for the distance r between two randomly chosen points in the unit square (with uniform distribution) is

$$p(r) = \begin{cases} 2\pi r - 8r^2 + 2r^3 & \text{if } 0 \le r \le 1 \\ 8r\arcsin\left(\frac{1}{r}\right) + 8r\sqrt{r^2 - 1} - 2r^3 - 2r(2 + \pi) & \text{if } 1 < r \le \sqrt{2} \end{cases}$$

Pg. 69

14. Explain how to construct an archetype method with an arbitrary number of classes.

Pg. 83

15. Explain why the encoding described in Chapter 10 using 64 domain basis vectors is not exact.

Pg. 209

Appendix C

Projects

Y. Fisher

This appendix contains "projects," that is, potential further work, both theoretical and applied. The projects consist of various suggestions and algorithms of varying value and difficulty which may (or may not) extend the ideas presented in this book. Like the exercises, these end with the pages numbers from which they are referred (when there is a reference). If you, gentle and ambitious reader, actually work out any of these problems, I would be interested in seeing your results.

C.1 Decoding by Matrix Inversion

Section 11.4 discusses how the fixed point of a PIFS can be found by matrix inversion. The matrix $I - S$ is very sparse, but its inverse is not, in general. Of course, the equation

$$f^\infty = |\tau| = (I - S)^{-1} O$$

is not solved by inverting $I - S$ but by partial pivoting. It should be possible to find a pivoting scheme that attempts to bring $I - S$ into upper triangular form with the smallest number of pivoting operations. Are there bounds on the number of operations required to pivot $I - S$ into this form ? A clever scheme can probably require only as many operations as the rapid pyramidal decoding method discussed in Chapter 5 or the pixel chasing scheme discussed below.

Pg. 227

C.2 Linear Combinations of Domains

Chapter 10 discusses ways to use linear combinations of domains. Doing a true search of all possible pairs (or larger numbers) of domains mapped to ranges is too computationally intensive.

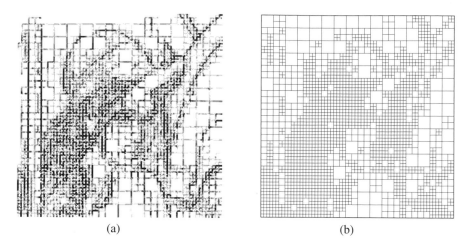

(a)   (b)

Figure C.1: (a) A (grey-level-histogram equalized) difference between a smoothed and un-smoothed encoding of Lenna. Darker portions correspond to a bigger difference. (b) A quadtree partition of Lenna in which the squares have roughly equal variance.

However, another possibility is to find the best domain and then find another domain to encode the difference between the transformed domain and range. That is, encode the collage error using another domain. The offset of this encoding can be combined with the offset from the first encoding of the range, so that this is equivalent to using a total of three coefficients.

Pg. 199

C.3 Postprocessing: Overlapping, Weighted Ranges, and Tilt

All of the schemes discussed in this book partition images into nonoverlapping pieces and then attempt to encode each piece independently. Artificial discontinuities at the range boundaries may arise because each range piece is matched with a domain independently of the other ranges.

One possible remedy is to select overlapping ranges. The overlap can then be averaged, hopefully resulting in a better image. Of course, each range must then contain more pixels, or more ranges must be used – either way, there is some cost. An alternative solution is to weigh the pixels at the boundary of a range more heavily than the internal pixels when seeking a domain-range fit. Doing this, however, will reduce the signal-to-noise ratio.

Figure C.1a shows the difference between a smoothed and unsmoothed image. Brighter portions in the image correspond to a bigger difference. This image shows that the schemes above should really be adaptive. It is clear that regions of high complexity required more adjustment on the range boundary than relatively smooth regions. Figure C.1b is a quadtree partition of Lenna in which each square has a roughly equal variance. This image shows that the variance is a reasonable measure of the image complexity and can be used to predict which squares will require special treatment.

Another postprocessing method consists of adding "tilt" to the blocks after they are rendered.

The tilt is added to each pixel in the block – it is a term of the form $a\,x + b\,y + c$, where x and y range over the block. The a, b and c coefficients can be computed using least squares to minimize the error along the block boundary. Since both blocks along a boundary contribute to this correction, the tilt for each can be computed to minimize half the error.

C.4 Encoding Optimization

Figure C.1b also shows that it is quite reasonable to base the partition on the variance and not on how well a candidate range is matched with a domain. Doing this seems to miss some of the larger squares, but it would result in a significant reduction in encoding time because none of the previously rejected ranges would have to be compared to domains.

C.5 Theoretical Modeling for Continuous Images

The other schemes discussed in this book do not give a bounded error in the value of the decoded image at any point. In some applications, for example in encoding of elevation data, it may be useful to specify the value of the encoded function exactly at some set of points. For such applications, a different theoretical model can be developed along the following lines.

As before, we find a contractive transformation composed of a union of affine maps in a space of images. In this case, we can take the space to consist of all functions over the unit square. However, the affine maps are defined differently. Given a function $f : I^2 \to \mathbb{R}$, we can break up f over squares (or regions with other shapes). We can then define a map between a square R_1 with coordinates (x_i, y_i), $i = 1, \ldots, 4$, and a square R_2 with coordinates (u_i, v_i), $i = 1, \ldots, 4$. We can then use a general affine transformation

$$w \begin{bmatrix} x \\ y \\ z \end{bmatrix} = \begin{bmatrix} a_{11} & a_{12}b & a_{13} \\ a_{21} & a_{22} & a_{23} \\ a_{31} & a_{32} & a_{33} \end{bmatrix} \begin{bmatrix} x \\ y \\ z \end{bmatrix} + \begin{bmatrix} b_1 \\ b_2 \\ b_3 \end{bmatrix}$$

and require that

$$w(x_i, y_i, f(x_i, y_i)) = (u_i, v_i, f(u_i, v_i)), \quad i = 1, 2, 3, 4.$$

This defines all 12 coefficients. As before, a w that minimizes some metric can be found by a search procedure, and the collection of all w defines a transformation W. For a slightly more general model, w can be taken to be an affine map in $\mathbb{R}^4$. In this case, there are four degrees of freedom that can be used to minimize the distance (in some metric) between the values of f above R_1 and the values of f above R_2. For example, a central point can be also be matched or positioned to minimize the error.

Since the functional values over the vertices of the R_i remain fixed, they remain invariant under iteration of W and must therefore belong to the fixed point. Thus, these values can be encoded exactly. The proper encoding and quantization of the mappings remain to be worked out, as does a solid theoretical foundation.

C.6   Scan-line Fractal Encoding

Consider the image as a collection of scan lines f_i, where each $f_i : \mathbb{R} \to \mathbb{R}$ is a function determining the grey level along the scan line. The first function f_1 can be encoded in one dimension using a scheme similar to the others presented in this book. For $i > 1$, f_i can be encoded using domains from the decoded f_{i-1}. This is a kind of 1-dimensional video encoding.

C.7   Video Encoding

The obvious generalization to video encoding is to use 3-dimensional range blocks, with two spatial dimensions and a temporal dimension. This would lead to infinite frame rate resolution, which might be interesting. Would motion smoothing occur? However, this would require buffering of the decoded signal, and this, coupled with decoding that would no longer be particularly computationally simple, may leave this approach on the weak side.

Current methods of video encoding attempt to use previous (and future) frames to compose the current frame. This is called *motion compensation*. When parts of a frame are copied from a previous frame, the memory required to store the data is greatly reduced. This falls very naturally into the fractal scheme.

Initial frames (I frames, in MPEG parlance) can be stored using some fractal (that is, self-referential) scheme. Successive frames are then built up from the decoded version of the previous frame. Motion compensation corresponds to having no contraction in the spatial directions. The parts of the frame that can't be encoded using motion compensation are encoded using some partition and spatially contracting transformations from the previous frame. Since this is not always possible (for example, if the previous frame is monotone), when the error becomes too high, the frames can be encoded self-referentially again.

Decoding of I frames is standard. Decoding of the other frames consists of simply copying the proper piece of the previous frame into the current frame, with any decimation that might be necessary. Since non-I frames are not self-referenced, the decoding for each requires just one iteration. When encoding the frame after the I frame, it is important to use the decoded (as opposed to the original) frame, since this will avoid some error. See [30].

C.8   Single Encoding of Several Frames

Consider a short sequence of several images $f_1, f_2, \ldots, f_N$. It would be interesting to devise a method for finding an encoding τ (using the notation of Chapter 11) that would optimally encode f_{n+1} in terms of f_n, for each $n = 1, \ldots, N$. That is, we seek $g_1 \approx f_1, g_2 \approx f_2, \ldots, g_N \approx f_N$ such that $\tau(g_n) = g_{n+1}$. If the original sequence shows some periodic motion, for example a ferris wheel turning, then we can set $f_{N+1} = f_1$ and repeated application of τ will lead to a short video sequence which may be very highly compressed. Such a scheme may also lead to a method for adding compression to video encodings (while adding to the encoding time), by encoding more than one frame at a time.

C.9   Edge-based Partitioning

Most, if not all, of the fractal schemes described in this book have most of their errors concentrated at the edges in an image. This is very reasonable because we don't expect to find exact local self-similarity at different scales. Moreover, the metrics used to measure the local self-similarity are insensitive to edges.

Nothing can be done about the lack of self-similarity at different scales, but we can try to fix the metrics. Even better, we can structure the algorithm to be sensitive to edges. The algorithm suggested below should encode image edges very sharply.

- Find the edges in the image.

- Vectorize them and form chains of points.

- Use a 2-dimensional fractal method to encode the edge chains, as in [43], for example. This will map arcs of a curve to arcs of a curve, but these can be turned into domains and ranges by "boxing" the domain arc as shown in Figure C.2.

- Over each box, compute the optimal grey-level scaling and offset.

- The portion of the image that is not in any range must then be partitioned and fit using some other scheme, such as the quadtree scheme. This region will contain no edges, so it may be possible to encode it well using a simple low-order polygonal fit of the grey shades.

The decoding algorithm must make some decision about precedence or averaging of overlapping ranges. Also, there is no longer any implicit ordering to the transformations, so that storing them will require more memory.

C.10   Classification Schemes

Chapters 4 and 9 are devoted to optimizing the classification of domains and ranges. Other approaches are possible. For example:

- Use variance (or some other method) to classify sub-images into "texture" or "non-texture" blocks. This is an easy extension of the code in Appendix A.

- Moments: The first moment is the average, which is not relevant. The second moment is the variance, which is what the scheme discussed in Chapter 3 attempts to use. Higher moments may yield better results.

- Large number of bins: A large number of bins may be created by subsampling or averaging a sub-image to a (relatively) small number of values, considered as a vector. These vectors can be classified by solid angle.

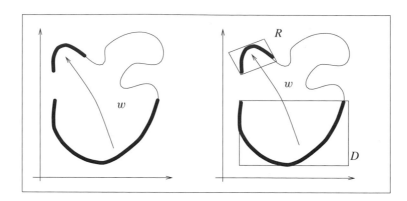

Figure C.2: The 2-dimensional transform can be "boxed" to yield a domain D and a range R. The curve shows an image edge and a transformation w derived from a 2-dimensional fractal encoding. The domain curve is surrounded by a rectangle, which becomes the domain block D. The range block R is the image of the domain block.

- Wavelet or Fourier methods: It is possible that some other representation of the sub-image may yield a small data set which can be used to gauge "similarity" between image pieces.

- One-dimensional relations: A sub-image is mapped onto the real line (using some scheme such as the one presented in the following project). The domain pool can then be chosen to be all the domains that fall in some interval about the range position on the real line.

Pg. 57

C.11   From Classification to Multi-dimensional Keys

D. Saupe

Fractal image compression allows fast decoding but tends to suffer from long encoding times. This project introduces a new twist for the encoding process. During encoding a large domain pool has to be repeatedly searched, which by far dominates all other computations in the encoding process. If the number of domains in the pool is N, then the time spent for each search is linear in N, $O(N)$. Previous attempts to reduce the computation times employ classification schemes for the domains based on image features such as edges or bright spots (from 3 in [48] up to 72 in [27]). Thus, in each search only domains from a particular class need to be examined. However, this approach reduces only the factor of proportionality in the $O(N)$ complexity.

In this project, we replace the domain classification by a small set of real-valued multi-dimensional keys for each domain. These keys are carefully constructed such that the domain pool search can be restricted to the nearest neighbors of a query point. Thus, we may substitute the sequential search in the domain pool (or in one of its classes) by multi-dimensional nearest neighbor searching. There are well known data structures and algorithms for this task which operate in logarithmic time $O(\log N)$, a definite advantage over the $O(N)$ complexity of the sequential search. These time savings may provide a considerable acceleration of the encoding and, moreover, facilitate an enlargement of the domain pool, potentially yielding improved image fidelity.

For simplicity, we present the approach for a special one-dimensional case and generalize later. We consider a set of N vectors $x^{(1)}, \ldots, x^{(N)} \in \mathbb{R}^d$ (representing d pixel values in each of the N domains of the pool) and a point $z \in \mathbb{R}^d$ (representing a range with d pixels). We let $E(x^{(i)}, z)$ denote the smallest possible least squares error of an approximation of the range data z by an affine transformation of the domain data $x^{(i)}$. In terms of a formula, this is $E(x^{(i)}, z) = \min_{a,b \in \mathbb{R}} ||z - (ae + bx^{(i)})||^2$, where $e = \frac{1}{\sqrt{d}}(1, \ldots, 1) \in \mathbb{R}^d$ is a unit length vector with equal components. Computing the optimal a, b and the error $E(x^{(i)}, z)$ is a costly procedure, performed for all of the domain vectors $x^{(1)}, \ldots, x^{(N)}$ in order to arrive at the minimum error $\min_{1 \leq i \leq N} E(x^{(i)}, z)$. This staggered minimization problem needs to be solved for many query points z in the encoding process (i.e., for all ranges). We now show that this search can be done in logarithmic time $O(\log N)$. The following lemma provides the mathematical foundation for the solution. We use the notation $d(\cdot, \cdot)$ for the Euclidean distance and $\langle \cdot, \cdot \rangle$ for the inner product in $\mathbb{R}^d$, thus, $||x|| = d(x, 0) = \sqrt{\langle x, x \rangle}$.

Lemma C.1 *Let $d \geq 2$, $e = \frac{1}{\sqrt{d}}(1, \ldots, 1) \in \mathbb{R}^d$ and $X = \mathbb{R}^d \backslash \{re \mid r \in \mathbb{R}\}$. Define the normalized projection operator $\phi : X \to X$ and the function $D : X \times X \to [0, \sqrt{2}]$ by*

$$\phi(x) = \frac{x - \langle x, e \rangle e}{||x - \langle x, e \rangle e||} \quad and \quad D(x, z) = \min(d(\phi(x), \phi(z)), d(-\phi(x), \phi(z))).$$

For $x, z \in X$ the least squares error $E(x, z) = \min_{a,b \in \mathbb{R}} ||z - (ae + bx)||^2$ is given by

$$E(x, z) = \langle z, \phi(z) \rangle^2 g(D(x, z)) \quad where \quad g(D) = D^2(1 - D^2/4).$$

Proof: The least squares approximation of z by a vector of the form $ae + bx$ (resp. $ae + b\phi(x)$) is given by the projection

$$\text{Proj}(z) = \langle z, e \rangle e + \langle z, \phi(x) \rangle \phi(x) = \langle z, e \rangle e + \langle z, \phi(z) \rangle \langle \phi(x), \phi(z) \rangle \phi(x),$$

where the last equation is derived from $z = \langle z, e \rangle e + \langle z, \phi(z) \rangle \phi(z)$ (see Figure C.3). The least squares error in this approximation is calculated as

$$E(x, z) = ||z - \text{Proj}(z)||^2 = \langle z, \phi(z) \rangle^2 (1 - \langle \phi(x), \phi(z) \rangle^2).$$

Since $d(\pm\phi(x), \phi(z)) = \sqrt{2(1 \mp \langle \phi(x), \phi(z) \rangle)}$ we have $D(x, z) = \sqrt{2(1 - |\langle \phi(x), \phi(z) \rangle|)}$. Solving the last equation for $|\langle \phi(x), \phi(z) \rangle|$ and inserting the square of the result in the formula for $E(x, z)$ completes the proof. ∎

The lemma states that the least squares error $E(x, z)$ is proportional to the simple function g of the Euclidean distance D between the projections $\phi(x)$ and $\phi(z)$ (or $-\phi(x)$ and $\phi(z)$). Since $g(D)$ is a monotonically increasing function for $0 \leq D \leq \sqrt{2}$ we conclude that *the minimization*

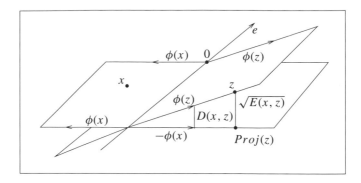

Figure C.3: Illustration of the geometry underlying the lemma.

of the least squares errors $E(x^{(i)}, z)$ for $i = 1, \ldots, N$ is equivalent to the minimization of the distance expressions $D(x^{(i)}, z)$. Thus, we may replace the computation and minimization of N least squares errors $E(x^{(i)}, z)$ by the search for the nearest neighbor of $\phi(z) \in \mathbb{R}^d$ in the set of $2N$ vectors $\pm\phi(x^{(i)}) \in \mathbb{R}^d$.

The problem of finding closest neighbors in Euclidean spaces has been thoroughly studied in computer science. For example, a method using k-d trees that runs in expected logarithmic time is presented in [31] together with pseudo code. After a preprocessing step to set up the required k-d tree, which takes $O(N \log N)$ steps, any search for the nearest neighbors of query points can be completed in expected logarithmic time $O(\log N)$. An even more efficient method that produces a list of so-called approximate nearest neighbors is presented in [1].

We conclude with some remarks on generalizations and implications.

1. In the above we have made the assumption that the number of components in all vectors is the same, namely d. This is not the case in practice, where small and large ranges need to be covered by domains that can be of a variety of sizes. To cope with this difficulty, we settle for a compromise and proceed as follows. We down-filter all ranges and domains to some prescribed dimension of moderate size, for example, $d = 8$ or 16. This allows the processing of arbitrary domains and ranges, however, with the implication that the formula of the lemma is no longer exact but only approximate.

2. For a given range, not all domains from the pool are admissible. There are restrictions on the resulting number b, the contraction factor of the affine transformation. This is necessary in order to ensure convergence of the iteration in the image decoding. Also, one imposes bounds on the size of the domain. To take that into consideration, we search the *entire* domain pool not only for the nearest neighbor of the given query point but also for, say, the next 10 or 20 nearest neighbors (this is called the all-nearest-neighbors problem and can still be solved in logarithmic time using a priority queue). The non-admissible domains are discarded from this set, and the remaining domains are compared using the ordinary least squares approach. This also alleviates the problem mentioned in the previous remark: that the estimate by the lemma is only approximate. Some preliminary empirical tests (using a 256×256 image) have shown that although the best domain for a given range is often not the first entry in the priority queue, it usually is among the first six or so. Thus, the result is identical to a search through the *entire*

domain pool.

3. Our technique for encoding one-dimensional image data readily carries over to (two dimensional) images. There are two possible approaches. The first is simply to represent an image by a one-dimensional scan, for example, by the common scan-line method or (better) using a Hilbert or Peano-scan. However, in most implementations, image domains and ranges are squares or rectangles. In that case image data from a rectangle, for example, can first be transformed to a square, then down-filtered to, say, an array of 3×3 or 4×4 intensity values. After applying the normalized projection operator ϕ we can use the result as a multi-dimensional key in the exact same fashion as described before.

4. We make two technical remarks concerning memory requirements for the k-d tree. First, it is not necessary to create the tree for the full set of $2N$ keys in the domain pool. We need to keep only one multi-dimensional key per domain if we require that this key has a non-negative first component (multiply the key by -1 if necessary). In this set-up a k-d tree of all $2N$ vectors has two symmetric main branches (separated by a coordinate hyperplane). Thus, it suffices to store only one of them. Second, there is some freedom in the choice of the geometric transformation that maps a domain onto a range. A square, for example, may undergo any of the 8 transformations of its symmetry group. This will create 7 additional entries in the k-d tree, enlarging the size of the tree. However, we can get away without this tree expansion. To see this, just note that we may instead consider the 8 transformations of the range (or just some of them) and search the original tree for nearest neighbors of each one of them.

5. The $O(N \log N)$ preprocessing time required to create the data structure for the multi-dimensional search is not a limitation of the method. To see that, observe that the number of ranges to be considered is of the same order $O(N)$ as the number of domains. Thus, the sum of all search times, including the preprocessing, is $O(N \log N)$, to be compared with $O(N^2)$ for the method using sequential search.

6. Another approach that uses several real numbers as keys for domains and ranges is given in terms of so-called Rademacher labellings in [9]. These labels are used to make a pre-selection of the domain blocks much in the spirit of the usual classification. Another concept using a similarity measure related to the above lemma is presented in Section 9.3.

C.12   Polygonal Partitioning

The scheme described in Chapter 6 partitions ranges into rectangles. It is possible to generalize this to a method that is much more versatile, but not particularly memory intensive. Rather than being partitioned only horizontally or vertically, the image may be partitioned along lines that run at angles of 0, 45, 90, or 135 degrees, as in [83]. This still leads to a tree-structured partitioning of the image, and only one extra bit is required per node to specify the orientation of the partition.

C.13   Decoding by Pixel Chasing

Consider a modification of the notation of Chapter 11 in which $x \in \mathbb{Z}^2$ represents a coordinate in a matrix representing an image. Also, let $f(x)$ represent the value of the matrix element, or pixel, at position x. When the domain decimation consists of subsampling, a decoding iteration consists of computing

$$f(x) = s(x)f(m(x)) + o(x)$$

for each x. Note that when this equation is satisfied, then we have found the fixed point. Rather than computing this equation once for each pixel and repeating, we can compute it repeatedly for each pixel. That is, fix x, and choose some random initial image f_0. Now "chase" the pixel values through the action of m. Compute

$$f_1(x) = s(x)f_0(m(x)) + o(x).$$

We really want to know the fixed point f_∞, but if $m(x) = y$ and $f_1(y)$ is known, then we can write the fixed point equation

$$f_1(x) = s(x)f_1(y) + o(x).$$

We are then done for $f(x)$, and must go on to the next value of x. If we don't know $f_1(m(x))$, which is the case initially, we compute

$$f_1(y) = f_1(m(x)) = s(m(x))f_0(m(m(x))) + o(m(x)),$$

so that

$$
\begin{aligned}
f_2(x) &= s(x)f_1(y) + o(x) \\
&= s(x)\left[s(m(x))f_0(m(m(x))) + o(m(x))\right] + o(x) \\
&= s(x)s(m(x))f_0(m(m(x))) + s(x)o(m(x)) + o(x),
\end{aligned}
$$

and so on. Again, if we know $f_2(m(m(x)))$, we can substitute this for $f_0(m(m(x)))$ and we will be done. (Eventually we will find a loop, leading to a system that can be solved, but this is not the point here. This is discussed, however, in [59].) Even though each $f_0(m^{\circ n}(x))$ is not well known, we can find an excellent estimate of the fixed point $f_\infty(x)$. We know (empirically, for example) that 10 iterations are sufficient to decode the image to a close approximation of the fixed point, so that $f_{10}(x)$ will be nearly equal to $f_\infty(x)$. As the algorithm progresses, the first pixels will require computation of $f_{10}(x)$, but eventually, the image will get filled up so that it will be sufficient to compute $f_n(x)$ for $n < 10$. On average, for example, when half the image is computed, there is a 50% chance that the $f_1(x)$ will be nearly exact for an x which is not yet computed. This means that there is some reduction in the number of computations that must be carried out.

We can compute how many operations this will take. Assume that the proportion of un-decided pixels is p and that the action of m is random. Then the proportion of pixels whose value will be determined through one reference of m is p. The proportion that will require two references through m is $p(1 - p)$; three references requires $p(1 - p)^2$, and so on. But since we allow a maximum of 10 references, the proportion of the pixels requiring 10 references is $(1-p)^9$. So the average pixel will require $10(1-p)^9 \sum_{n=1}^{9} n\,p(1-p)^{n-1}$. Since the proportion ranges from 0 to 1, we can estimate the average number of references by

$$\int_0^1 \sum_{n=1}^{10} n\,p^n \approx 3.$$

That means that on average, each pixel will require 3 references, not 10. Each reference is equivalent to an iteration, so that the computational burden of this scheme is comparable to doing just 3 iterations in a normal iterative decoding method. However, this model just accounts for two sorts of pixels – "known," with 10 references or more, and "unknown," whose number of references is considered 0. The number of operations can be significantly reduced if we keep track of populations of pixels with 0, 1, ...,9 ,10 computed references. With these improvements, the decoding scheme should be significantly improved. It should also be pointed out that most of the pixels will have values that are equivalent to doing more than 10 references, since they will be iterates of pixels that have a minimum of 10 references.

Pg. 59

C.14 Second Iterate Collaging

Given an image f, the Collage Theorem motivated the minimization of $d(f, W(f))$. However, what we really want to minimize is $d(f, x_W)$, where x_W is the fixed point of W. Unfortunately, this is not usually possible. However, it may be possible to minimize $d(f, W(W(f)))$. Since $W(W(f))$ is closer to the fixed point than $W(f)$, this should improve the encoding. One approach might be to find proper domains and ranges using the collage minimization, but to find the optimal scaling and offset values using the second iterate condition.

C.15 Rectangular IFS Partitioning

Figure C.4: An encoding of 256×256 Lenna using a simple IFS on 8×8 blocks.

Figure C.4 shows an encoding of 256×256 Lenna using a simplification of Dudbridge's scheme in [21]. The original image was partitioned into 8×8 blocks, each of which was encoded using a simple square IFS as in Figure 12.1 with an affine transformation of the grey levels. This is equivalent to finding a fixed point for the operator

$$\tau(f) = s(x)f(m(x)) + o(x),$$

as in Chapter 11, with $s(x)$ and $o(x)$ piecewise constant over the quadrants of the block and with $m(x) = 2x \bmod 1$. In this notation, $x = (u, v)$ represents a point in $\mathbb{R}^2$, so that

$$m \begin{bmatrix} u \\ v \end{bmatrix} = \begin{bmatrix} 2u \bmod 1 \\ 2v \bmod 1 \end{bmatrix}.$$

Notice that edges in the image that run at 0, 45, and 90 degrees are well encoded, but that the other edges are not. This is because the map $m(x) = 2x \bmod 1$ has a natural affinity for such things. For example, if two of the diagonal $s(x)$ values are zero, then the attractor for the set will naturally be a line along the other diagonal of the block. If edges at other angles are also well encoded, the scheme may be viable.

To do this, we can consider a more general $m(x)$. For example, if

$$m \begin{bmatrix} u \\ v \end{bmatrix} = \begin{bmatrix} 3u \bmod 1 \\ 2v \bmod 1 \end{bmatrix},$$

then we should be able to encode edges with slopes of $-2/3$ and $2/3$ well. In this case, $s(x)$ and $o(x)$ would be piecewise constant over six rectangles, not four as before. To encode an image, a small set of possible $m(x)$-functions can be tested, with the optimal one used for the encoding.

C.16 Hexagonal Partitioning

Consider the following process, shown in Figure C.5, which constructs a hexagonal tiling of the plane. Begin with a subset of the plane A_0 which is a filled hexagon; surround A_0 with six copies of itself, laid adjacent to A_0; scale the new subset to exactly fit in A_0 and call this A_1. Now repeat. The limit is a fractal that forms a tiling of the plane and which is built up of seven copies of itself.

Just as the square is a basic tile for the quadtree partition, we can make a hept-tree partition based on this shape. As shown in Figure C.6, if the pixels on every other row are shifted one half of a pixel width, the centers of the pixels now lie on a roughly hexagonal tile; the aspect ratio is not 1, but the connectivity is the same. There are several obvious difficulties; it is easier to think of pixels as squares which fit nicely to form larger squares than it is to think of them as hexagonal tiles.

The potential advantages of this scheme are:

- The overlap between ranges can be used to minimize artifacts along range boundaries. (This follows the long tradition of claiming bugs as features.)

- Since the range boundaries are not straight edges, artifacts will be less distracting.

- The domain-range map can take on 12 orientations. This breaks away from the strict vertical and horizontal self-similarity of the quadtree scheme.

Pg. 56

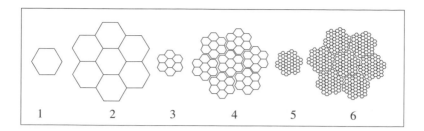

Figure C.5: The first steps in building a hexagonal tile of the plane. The tiles are displayed slightly separated to show how they fit together. The steps are: (1) an initial hexagon A_0, (2) seven hexagons, (3) the seven hexagons reduced to form A_1, (4) seven copies of A_1, (5) the seven copies of A_1 reduced to form A_2, and (6) seven copies of A_2.

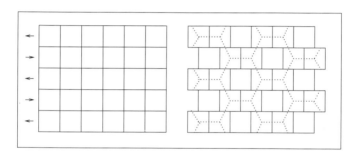

Figure C.6: A way of generating hexagonal tiling on a square lattice.

C.17 Parallel Processing

While computationally demanding and not well suited to vectorization, the compression methods discussed in this book can fortunately be readily parallelized: they can be implemented in parallel on separate processors sharing the same memory. The most straightforward method simply assigns a separate piece of the image to each processor.

A better approach might incorporate some form of pipelining. For example, a master processor could determine ranges and deal with input/output, while each of several sub-processors compares a range to some subset of the domain pool. Each subprocessor would find the best domain in its subset, and return it to the master processor, which would find the optimal domain and send the next range for matching.

C.18 Non-contractive IFSs

The image in Figure C.7a shows an expansive IFS which is eventually contractive. One of the transformations, shown by the larger rectangle, is expansive. Therefore, the W map is not

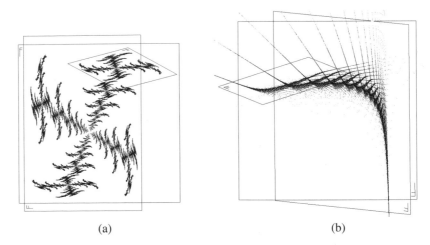

(a)   (b)

Figure C.7: (a) An expansive IFS which is eventually contractive. (b) An expansive IFS which is not eventually contractive, but still has an interesting attractor. Both IFS's are shown as the image of a square (marked to indicate orientation).

contractive in the Hausdorff metric. However, the second iterate of W

$$W^2 = w_1 \circ w_1 \cup w_1 \circ w_2 \cup w_2 \circ w_1 \cup w_2 \circ w_2$$

is contractive.

Figure C.7b shows another expansive IFS. In this case, the attractor is not bounded (so the figure shows only a part of the attractor, and not well). Nevertheless, the attractor exists and has interesting properties.

There should be interesting theory of non-contractive IFS's.

Pg. 37

Appendix D

Comparison of Results

Y. Fisher

In this appendix we briefly compare the results in this book with other results, such as wavelets, JPEG, and Iterated Systems, Inc. software. Some of these comparisons appear also in [29]. The results should be taken with a grain, or more aptly a large mountain, of salt for the reasons listed below.

- PSNR is not a measure of perceived quality.

- Encoding time is not considered as a factor.

- Almost none of the code has been written with optimal bit allocation. The degree of optimization varies greatly and may overwhelm any potential strength of the compared method.

- The Iterated Systems, Inc. code is written to compress color images, and therefore its performance on grey-scale images must be estimated.

- While the quadtree, HV, wavelet, and JPEG results were done on the same images, the other results used the same images in name but not necessarily in content. Even the same scheme will give varying results if the image is slightly modified, for example, by changing its contrast even slightly.

The compared methods consist of the results from this book, as listed in the Table D.1, and the following:

JPEG. The Independent Group's public domain JPEG routines were used. This uses the JFIF standard (which has the misfortune of sounding like, but being very different from, the FIF format of Iterated Systems, Inc.) This code was written by Tom Lane, Philip Gladstone, Luis Ortiz, Lee Crocker, Ge' Weijers, and other members of the Independent

311

JPEG Group. Their disclaimers state that "THIS SOFTWARE IS NOT COMPLETE NOR FULLY DEBUGGED. It is not guaranteed to be useful for anything, nor to be compatible with subsequent releases, nor to be an accurate implementation of the JPEG standard." (The uninitiated to the world of disclaimers should not view this as meaning that the code is not excellent).

Wavelets. The wavelet compression results used the public domain EPIC (Efficient Pyramid Image Coder) routines. The compression algorithms are based on a biorthogonal critically sampled wavelet, subband decomposition, and a combined run-length/Huffman entropy coder. This code was designed by Eero P. Simoncelli and Edward H. Adelson, and was written by Eero P. Simoncelli at the Massachusetts Institute of Technology Media Laboratory, in the Vision Science Group.

FIF. Iterated Systems Fractal Image Format (FIF) compression method was also used for comparison (Images Inc. Version 3.10, "best quality" compression). This code encodes color images only, and so the results must be interpreted somewhat. Since their routines are proprietary, the interpretation is essentially a "best guess" of their color encoding method. In the comparison, it is assumes that the color data is converted to YIQ or YUV representation and that the chrominance signals are decimated by a factor of 2 or 4 and compressed to a bit rate in the range of one-third to one times the bit rate of the luminance channel. This leads to a range of compression values for the Y signal. The data is shown with error bars along the compression axis. The error bars represent the range of 0% to 33% of the compressed image size used for color information. The data point is at 16%.

We present results for 512×512 Lenna and for 512×512 Boat. The results span a wide range of compression, with fidelities falling far below the useful range. Nevertheless, the strengths and weaknesses of each scheme can be seen at the extremes.

Figure D.1 shows PSNR versus compression results for 512×512 Lenna using the compression methods discussed above. The quadtree results are the best results from Chapter 3. It is possible to slightly improve these results by comparing all the domains in a very large lattice, but this is pointless, since the computation time would be enormous, if not prohibitive. The quadtree results are worse than both the JPEG and wavelet results, at least over the range of fidelities that are useful. The HV-partitioned results, however, are better than, or comparable to, all the other results, except at low compression ratios. The wavelet method has the potential advantage of perfect reconstruction at a compression ration of 1, something that JPEG and the fractal methods cannot do.

Figure D.2 is very similar. Figure D.3 shows sample results from encoding the Boat image using the above methods. Again, the HV scheme performs well, especially given the crude nature of its bit allocation and quantization.

Acknowledgment

Dr. Fisher would like to thank Joel Goldberger of Infomagic, Inc. for providing the FIF data.

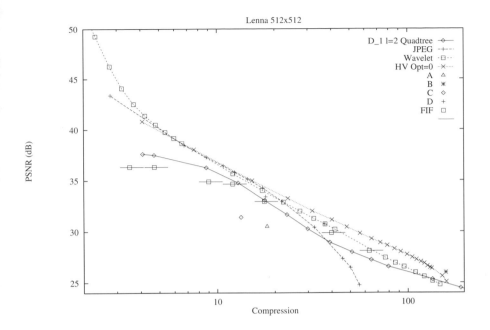

Figure D.1: Comparison of results for 512×512 Lenna.

Table D.1: Key to symbols in Figures D.1 and D.2.

Symbol	Source
A	Chapter 10
B	Chapter 13
C	[49]
D	[27]
A1	Chapter 8

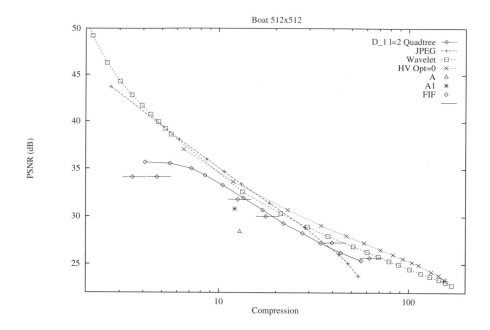

Figure D.2: Comparison of results for 512×512 Boat.

Quadtree Fractal Encoding
512x512 Boat Image
Compression = 57.17:1 (.140 bpp)
PSNR = 25.3dB

JPEG
512x512 Boat Image
Compression = 54.3:1 (0.147 bpp)
PSNR = 23.7dB

HV Fractal Encoding
512x512 Boat Image
Compression = 58.1:1 (0.137 bpp)
PSNR = 27.2dB

Epic Wavelet
512x512 Boat Image
Compression = 58.0:1 (0.138 bpp)
PSNR = 26.4dB

Figure D.3: The Boat image compressed using four methods.

Appendix E

Original Images

Figure E.1: 256 × 256 San Francisco.

Figure E.2: 256 × 256 Collie.

Figure E.3: 256 × 256 Tank Farm.

Figure E.4: 512 × 512 Lenna.

Figure E.5: 512 × 512 Mandrill.

Figure E.6: 512 × 512 Kiel Harbor.

Figure E.7: 512 × 512 Peppers.

Figure E.8: 512 × 512 Boat.

Figure E.9: 256 × 256 String Cheese.

Bibliography

[1] S. Arya, M. M. Mount, N.S. Netanyahu, R. Silverman, and A. Wu. An optimal algorithm for approximate nearest neighbor searching. In *Proc. 5th Annual ACM-SIAM Symposium on Discrete Algorithms*, pages 573–582, 1994.

[2] Z. Baharav, D. Malah, and E. Karnin. Hierarchical interpretation of fractal image coding and its application to fast decoding. In *Intl. Conf. on Digital Signal Processing*, Cyprus, July 1993.

[3] M.F. Barnsley. Fractal functions and interpolation. *Constr. Approx.*, 2:303–329, 1986.

[4] M.F. Barnsley. *Fractals Everywhere*. Academic Press, San Diego, 1988.

[5] M.F. Barnsley, J.H. Elton, and D.P. Hardin. Recurrent iterated function systems. *Constr. Approx.*, 5(1):3–48, 1989.

[6] M.F. Barnsley, V. Ervin, D. Hardin, and J. Lancaster. Solution of an inverse problem for fractals and other sets. *Proc. Natl. Acad. Sci. USA*, 83:1975–1977, April 1986.

[7] M.F. Barnsley and L.P. Hurd. *Fractal Image Compression*. AK Peters, Ltd., Wellesley, Ma., December 1992.

[8] M.F. Barnsley and A. Jacquin. Applications of recurrent iterated function systems to images. *SPIE Visual Communications and Image Processing*, 1001:122–131, 1988.

[9] T.J. Bedford, F.M. Dekking, and M.S. Keane. Fractal image coding techniques and contraction operators. *Nieuw Arch. Wisk. (4)*, 10(3):185–218, 1992.

[10] R.D. Boss and E.W. Jacobs. Fractal-based image compression. Technical Report 1315, Naval Ocean Systems Center, San Diego, CA, September 1989.

[11] R.D. Boss and E.W. Jacobs. Studies of iterated transform image compression, and its application to color and DTED. Technical Report 1468, Naval Ocean Systems Center, San Diego, CA, 1991.

[12] R. Boynton. *Human Color Vision*. Optical Society of America, 1992.

[13] P.J. Burt and E.H. Adelson. The Laplacian pyramid as a compact image code. *IEEE Trans. on Communications*, 31(4):532–540, April 1983.

[14] A. Buzo, A.H. Gray, R.M. Gray, and J.D. Markel. Speech coding based upon vector quantization. *IEEE Trans. on Acoustics, Speech and Signal Processing*, 28:562–574, October 1980.

[15] G. J. Chaitin. Algorithmic information theory. *IBM Journal of Research and Development*, 21:350–359, 1977.

[16] W. Chen and W.K. Pratt. Scene adaptive coder. *IEEE Transactions on Communications*, 32:225–232, 1984.

[17] W.H. Chen, C.H. Smith, and S.C. Fralick. A fast computational algorithm for the discrete cosine transform. *IEEE Trans. Commun.*, COM-25(9):1004–1009, September 1977.

[18] K.-M. Cheung and M. Shahshahani. A comparison of the fractal and JPEG algorithms. *TDA Progress Report 42-107*, pages 21–26, November 15, 1991.

[19] F. Cocurullo, F. Lavagetto, and M. Moresco. Optimal clustering for vector quantizer design. In J. Vandewalle, R. Boite, M. Moonen, and A. Oosterlinck, editors, *SIGNAL PROCESSING VI: Theories and Applications*. Elsevier Science Publishers, 1992.

[20] R.A. DeVore, B. Jawerth, and B.J. Lucier. Image compression through wavelet transform coding. *IEEE Transactions on Information Theory*, 38:719–746, 1992.

[21] F. Dudbridge. *Image approximation by self-affine fractals*. PhD thesis, University of London, 1992.

[22] H.M. Eikseth. l_1 / l_2-normen til en endelig-dimensjonal lineær avbildning (in norwegian). Diploma Thesis, Norges Tekniske Høgskole, 1992.

[23] K. Falconer. *Fractal Geometry – Mathematical Foundations and Applications*. John Wiley and Sons, Chichester, 1990.

[24] Y. Fisher. Fractal image compression. In *SIGGRAPH '92*, 1992. Fractal Course Notes.

[25] Y. Fisher. Fractal image compression. In *Chaos and Fractals: New Frontiers of Science*. Springer-Verlag, New York, 1992. Appendix A. Heinz-Otto Peitgen, Dietmar Saupe, and Hartmut Jürgens, Authors. See [69].

[26] Y. Fisher, E.W. Jacobs, and R.D. Boss. Fractal image compression using iterated transforms. Technical Report 1408, Naval Ocean Systems Center, San Diego, CA, 1991.

[27] Y. Fisher, E.W. Jacobs, and R.D. Boss. Fractal image compression using iterated transforms. In James A. Storer, editor, *Image and Text Compression*, pages 35–61. Kluwer Academic Publishers, Boston, MA, 1992.

[28] Y. Fisher and A. Lawrence. Fractal image encoding: S.B.I.R. Phase 1, Final Report, 1990. Technical report.

[29] Y. Fisher, T. P. Shen, and D. Rogovin. A comparison of fractal methods with dct (jpeg) and wavelets (epic). In *SPIE Procedings, Neural and Stochastic Methods in Image and Signal Processing III*, volume 2304-16, San Diego, CA, July 28-29 1994.

[30] Y. Fisher, T.P. Shen, and D. Rogovin. Fractal (self-vq) video encoding. In *Visual communicationa and image processing '94 (VCIP'94)*, volume 2304-16, Chicago, IL, September 28-29 1994.

[31] J.H. Friedman, J. L. Bentley, and R. A. R. A. Finkel. An algorithm for finding best matches in logarithmic expected time. *ACM Trans. Math. Software*, 3(3):209–226, 1977.

[32] F. Gantmacher. *The Theorey of Matrices*. Chelsea, New York, 1960.

[33] M. Gharavi-Alkhansari and T. Huang. A fractal-based image block-coding algorithm. In *Proc. ICASSP*, volume 5, pages 345–348, 1993.

[34] D.P. Hardin and P.R. Massopust. The capacity for a class of fractal functions. *Communications in Mathematical Physics*, 105:455–460, 1986.

[35] M.H. Hayes, G. Vines, and D.S. Mazel. Using fractals to model one-dimensional signals. In *Proceedings of the Thirteenth GRETSI Symposium*, volume 1, pages 197–200, September 1991.

[36] J. E. Hutchinson. Fractals and self-similarity. *Indiana University Mathamatics Journal*, 3(5):713–747, 1981.

[37] K. Culik II and S. Dube. Rational and affine expressions for image description. *Discrete Applied Mathematics*, 41:85–120, 1993.

[38] K. Culik II, S. Dube, and P. Rajcani. Efficient compression of wavelet coefficients for smooth and fractal-like data. In J.A. Storer and M. Cohn, editors, *Proceedings of the Data Compression Conference*, pages 234–243. IEEE Computer Society Press, 1993. CDD'93, Snowbird, Utah.

[39] K. Culik II and J. Karhumäki. Automata computing real functions. *SIAM J. on Computing*, to appear. Tech. Report TR 9105, University of South Carolina, Columbia (1991).

[40] K. Culik II and J. Kari. Computational fractal geometry with wfa, 1993. manuscript.

[41] K. Culik II and J. Kari. Image compression using weighted finite automata. *Computer and Graphics*, 17(3):305–313, 1993.

[42] K. Culik II and J. Kari. Image-data compression using edge-optimizing algorithm for wfa inference. *Journal of Information Processing and Management*, To Appear.

[43] E.W. Jacobs, R.D. Boss, and Y. Fisher. Fractal-based image compression ii. Technical Report 1362, Naval Ocean Systems Center, San Diego, CA, June 1990.

[44] E.W. Jacobs, Y. Fisher, and R.D. Boss. Image compression: A study of the iterated transform method. *Signal Processing*, 29:251–263, 1992.

[45] A. Jacquin. *A Fractal Theory of Iterated Markov Operators with Applications to Digital Image Coding*. PhD thesis, Georgia Institute of Technology, August 1989.

[46] A. Jacquin. Fractal image coding based on a theory of iterated contractive image transformations. In *Proc. SPIE's Visual Communications and Image Processing*, pages 227–239, 1990.

[47] A. Jacquin. A novel fractal block-coding technique for digital images. *IEEE ICASSP Proc.*, 4:2225–2228, 1990.

[48] A. Jacquin. Image coding based on a fractal theory of iterated contractive image transformations. *IEEE Transactions on Image Processing*, 1(1):18–30, January 1992.

[49] A. Jacquin. Fractal image coding: A review. *Proceedings of the IEEE*, 81(10):1451–1465, October 1993.

[50] N.S. Jayant and P. Noll. *Digital Coding of Waveforms – Principles and Applications to Speech and Video*. Prentice-Hall, Englewood Cliffs, NJ, 1984.

[51] E. Kreyszig. *Introduction to Functional Analysis with Applications*. Robert E. Krieger Publishing Company, Malabar, FL, 1989 (reprint edition).

[52] S. Lepsøy. Removal of blocking effects by FIR filtering, 1991. Internal note (in Norwegian), The Norwegian Institute of Technology.

[53] S. Lepsøy. *Attractor Image Compression – Fast Algorithms and Comparisons to Related Techniques*. PhD thesis, Norwegian Institute of Technology, Trondheim, Norway, June 1993.

[54] S. Lepsøy, G.E. Øien, and T.A. Ramstad. Attractor image compression with a fast non-iterative decoding algorithm. In *Proc. ICASSP*, pages 5:337–340, 1993.

[55] Y. Linde, A. Buzo, and R. M. Gray. An algorithm for vector quantizer design. *IEEE Trans. on Communications*, COM-28(1):84–95, January 1980.

[56] S.P. Lloyd. Least squares quantization in PCM. *IEEE Trans. Information Theory*, IT-28:127–135, March 1982.

[57] L. Lundheim. An approach to fractal coding of one-dimensional signals. In *KOMPRESJON-89*, pages 11–25, Oslo, 1989.

[58] L. Lundheim. Filters with fractal root signals suitable for signal modelling and coding. In *The 1990 Digital Signal Processing Workshop*, pages 1.3.1–1.3.2, New Palz, 1990. IEEE.

[59] L. Lundheim. *Fractal Signal Modelling for Source Coding*. PhD thesis, The Norwegian Institute of Technology, September 1992.

[60] B.B. Mandelbrot. *The Fractal Geometry of Nature*. W. H. Freeman and company, New York, 1983.

[61] R. Mañè. *Ergodic Theory and Differentiable Dynamics*. Springer-Verlag, New York, 1987.

[62] J. Max. Quantizing for minimum distortion. *IEEE Trans. Information Theory*, 6:7–12, March 1960.

[63] D.S. Mazel and M. H. Hayes. Fractal modeling of time-series data. In *Proceedings of the Twenty-Third Asilomar Conference of Signals, Systems and Computers*, 1989.

[64] G.E. Øien. *L_2-Optimal Attractor Image Coding with Fast Decoder Convergence*. PhD thesis, Norwegian Institute of Technology, Trondheim, Norway, June 1993.

[65] G.E. Øien, Z. Baharav, S. Lepsøy, E. Karnin, and D. Malah. A new improved collage theorem with applications to multiresolution fractal image coding. In *Proc. ICASSP*, 1994.

[66] G.E. Øien, S. Lepsøy, and T. Ramstad. An inner product space approach to image coding by contractive transformations. In *ICASSP-91*, pages 2773–2776. IEEE, 1991.

[67] H.-O. Peitgen and D. Saupe, editors. *The Science of Fractal Images*. Springer-Verlag, New York, 1989.

[68] H.-O. Peitgen, D. Saupe, and H. Jürgens. *Fractals for the Classroom*. Springer-Verlag, New York, 1991.

[69] H.-O. Peitgen, D. Saupe, and H. Jürgens. *Chaos and Fractals: New Frontiers of Science*. Springer-Verlag, New York, 1992.

[70] W.H. Press, B.P. Flannery, S.A. Teukolsky, and W.T. Vetterling. *Numerical Recipes in C*. Cambridge University Press, New York, 1988.

[71] M. Rabbani and P.W. Jones (ed.). *Digital Image Compression Techniques*. SPIE Press, Tutorial Texts in Optical Engineering, Vol. TT7, Bellingham, Washington, 1991.

[72] B. Ramamurthi and A. Gersho. Classified vector quantization of images. *IEEE Trans. Comm.*, COM-34:1105–1115, November 1986.

[73] A. Rosenfeld, editor. *Multiresolution Image Processing and Analysis*. Springer-Verlag, 1984.

[74] H. L. Royden. *Real Analysis*. Macmillan, New York, 1988.

[75] W. Rudin. *Real and Complex Analysis*. McGraw-Hill, New York, 1972.

[76] G. Strang. *Linear Algebra and Its Applications*. Harcourt Brace Jovanovich, San Diego, 3rd edition, 1988.

[77] J.T. Tou and R.C. Gonzalez. *Pattern Recognition Principles*. Addison-Wesley, Reading, MA, 1974.

[78] G. Vines. *Signal Modeling With Iterated Function Systems*. PhD thesis, Georgia Institute of Technology, Atlanta, GA, 1993.

[79] G. Vines and M.H. Hayes. Orthonormal basis approach to IFS image coding.

[80] G. Vines and M.H. Hayes. Nonlinear interpolation in a one-dimensional fractal model. In *Proceedings of the Fifth Digital Signal Processing Workshop*, pages 8.7.1–8.7.2, September 1992.

[81] G. Vines and M.H. Hayes. Nonlinear Address Maps in a One-Dimensional Fractal Model. *IEEE Trans. on Signal Processing*, 41(4):1721–1724, April 1993.

[82] P. Waldemar. Kompleksitetsreduksjon i enkoderdelen av en attraktorkoder for bilder (in norwegian). Term project, Norges Tekniske Høgskole, 1992.

[83] X. Wu and C. Yao. Image coding by adaptive tree-structured sementation. In *Data Compression Conference Proceedings, Snowbird Utah*, pages 73–82. IEEE Computer Society Press, Los Alamitos, CA, 1991.

[84] K. Zeger, J. Vaisey, and A. Gersho. Globally optimal vector quantizer design by stochastic relaxation. *IEEE Transactions on Signal Processing*, 40(2):310–322, February 1992.

Index